Photograph by William Tobey

"Throughout a long professional life, the late Professor Rüdenberg combined leadership in industrial research and design work with academic activities first in Germany and later in the United States ... Starting from physical principles and using classical mathematical methods, the author examines the wide subject of switching and surge-propagation phenomena in simple and complex transformer and transmission systems and in transformer and machine winding. Many of these investigations are based on original work that has since been accepted as a standard ... A bibliography of over 500 references is appended. The entire subject is given a crystal-clear treatment, which is aided by a large number of practical examples. The text can be highly recommended to practicing engineers and to students and teachers alike."

—*Journal of the Institute of Electrical Engineers*

The late Reinhold Rüdenberg was Professor Emeritus, Technical University, Berlin, and Gordon McKay Professor of Electrical Engineering, Harvard University. Prior to his consulting practice in the United States and Britain, he was chief engineer of Siemens-Schuckertwerke.

ELECTRICAL SHOCK WAVES IN POWER SYSTEMS

ELECTRICAL SHOCK WAVES IN POWER SYSTEMS

Traveling Waves in Lumped and Distributed Circuit Elements

Reinhold Rüdenberg

Late Gordon McKay Professor of Electrical Engineering, *Emeritus*
Harvard University

Harvard University Press
Cambridge, Massachusetts
1968

Translated by Hanns J. Wetzstein from the fourth, enlarged German edition of *Elektrische Wanderwellen: auf Leitungen und in Wicklungen von Starkstromanlagen* (Berlin: Springer, 1962)

This book is a sequel to the earlier work by the same author: *Transient Performance of Electric Power Systems* (New York: McGraw-Hill Book Company, Inc., 1950)

©

Distributed in Great Britain by
Oxford University Press
London

LIBRARY OF CONGRESS CATALOG CARD NUMBER 68–14272
PRINTED IN THE UNITED STATES OF AMERICA

PREFACE

This book on the performance of power lines and windings under shock-wave conditions is an enlargement of Part C of the third edition of *Transient Performance of Electric Power Systems*, which appeared in 1933. It examines the generation and propagation of shock waves in electric power systems. Parts A and B, dealing with switching phenomena in lumped circuits, appeared in a fourth edition in 1953 (in English in 1950). Because this book is thus an extension of the earlier work in the direction of distributed circuit elements and traveling-shock-wave phenomena, the earlier work, *Transient Performance of Electric Power Systems* (New York: McGraw-Hill, 1950), should be considered a companion volume and will be referred to frequently as Ref. 1.

Thus the prefaces and introductions of the earlier works apply also to this volume and are repeated here with some changes.

Since the first edition of this book appeared, in 1923, the transient performance of electric power systems has been recognized as a fundamental aspect of high-voltage technology. This is due to the fact that the transient performance is of equal and sometimes even greater importance in the construction and operation of lines, installations, and machinery than the steady-state performance. In fact, the transients or shock waves connected with short circuits, ground faults, and lightning strokes on lines, cables, and machines have become the governing consideration in the design of modern high-voltage and power systems and their components.

Such transients may be the result of planned or intended switching operations to change system performance or they may be caused accidentally by short-circuit or excess-voltage disturbances of the system.

While Ref. 1 treated mainly the performance of lumped circuit elements, this book extends the treatment to distributed circuit elements such as lines and windings. The distributed parameters and the time taken to propagate electromagnetic energy are given detailed examination. Since the last edition in 1933, considerable work has been done and published. Thus it was necessary to extend the treatment, though this has been done only in the direction of the most important new developments. A number of books on special subjects connected with the general performance, such as the impulse strength of insulation or excess voltages in networks, are listed in the references.

The areas covered, the detailed processes examined, and the method of analysis used are based on my experience in the practice and teaching of electrical engineering in several industrial countries. Many chapters contain in part lecture material that I used over a number of years at the Technical University of Berlin and at Harvard University. This covered electrical machinery, apparatus, and instruments, power transmission and distribution, switching transients, and traveling shock waves. I have tried to select typical examples that allow an insight into the fundamental physical behavior rather than details of a range of practical applications. Problems in the field of communications technology are not covered, although many of the analyses developed for pulse behavior can be considered applicable to it.

I have tried to arrange the material in each section and chapter in a manner appropriate to a teaching text, proceeding from the simpler to the more difficult cases and configurations. To facilitate still further the use of the book as a general reference work, the chapters are presented in fairly self-contained form and the equations and figures are numbered accordingly. Cross references between chapters are kept to a minimum consistent with the interdependence of the phenomena treated.

During my working life of more than thirty years as a practicing electrical engineer, and even longer as a teacher, I have found that students and engineers derive permanent value from the analysis of complicated problems only by means of the simplest possible treatment, which they can absorb and retain in their memory until such time as they encounter an opportunity to use it in their professional work. All treatments of a deeper and more complex nature are better suited to the needs of applied mathematicians and pure research workers than to those of practicing engineers, even when the latter are engaged in scientific research. For engineers, insight into the physical basis of a process and the simplicity of the resulting analysis are of far greater importance than mathematical rigor and generality of method in the solution obtained.

In this book, then, the mathematical method will be regarded merely as a tool. For each case the one most appropriate to it is used. This approach usually leads to the introduction of improved or new arrangements and design guidelines, as is necessary in practice.

Further, the engineer can use his general knowledge of dynamic processes to understand transient processes in lumped and distributed circuits on a broad general basis rather than as a separate subject accessible only to the practicing specialist. Thus almost every problem treated in this book can be grasped without preparation in a specialized mathematical treatment and the reader can proceed at once to a deeper technical discussion of the case under consideration.

In recent years a number of papers and books have appeared that deal with the application of operational and transform methods of calculation to the area of switching, transient, and shock-wave phenomena. These are included in the literature references and the reader who is specially interested will want to use them occasionally for the rigorous treatment of very complex transient phenomena. However, the aim of this book is to enlarge the reader's insight into the technical area of switching processes and transients in electric power systems and hence to stress the basic physical phenomena rather than the mathematical analytic tools.

Numerical examples are used freely to keep the implementation and practical use close to the analysis. They are mostly taken from actual cases and represent typical values encountered in practice. Thus they enable the reader to assess the practical implications of the analytical results with ease. The numerous oscillograms and measurements, largely based on my industrial work, further serve as a handy bridge between theory and practice.

References are not used in the text; instead a list of literature references is given, arranged by chapters and containing a number of important pertinent publications. In view of the large volume of material published in many countries pertaining to transients, this list cannot be considered complete. However, it will serve as a useful guide for the reader who wishes to pursue a particular subject further.

In particular, the first reference listed under Chapter 17, by P. A. Abetti, cites about 1000 publications pertinent to Chapters 17 to 22. This is indicative of the many approaches used in this important area of the transient behavior of windings. All are increasingly converging on the methods used here.

I acknowledge with thanks the help of several industrial concerns and colleagues in the form of illustrations for this book. My wife deserves special thanks for her help in the preparation of the manuscript of the additions and in the proofreading of the entire book. I also thank the publisher for the up-to-date format used for the book.

REINHOLD RÜDENBERG

TRANSLATOR'S PREFACE

As directed by the publisher, the translator has followed the original text of the author and his analytic developments in full detail. This presented only one problem in the use of words, namely, the term "traveling waves," which is a literal translation of the word *Wanderwellen* used in the title and throughout the original text. The translator felt that in technical English the following possible alternatives were available: "traveling wave," "shock wave," "pulse," "impulse," "surge," "wave train," "wave packet." In most places in the text he found that of all of these the word "pulse" is most descriptive; moreover, it is in such general use in other technical areas, such as communications, that it makes the work more accessible and useful to a larger readership. At the same time the concept of shock waves is most appropriate to the aim of the author in presenting the entire work pertaining to the proper strength in design of power systems when stressed by such shock waves. For this reason this term was used in the English title. Where appropriate or expedient in the text, one or more of these words are used if they serve either to identify the treatment with established areas or to indicate to the reader the similarity of concept with an established area. By this slight variation and variability the translator hopes to have made the author's excellent simple treatment of complex lumped and distributed structures accessible to the largest possible readership without deviating in any way from the rigor or practical application intended and achieved by the author.

It can then be hoped that, because of its physical insight, this book will be found useful when electrical or other shock-wave phenomena need to be analyzed in lumped or distributed structures in other technological areas where energy transfer between separate elements will become increasingly a problem as units of larger power ratings are built and interconnected, as was the case in electric power networks.

Some of these areas are:

Electrical distribution systems (lines and cables) of even higher voltage and power rating;

The use of semiconductor devices in such systems for control, switching, rectification, and inversion;

The introduction of magnetohydrodynamic power generators of high power rating (possibly in pulse form also);

Superconducting magnets for various applications (involving considerable energy storage);

Electrically pulsed excitation of laser and maser devices emitting electromagnetic radiation;

Plasma devices to explore and perhaps later to exploit fusion phenomena.

While most of these applications will involve configurations substantially different from those treated in this book, the generalized simple approach developed is likely to be of considerable use in these areas as they enter the realm of practical engineering. In the areas of particle-accelerator power engineering and of production of high temperatures by exploding wires, the translator is aware of some cases in which the author's work in this book and in Ref. 1 was directly applicable.

It will probably be many years before fundamental texts of the scope and general simplicity of this book become available for these emerging technologies, which treat the transient phenomena connected with the propagation of shock waves in the energy-transfer systems including the many novel energy-transformation devices in the process of evolution. In the meantime this book, because of its simplicity and elegance of treatment, may well serve as a valuable case-reference book.

As a former graduate student of the author's, the translator can also attest to the fact that he has been able to use to great advantage the insight gained from the study of power-system cases in many almost entirely different areas. In preparing the translation he had available, by courtesy of the author's widow, most of the author's original material, including papers published long before 1933. The translator found that these early papers contained an approach that was sufficiently basic to allow the author to add refinements over a number of decades as these became necessary owing to system demand and possible through better instrumental observation.

The translator also wishes to place on record the wonderful experience it was, during the author's last teaching years, to listen to his lectures, which were presented always with a strong dynamism. Surely none of his students will ever forget the swinging of the ax with which the workman short-circuits the cable and which evaporates in the arc. But beyond this the author also imbued his students with the social and economic importance of engineering to harness the forces available from nature productively. To the translator he also communicated on a few occasions his concern over the concentration of government support in the technical and scientific faculties of universities, as contrasted with the lack of such support in the social, medical, and humanistic faculties.

The translator felt it appropriate to make these few personal observations as a mark of gratitude and respect for the author which may be of interest to the reader.

Cochituate, Massachusetts   HANNS J. WETZSTEIN

CONTENTS

CHARTS

TABLES

LETTER SYMBOLS MOST FREQUENTLY USED

LATIN AND GERMAN SYMBOLS

A	work done; width; current density
a	spacing; distance; width; length
B	amplitude
B, 𝕭	magnetic induction
b	distance; breadth
C, c	capacitance
D, d	distance; thickness; diameter
E	voltage; amplitude of voltage; electromotive force
𝕰	electric field strength; voltage of adjacent conductor
e, 𝖊	instantaneous voltage
$_e$	excess voltage (subscript)
f	frequency (cycles per second)
G, g	mutual electrostatic coupling
g	distance to ground
H	magnetic field strength; pulse-front length
h	height; thickness
I	current amplitude
𝕴, 𝖎	displacement current in windings
i	instantaneous current; current density; insulation thickness
j	$\sqrt{-1}$, imaginary unit
k	height; length; pulse-front length
L, l	self-inductance; loss
l	length
M, m	mutual inductance
m	number of pulses or waves
N	number of turns; number of elements
n	number of order, mode; number of conductors
P, p	electrostatic potential; insulation resistance; parallel resistance
Q	electric charge
q	electrostatic self-influence
R, r	ohmic resistance; radius
r	distance; ratio
$_r$	retrograde, rearward (subscript)
S	conductor spacing; stray or leakage inductance of windings
s	distance; length; specific resistance or resistivity; slope or pitch of windings
T	time constant
t	time
V	heat loss
v	velocity; velocity of light
$_v$	forward (subscript)
W	work; power; energy
w	velocity of pulse front; winding length
X	reactance; spatial decay constant
x	length coordinate; distance
y	length coordinate; distance
Z, 𝖅	surge impedance
z	mutual surge impedance

GREEK SYMBOLS

α	exponential coefficients; density of waves
β	refraction coefficient; exponential coefficient; angle subtended
γ	winding capacitance
Δ, δ	difference
δ	length; exponent of damping; diameter
ε	dielectric constant; protective effect; 2.718, base of natural logarithms
η	shielding effect; efficiency
ζ	capacitance to line
θ	damping constant
κ	electrostatic self-influence
Λ, λ	length; wavelength
λ	self-inductance
μ	permeability; mutual inductance; wave number
ν	natural frequency; critical frequency
ξ	length
ρ	radius; distance; damping coefficient; reflection coefficient; resistance
τ	duration; period; time constant
ϕ	magnetic flux
ω	angular frequency

ELECTRICAL SHOCK WAVES IN POWER SYSTEMS

INTRODUCTION

In the early years of electrical engineering, power systems were designed according to the requirements of regular sustained operation. By thorough investigation of the materials used and by study of the properties of the machines, apparatus, and lines in steady-state operation, remarkable success was attained with respect to the magnitude of the power, the level of the voltage, and the distance over which the electrical energy was transmitted. However, experience showed that with switching processes in the circuits and under similar intentional or accidental conditions peculiar phenomena appeared which could greatly disturb the regular operation of the system. Since then numerous investigations have endeavored to clarify scientifically these phenomena and to devise methods to prevent damage in the networks.

Hence, in the operation of every electrical system, we must distinguish between the stationary phenomena existing in the steady state and the transient performance resulting from any switching process or a similar change in the circuit. Mostly, the latter initiate disturbances created by the charge or discharge of the local energies connected with every electrical circuit. The increasing magnitude and density of the energy in all parts of modern power systems cause the electrical and mechanical transient phenomena to play an ever-increasing role. Nowadays in every electrical system the control of these transient phenomena is equal in importance to the command of all the phenomena of the stationary state.

In addition to intentional switching processes, consisting of the correct manipulation of the controlling apparatus, there frequently occur, particularly in high-voltage and long-distance systems, unintentional transient phenomena; these may consist of ground faults, short circuits, conductor breaks, lightning strokes, or even erroneous operation of circuit breakers and similar inadvertent processes. Such accidents nearly always lead to severe disturbances throughout the electrical system, giving rise to high excess currents or voltages and sometimes to currents of substantially different frequency or entirely distorted wave shape. Related to such phenomena are disturbances produced in the steady state by higher harmonics or by resonances in certain parts of the circuits, both causing parasitic oscillations in the network.

The voltage or current in connecting lines or the speed of machines connected to them will be changed by any switching process, and at the same time the energy linked with the circuit will also vary. If this energy is concentrated at a certain place, for example in the magnetic field of a generator, in the electric field of a condenser, or in the inertial mass of a motor armature, the change in or the transition of the amounts of energy after the switching will take place simultaneously in the entire circuit. In general such transients decay relatively slowly, within times of the order of magnitude of 1 sec. In this case the phenomena are said to be slow, or quasi-stationary, since voltage and current are distributed over the conductors in a manner very similar to the steady state.

However, it is well known that all electrical phenomena actually spread out with finite, though very high, speed, namely with the velocity of propagation of light, which in a vacuum is 300,000 km/sec. Electric currents in wire conductors also cannot exceed this velocity. Hence a change of current or voltage after a switching process will not occur exactly simultaneously in all parts of the circuit. Rather, voltage and current spread out in the form of traveling waves of enormous velocity from the switching place over the entire network. Thus the change of the electric and magnetic fields existing at any point of the conductors occurs with extreme rapidity. For the most part, these phenomena have entirely decayed when the slow transient process mentioned before just comes into play. Nevertheless, severe disturbances may be

1

caused by the wavelike distribution of voltage and current. In this case the phenomena are designated rapid transients produced by traveling waves (or pulses or shock waves) on the conductors. The behavior of such traveling waves on open lines or cables and in the windings of machines and transformers is treated in the present volume.

We shall attempt to give a survey of typical behavior of traveling waves in high-voltage power systems. As far as possible we shall select the most essential aspects of this behavior, using simple and easily comprehended physical relations. From these simple examples we shall extrapolate conclusions to other and more complicated cases that need to be treated in practice. However, we can neither treat all cases for which solutions are known nor obtain all the details of the sometimes extremely complex phenomena that occur in practice. By carefully chosen approximations we shall be able to put the analytic results into easily usable form. The practical examples and the interspersed oscillograms will serve to show the utility of the form and will also give a feel for representative values of the parameters over a range of practical interest.

The fundamental regularity of our solutions will lead to many design rules and constraints for the construction of high-current and high-voltage installations and their components. However, we do not here have space to treat conventional and purely structural details of design.

In the actual treatment of the various transient phenomena, we must use the mathematical method, since this alone enables us to describe clearly and concisely complex interrelations among numerous quantities. We shall attempt to use the simplest possible mathematical tools compatible with a particular problem; these are essentially the elements of the differential and integral calculus. This discipline is unavoidable, since transient processes consist of phenomena that vary with time and such changes can be treated exactly only by the calculus. In a few chapters dealing with more intricate phenomena, particularly those due to spatial as well as temporal changes, the elements of partial differential equations and finite-difference equations are used. Still higher mathematical disciplines

need not necessarily be applied in order to classify most of the transient and traveling-wave phenomena that occur in practice in power networks.

As a basis for the analysis of the transient behavior of a dc or an ac system it is necessary to assume that its steady-state performance is known. Similarly, some of the results pertaining to lumped circuit elements developed in the earlier book, Ref. 1, will be used here and specific reference will be made to them where appropriate.

In the investigation of phenomena of oscillation as described by sine and cosine functions, the formal work of calculation is greatly simplified by the use of complex quantities. Designating the imaginary unit by

$$j = \sqrt{-1} \qquad (1)$$

and the base of natural logarithms by

$$\varepsilon = 2.718, \qquad (2)$$

we have the following mathematical relation:

$$\varepsilon^{j\omega t} = \cos \omega t + j \sin \omega t. \qquad (3)$$

Differentiation and integration of the exponential function in the left-hand member of Eq. (3) are very much simpler than the same operations on the right-hand member, since the exponential function remains unchanged by these operations. Therefore, it is useful to express an alternating current of frequency ω, changing in accordance with a cosine or sine function of the time t, not as

$$i = I \cos \omega t \quad \text{or} \quad i = I \sin \omega t \qquad (4)$$

but rather as

$$i = I\varepsilon^{j\omega t}. \qquad (5)$$

For by Eq. (3), the last expression contains in summary the two functions of Eq. (4) and thus will reduce every formal calculation so greatly that it often shrinks to a few lines. In the final result, the exponential quantities must be resolved by Eq. (3) into their real and imaginary components. Then the real part yields the cosine component and the imaginary part the sine component of the solution. We call these exponential functions, as well as the cosine and sine functions, harmonic functions.

Though this complex notation is already very useful for the simple solution of the wave equation, for example, its greatest utility is seen in the treatment of more difficult problems such as the interaction of oscillatory and traveling-wave phenomena in the windings of machines and transformers.

The temporally and spatially changing processes in our electric circuits can be described either as stationary oscillations or as traveling waves. Both viewpoints are physically correct and mathematically equivalent, as shown in Chapter 1, Sec. (*d*), owing to the equivalence of certain elementary trigonometric functions. However, when it comes to utility in actual analysis, the two viewpoints differ considerably in merit. Stationary oscilla- tions lead to solutions in the form of slowly converging series. Traveling waves (or pulses), on the other hand, lead in most cases of practical interest to solutions in almost closed form, so that the desired physical and technical insight can be derived from the analysis. Thus, this viewpoint is strongly favored in this book.

The units used in all the derivations and examples are taken either from the meter-kilogram-second (mks) system, or from the centimeter-gram-second (cgs) system, and are electromagnetic or electrostatic as best suits the special case. In this way great simplicity is attained in all the numerical calculations.

PART I

HOMOGENEOUS LINES

CHAPTER 1

LAWS OF PROPAGATION OF TRAVELING WAVES

All electrical phenomena propagate in free space or along the surfaces of metallic conductors with the high but finite velocity of light, about 300,000 km/sec. Thus to traverse a distance of 2,500 km an electric wave requires 1/120 sec. This is a half period of the usual 60-cy/sec alternating-current frequency. The voltage at the far end of the line will then lag in phase 180° behind that at the near end. In cables with oil-impregnated paper insulation the velocity of propagation is only about half the value in air, owing to the higher dielectric constant.

TABLE 1.1. Phase angle for 60-cy/sec a.c. lines.

Distance traveled (km)		Phase difference	
Open line	Cable	Degrees	Period
5000	2500	360	1
2500	1250	180	1/2
1250	625	90	1/4
834	416	60	1/6
416	208	30	1/12

Table 1.1 shows phase differences for lines of different lengths. From this table it is seen that the assumption of constant phase applies only along overhead lines considerably shorter than 400 km or cables considerably shorter than 200 km. Such short lines can be treated as quasi-stationary circuits.

For longer transmission lines, as shown in Table 1.1, the finite velocity of propagation produces effects that disturb the normal performance of the lines as well as of machines and other equipment connected to them. In particular, lengths of lines giving rise to phase differences of 90° or multiples thereof can produce very dangerous resonance phenomena.

Changes of state imposed on the circuit by switching action also propagate along the lines with the velocity of light.

Abrupt changes such as the switching of voltage onto a line will thus produce a voltage at the far end only after a definite time delay. Figure 1.1 shows the example of an overhead line grounded at both ends. This grounding is not sufficient protection at points along the line against an incident atmospheric discharge, which can produce a high potential with respect to the ground before it propagates

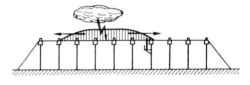

Fig. 1.1

with the velocity of light to discharge itself at the ends.

In windings of generators, motors, transformers, and other pieces of equipment, high voltages can be produced between some conducting surfaces owing to their proximity and the finite propagation time required within the complex structures. Such voltages, though of short duration, may be far above the normal steady-state operating ones and thus can cause failures in the insulation.

All these phenomena are clearly due to the finite velocity of propagation.

(a) *Propagation of current and voltage.* The usual case in practice is a circuit in which the

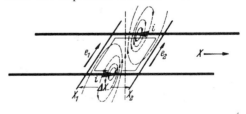

Fig. 1.2

conductors of current to and from the load are closely spaced and parallel. If at first we neglect the losses in the conductors and in the insulation between them, the distribution of current and voltage will be determined by the inductive and capacitive laws applicable.

7

Figure 1.2 shows a line element of length Δx and a rectangle formed by the two currents i and voltages e_1 and e_2. This rectangle encloses a part of the magnetic flux surrounding the line. The change with time of this flux produces a voltage in the rectangle and thus an increase in voltage along the line given by the equation

$$e_1 - e_2 = -\Delta e = l\Delta x \frac{\partial i}{\partial t}, \qquad (1.1)$$

in which l is the self-inductance of the line per unit length.

Equation (1.1) leads directly to the differential form:

$$\frac{\partial e}{\partial x} = -l \frac{\partial i}{\partial t}. \qquad (1.2)$$

A change with time of the voltage e between the two conductors shown in Fig. 1.3 produces

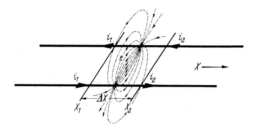

Fig. 1.3

a displacement current proportional to the rate of change of voltage with respect to time. Thus in the given rectangle a part of the entering current flows as a displacement current across to the opposite conductor. Hence in the element Δx of the line there is an increase of current given by

$$i_1 - i_2 = -\Delta i = c\Delta x \frac{\partial e}{\partial t}, \qquad (1.3)$$

in which c is the capacitance per unit length of line.

Equation (1.3) leads directly to the differential form:

$$\frac{\partial i}{\partial x} = -c \frac{\partial e}{\partial t}. \qquad (1.4)$$

Equations (1.2) and (1.4) show that a change of current with time produces a change of voltage with position along the line and that a change of voltage with time produces a change of current with position

along the line. The changes in position along the line are proportional to the changes with time and to the self-inductance and capacitance per unit length of line.

We wish to examine the currents and voltages along the line in the transient modes caused by switching or other change-of-state processes. The distribution in space and time of these transient currents and voltages must satisfy Eqs. (1.2) and (1.4) and the boundary conditions of a given problem.

To solve these equations we simply substitute, by analogy with Ohm's law,

$$e = Zi, \qquad (1.5)$$

in which Z is a constant. To see if this relation between voltage and current satisfies Eqs. (1.2) and (1.4) we obtain by insertion of (1.5) in (1.2)

$$Z \frac{\partial i}{\partial x} = -l \frac{\partial i}{\partial t} \qquad (1.6)$$

and in (1.4)

$$\frac{\partial i}{\partial x} = -cZ \frac{\partial i}{\partial t}. \qquad (1.7)$$

Upon division of (1.6) by (1.7) the time and space variations drop out and we obtain

$$Z = \frac{l}{cZ}. \qquad (1.8)$$

Thus the trial solution (1.5) proves valid if Z is given by

$$Z = \pm \sqrt{\frac{l}{c}}. \qquad (1.9)$$

By analogy with the direct-current resistance, let us here call Z the surge impedance of the line, which is also measured in ohms. Equation (1.5) as the solution then shows that, regardless of time and space variations, the current and voltage are always related simply by

$$e = \pm \sqrt{\frac{l}{c}} i, \qquad (1.10)$$

Thus the shape of voltage pulses is always the same as that of current pulses.

Whereas in direct-current circuits a given voltage can produce only a single current, we see that in the transient mode there are two solutions, corresponding to the plus or minus sign in Eq. (1.10).

Take a voltage-pulse distribution between two conductors as shown in Fig. 1.4; by Eq. (1.10), this can correspond to a positive current wave or pulse as shown in Fig. 1.4*a* or to a negative current wave or pulse as shown in Fig. 1.4*b*. Both solutions satisfy the differential equations and can thus appear. They can be distinguished when the progress in time and space of these pulses is taken into account.

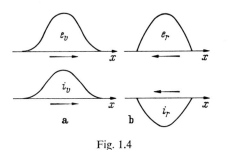

Fig. 1.4

From Eq. (1.6) we obtain the change of current with time by insertion of the surge impedance given by Eq. (1.9) as

$$-\frac{\partial i}{\partial t} = \frac{Z}{l}\frac{\partial i}{\partial x} = \pm\frac{1}{\sqrt{lc}}\frac{\partial i}{\partial x}. \qquad (1.11)$$

At any point along the line the change of current with time is thus proportional to the spatial gradient at that point. An increase of current along the *x*-axis at any point in Fig. 1.5

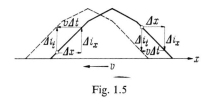

Fig. 1.5

thus results in a proportional increase with time in current at this same point. For a decrease in current along *x* there is correspondingly a proportional decrease with time in current. Thus after a short time Δt every point of the current-pulse envelope curve is displaced slightly toward the left with a velocity given by the quotient of the space and time elements and with a negative sign due to the direction of progress along the *x*-axis. Thus

$$-v = \frac{\Delta x}{\Delta t} = \frac{1/\Delta t}{1/\Delta x} = \frac{\partial i/\partial t}{\partial i/\partial x}. \qquad (1.12)$$

The last step in Eq. (1.12) simply substitutes the quotient of the time and space differentials of current for the quotient of the space and time elements, so that from Eq. (1.11) we obtain

$$v = \pm\frac{1}{\sqrt{lc}}. \qquad (1.13)$$

In this equation, *v* is the velocity with which all points in a current-pulse envelope propagate along the line. Because voltage is proportional to current, this velocity applies also to voltage-pulse envelopes. The velocity *v* is independent of the magnitude of current and voltage and is determined, according to Eq. (1.13), only by the constants *l* and *c* of the line. Thus all points on the current- and voltage-pulse envelopes move with the same velocity and the spatial shape of these pulse envelopes is propagated without distortion.

The plus and minus signs in Eq. (1.10) lead via Eq. (1.13) to two velocities of propagation that are equal but in opposite directions. When voltage- and current-pulse envelopes are of the same sign, as shown in Fig. 1.4*a*, this corresponds to the plus sign in Eq. (1.13) and hence to propagation in the positive direction along the *x*-axis. This will be regarded as the forward or advancing (*vowärts*) direction of propagation and is indicated by the subscript *v* in Fig. 1.4*a*. When voltage- and current-pulse envelopes are of opposite sign, as shown in Fig. 1.4*b*, this corresponds to the minus sign in Eq. (1.13) and hence to propagation in the negative direction along the *x*-axis. This will be regarded as the rearward or retrograde (*rückwärts*) direction of propagation and is indicated by the subscript *r* in Fig. 1.4*b*. The shape of the voltage- or current-pulse envelope is in no way restricted except for the proportionality between them imposed by Eq. (1.10).

Both the advancing and retrograde pulses (pulse envelopes or waves) have the absolute velocity

$$v = \frac{1}{\sqrt{lc}}. \qquad (1.14)$$

The direction is distinguished by the subscripts *v* and *r* so that in Eq. (1.9) we can omit the signs and simply obtain the surge impedance as

$$Z = \sqrt{\frac{l}{c}} \cdot \tag{1.15}$$

The forward or advancing pulse or wave is then

$$e_v = + Z i_v, \tag{1.16}$$

and the backward or retrograde pulse or wave is

$$e_r = - Z i_r. \tag{1.17}$$

The shape of the current- or voltage-pulse envelope or wave is still to be determined from the boundary conditions of a given problem. The surge impedance is determined only by the characteristic inductance and capacitance of the line, as shown in Eq. (1.15), and thus is sometimes called the characteristic or surge impedance of the line.

The voltages of the pulses traveling in the two directions are proportional only to their current counterparts traveling in the same direction and are independent of their current counterparts traveling in the opposite direction. Thus they can occur in simple superposition, leading to a total voltage on the line

$$e = e_v + e_r. \tag{1.18}$$

Similarly, the two current pulses give by superposition the total current on the line

$$i = i_v + i_r. \tag{1.19}$$

Thus the electrical state of the line can always be represented by the superposition of these two pulse systems traveling in opposite directions.

For lines with conductors of diameter 8 mm (hence a cross-sectional area of 50 mm²) and heights above the ground that are customary in practice, the values in Table 1.2 are observed. From this table and Eq. (1.15) the surge impedance of an air-insulated line

TABLE 1.2. Line constants.

Construction	Self-inductance (mH/km)	Capacitance (μF/km)
Open line	1.67	0.0067
Cable	0.33	0.133

is thus about

$$Z = \sqrt{\frac{1.67 \cdot 10^{-3}}{0.0067 \cdot 10^{-6}}} = 500 \ \Omega$$

and the velocity of pulse propagation from Eq. (1.14) is about

$$v = \frac{1}{\sqrt{1.67 \cdot 10^{-3} \cdot 0.0067 \cdot 10^{-6}}}$$
$$= 300,000 \text{ km/sec.}$$

For an oil-impregnated cable a typical value of surge impedance is

$$Z = \sqrt{\frac{0.33 \cdot 10^{-3}}{0.133 \cdot 10^{-6}}} = 50 \ \Omega$$

and of velocity of pulse propagation,

$$v = \frac{1}{\sqrt{0.33 \cdot 10^{-3} \cdot 0.133 \cdot 10^{-6}}}$$
$$= 150,000 \text{ km/sec.}$$

Any line can be characterized by a pair of numbers, either the inductance l and capacitance c per unit length or the surge impedance Z and velocity of propagation v. The first pair vary considerably with changes in dimensions of the line. In contrast to this, v is a specific constant of the insulating material used and thus has for air or oil-impregnated paper the fixed values just derived. Deviations in other insulating materials are small and easily obtained. The surge impedance Z is dependent on dimensions of the line but in a rather moderate way so that it stays close to the order of magnitude of the values derived above for lines and cables of different dimensional proportions.

The relations between the characteristic constants in Eqs. (1.14) and (1.15) can be expressed very simply:

$$lv = Z \tag{1.20}$$

and

$$cv = \frac{1}{Z}. \tag{1.21}$$

Hence l and c can be easily calculated from a knowledge of Z and v. As v is usually well known as a basic physical constant, it is possible to determine Z from a single measurement of l or c on the actual line.

(b) *Energy and damping.* A traveling pulse has an energy content A that is the sum of the electrostatic and electromagnetic energies stored. The electrostatic component is given by the voltage and capacitance per unit length as

$$a_s = \tfrac{1}{2}ce^2. \qquad (1.22)$$

The electromagnetic component is given by the current and inductance per unit length as

$$a_m = \tfrac{1}{2}li^2. \qquad (1.23)$$

As Eq. (1.10) gives the relation

$$\frac{e^2}{i^2} = \frac{l}{c} = Z^2 \qquad (1.24)$$

the two components of energy storage are equal and the total energy stored per unit length is

$$a = 2a_s = 2a_m. \qquad (1.25)$$

The total energy content is then obtained by integration over the length of the line:

$$A = \int a\,dx = c\int e^2 dx = l\int i^2 dx. \qquad (1.26)$$

Thus A is dependent on the shape of the current and voltage pulses and can be calculated from Fig. 1.6.

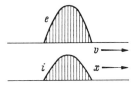

Fig. 1.6

The power of the pulses that pass along the line through a given cross section is equal to the product of energy content and velocity of propagation:

$$W = av = i^2 l \frac{1}{\sqrt{lc}} = i^2 Z = \frac{e^2}{Z}. \qquad (1.27)$$

For equal voltages the power is thus far higher in cables than in open lines, owing to the smaller surge impedance of the former. All of these relations are applied separately to the pulses traveling in the forward and backward directions.

In an open line with $Z = 500\ \Omega$ a pulse of $e = 100,000$ V has a power

$$W = \frac{100,000^2}{500} \cdot 10^{-3} = 20,000 \text{ kW}.$$

By Eq. (1.16) the current is

$$i = \frac{100,000}{500} = 200 \text{ A}.$$

In a cable with $Z = 50\ \Omega$ a pulse of $e = 10,000$ V already develops a power

$$W = \frac{10,000^2}{50} \cdot 10^{-3} = 2,000 \text{ kW},$$

while the current is still

$$i = \frac{10,000}{50} = 200 \text{ A}.$$

These large powers accompanying traveling pulses can produce strong effects in high-voltage networks. These effects are limited only by the high velocity of propagation, which limits the duration of the effects at any given point.

So far the ohmic resistance of the conductors and insulation in the lines has been neglected. The power W contained in the pulse is then propagated along the line without loss or distortion in shape. However, in real lines the resistance losses produce a power loss in the pulse which is dissipated as heat in both the conductor resistance r (per unit length) and insulation resistance p (per unit length). The total heat loss for an element dx of line is

$$L = i^2 r\, dx + \frac{e^2}{p}\, dx \qquad (1.28)$$

or by substitution from Eq. (1.24)

$$L = i^2 \left(r + \frac{Z^2}{p}\right) dx. \qquad (1.29)$$

According to Eq. (1.27), the pulse power entering the line element is, for the forward direction of pulse travel,

$$W = i^2 Z. \qquad (1.30)$$

However, this is valid only if the resistances are so small that the relation between voltage and current, shown by Eq. (1.24) as directly dependent on inductance and capacitance, is not significantly altered.

Differentiation of Eq. (1.30) gives the power loss of the pulse for each element of line,

$$-dW = -2i\, di\, Z. \qquad (1.31)$$

Equating this to the heat losses given by (1.29) gives

$$i^2 \left(r + \frac{Z^2}{p}\right) dx = -2Zi\, di. \qquad (1.32)$$

This equation contains only the variables

i and x, which when separated give the differential equation

$$\frac{di}{i} = -\frac{1}{2}\left(\frac{r}{Z} + \frac{Z}{p}\right) dx. \qquad (1.33)$$

By integration this gives the change in current along the line,

$$i = i_0 \exp\left[-\frac{1}{2}\left(\frac{r}{Z} + \frac{Z}{p}\right)x\right]. \qquad (1.34)$$

Here i_0 is the amplitude of the current pulse at any arbitrarily selected origin or starting point. For the voltage pulse the corresponding change along the line is

$$e = e_0 \exp\left[-\frac{1}{2}\left(\frac{r}{Z} + \frac{Z}{p}\right)x\right]. \qquad (1.35)$$

Thus the resistive losses in the conductors and insulation of the line produce an exponential damping or attenuation of both the current and the voltage amplitudes of the pulse. This is shown in Fig. 1.7 for a pulse

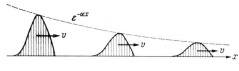

Fig. 1.7

traveling in the forward direction. The exponential coefficient of attenuation contains only the specific values of surge impedance, resistance of conductors, and resistance of insulation. As these are all constants of the line, attenuation is independent of the pulse amplitude and is determined only by the distance traveled by the pulse. The pulses thus retain their shape without distortion, only suffering a gradual attenuation in amplitude. The resistances do not affect the velocity of propagation of the pulses.

High-voltage lines usually have a very high value of p; hence insulation resistance losses can be neglected in comparison with resistance losses in the conductors. Thus it is almost always sufficient to consider only the first term in the attenuation coefficient as given in Eqs. (1.34) and (1.35) and to put

$$\frac{e}{e_0} = \frac{i}{i_0} = \varepsilon^{-rx/2Z}. \qquad (1.36)$$

This shows that damping is directly proportional to the conductor resistance and

inversely proportional to the surge impedance of the line. Thus it is much higher in cables than in open-air lines.

For the previously considered lines of 50-mm^2 cross-sectional area (Table 1.2), the direct-current resistance for the two-conductor loop r as measured was 0.7 Ω/km. From the previously obtained values of surge impedance it is seen that the amplitudes of the current and voltage pulses are reduced to half their value in an open-air line of length 1,000 km, as shown by

$$\varepsilon^{-0.7 \cdot 1000/2 \cdot 500} = \varepsilon^{-0.7} = 0.5,$$

and in a cable of length 100 km, as shown by

$$\varepsilon^{-0.7 \cdot 100/2 \cdot 50} = \varepsilon^{-0.7} = 0.5.$$

However, for the rapidly propagating pulses considered here, the resistance of the conductors is increased considerably above the direct-current values measured, owing to the skin effect. This increase of resistance due to the skin effect for arbitrary pulse shapes treated here may lead to a severalfold increase in the conductor resistance.

Thus the actual lengths of lines that will produce amplitude reductions to half values may be only fractions of the values just derived. However, even a tenfold increase in conductor resistance due to skin effect would still produce a halving of amplitudes in open-air lines only 100 km long and in cables 10 km long. Investigations of pulse-travel phenomena often deal with shorter lengths of lines. In such investigations it is then possible to neglect in a first approximation the effect of conductor resistance on the initial development of pulse phenomena. Longer lines, however, produce a considerable attenuation of the traveling pulses.

If one of the line conductors is replaced by a ground return, then of course the resistance of the remaining conductor and the earth surface layers involved govern the damping. In the ground the pulse current occupies a rather wide zone but only a thin surface layer, so that the ground-return resistance is usually several times that of the conductor, since the specific resistance of the surface soil involved is usually rather high.

Because of the high velocity of propagation, pulses traverse transmission lines having lengths in general use in very short times. Let us calculate the time taken by the pulse to be completely damped out. Substitute in Eq. (1.36) the time t required to travel the distance x according to the relation

$$x = vt. \qquad (1.37)$$

Using only damping due to line resistance and Eq. (1.20) we obtain

$$\frac{e}{e_0} = \frac{i}{i_0} = \varepsilon^{-rvt/2Z} = \varepsilon^{-(r/2l)t}. \qquad (1.38)$$

The exponent contains the quotient of the line resistance to the self-inductance, which is the reciprocal of the time constant of the line. Damping of pulses as a function of time is thus governed by the general relations applicable to transient processes in lumped oscillatory circuits, as developed, for example, in Ref. 1, Chapter 5.

In an interval equal to three time constants the pulses (or other phenomena) are damped to 5 percent of their initial value. If we again assume a tenfold increase in the dc resistance of the line previously examined, so that $r = 7\,\Omega/\text{km}$ for traveling pulses, and the line constants in Table 1.2, then the pulses are essentially damped out in open-air lines in

$$3\frac{2l}{r} = 3\frac{2 \cdot 1.67 \cdot 10^{-3}}{7} = 1.43 \cdot 10^{-3}\ \text{sec}$$

and in cables in

$$3\frac{2l}{r} = 3\frac{2 \cdot 0.33 \cdot 10^{-3}}{7} = 0.28 \cdot 10^{-3}\ \text{sec}.$$

This shows that all phenomena caused by the propagation of pulses occur only during a very short time interval, of the order of 1 msec, following a switching process. Thus they are essentially damped out in less than a half period (about 8.3 msec) of the 60-cy/sec power systems.

To obtain a simple picture of traveling-pulse phenomena, we shall from now on neglect the effect of resistance.

We can, however, always introduce it in retrospect by exponential damping according to the values just calculated.

(c) *Surge impedances.* As the surge impedance of lines plays a dominant role in our evaluations, it is important to determine its magnitude for a few representative line-conductor arrangements in general use.

For a single conductor of diameter d and height above ground h (Fig. 1.8), the capacitance per unit length, as derived in Ref. 1, p. 377, Eq. (30), is

$$c = \frac{1}{2v^2 \ln (4h/d)}. \qquad (1.39)$$

For the self-inductance we must observe that the pulse currents travel only in the top layer of the ground. To do so, Eq. (48), p. 407, of Ref. 1 was developed, and gives us here

$$l = 2 \ln \left(\frac{4h}{d}\right) + \frac{\sqrt{s/f}}{\sqrt{2\pi h}}. \qquad (1.40)$$

The second term of Eq. (1.40) is due to the magnetic field of the earth currents. Because it is only a few percent of the value of the

Fig. 1.8

first term for the fast-traveling pulses, we shall neglect it here entirely. Then we obtain for the surge impedance

$$Z' = \sqrt{\frac{l}{c}} = 2v \ln \left(\frac{4h}{d}\right) = 60 \ln \left(\frac{4h}{d}\right)\ \Omega. \qquad (1.41)$$

In the last of these equations a factor of 10^{-9} has been introduced to make possible practical calculations directly in ohms.

For a conductor of 8-mm diameter and 10-m average height above the ground this gives

$$Z' = 60 \ln \left(\frac{4 \cdot 10^3}{0.8}\right) = 60 \cdot 8.5 = 510\ \Omega.$$

This value of about 500 Ω will be used frequently in examples.

Pulses traveling between two conductors (Fig. 1.9) are subject to a smaller inter-conductor capacitance. We can deduce this capacitance from the previous example if we introduce a plane of symmetry between the two conductors as shown in Fig. 1.9 and neglect the small influence (end effect) of the actual ground plane. We then substitute the distance $\frac{1}{2}s$ for the height h in Eq. (1.39) and

Fig. 1.9

divide by 2 on account of the series connection of the two capacitances. We obtain

$$c = \frac{1}{4v^2 \ln (2s/d)}. \qquad (1.42)$$

The surge impedance is now simply given by Eq. (1.21) as

$$Z'' = \frac{1}{cv} = 4v \ln \left(\frac{2s}{d}\right) = 120 \ln \left(\frac{2s}{d}\right) \Omega.$$
$$(1.43)$$

For the 8-mm conductor with 2-m spacing s, this gives

$$Z'' = 120 \ln \left(\frac{2 \cdot 200}{0.8}\right) = 120 \cdot 6.2 = 750 \ \Omega.$$

The three-phase line arrangement is shown in Fig. 1.10. If the pulses travel along one

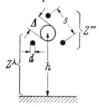

Fig. 1.10

conductor for forward current and the other two as returns, it is possible to show that the capacitance increases by one-third by superposition of the line potential fields. The surge impedance is thus three-quarters of that in Eq. (1.43); hence

$$Z''' = 3v \ln \left(\frac{2s}{d}\right) = 90 \ln \left(\frac{2s}{d}\right) \Omega, \qquad (1.44)$$

and in our numerical example $Z''' = 560 \ \Omega$.

All three conductors together, however, have a higher capacitance to ground, given by an equivalent conductor of diameter Δ shown in Fig. 1.10. In Ref. 1, p. 390, Eq. (24), Δ is derived as

$$\Delta = \sqrt[3]{4ds^2} = d \sqrt[3]{4 \left(\frac{s}{d}\right)^2}. \qquad (1.45)$$

For our example this gives

$$\Delta = 0.8 \sqrt[3]{4 \left(\frac{200}{0.8}\right)^2} = 50.5 \text{ cm}.$$

The surge impedance of the three-conductor line to ground is thus, according to Eq. (1.41),

$$Z^\lambda = 2v \ln \left(\frac{4h}{\Delta}\right) = 60 \ln \left(\frac{4h}{\Delta}\right) \Omega. \qquad (1.46)$$

For our example this gives

$$Z^\lambda = 60 \ln \left(\frac{4 \cdot 10^3}{50.5}\right) = 60 \cdot 4.38 = 260 \ \Omega,$$

which is only about one-half the value for the single-conductor line.

(*d*) *Other methods of solution.* It is possible to derive the relations governing the propagation of pulses along double-conductor lines in a substantially different manner for use in calculations. We wish to assemble here briefly these other, more complicated, forms of solution in order to have them available later for occasional use. If we differentiate Eq. (1.2) with respect to x and Eq. (1.4) with respect to t, multiply the latter by l, and subtract it from the former, we obtain the partial differential equation

$$\frac{\partial^2 e}{\partial x^2} = lc \frac{\partial^2 e}{\partial t^2}. \qquad (1.47)$$

This wave equation for the voltage e has an equivalent form for the current i.

A solution of this equation is given by the trial substitution

$$e = E \sin vt \sin \mu x. \qquad (1.48)$$

By repeated differentiation and substitution in Eq. (1.47) we obtain the relation

$$\mu = \pm \sqrt{lcv} \qquad (1.49)$$

between the frequency v in 2π sec and the wave number μ along 2π cm. Amplitude and frequency are still available as arbitrary constants. Figure 1.11 shows this Bernoulli solution, which exhibits a standing wave.

This form of solution is useful if one wishes to satisfy some local boundary conditions

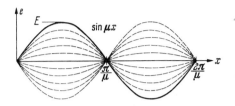

Fig. 1.11

along the line. For example, according to Eq. (1.48)

$$e = 0 \quad \text{for} \quad \mu x = 0, \pi, 2\pi, 3\pi, \ldots$$

This leads to nodal points along the line and in turn, from Eq. (1.49) to particular frequencies

$$v = \frac{\pi}{x\sqrt{lc}}, \frac{2\pi}{x\sqrt{lc}}, \frac{3\pi}{x\sqrt{lc}} \cdots, \quad (1.50)$$

which represent the self-oscillatory frequencies of the line as a function of its length x. This

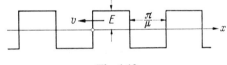

Fig. 1.12

formulation then describes the many partial oscillations that can take place in a line owing to its many degrees of freedom.

A variation of Eq. (1.48) in complex notation is

$$e = E\varepsilon^{jvt}\varepsilon^{j\mu x} = E\varepsilon^{j(vt+\mu x)} = E\varepsilon^{jv(t+x/v)}.$$
$$(1.51)$$

This is also a solution of the wave equation (1.47) and gives, according to Eq. (1.49),

$$v = \frac{v}{\mu} = \pm \frac{1}{\sqrt{lc}} \quad (1.52)$$

in which v is the velocity of propagation previously introduced. Equation (1.51) shows

that voltage wave packet or pulse components of arbitrary amplitude and frequency can travel in both directions along the line. The velocities of all these components are equal because $x = -vt$.

It is possible to use a number of harmonically related components, with fundamental and harmonics as given in Eq. (1.50), and build up a wide range of pulse or wave shapes. For example, the Fourier series

$$e = \frac{4E}{\pi}\left[\sin v_0\left(t + \frac{x}{v}\right) + \frac{1}{3}\sin 3v_0\left(t + \frac{x}{v}\right)\right.$$
$$\left. + \frac{1}{5}\sin 5v_0\left(t + \frac{x}{v}\right) + \cdots\right]$$
$$= \frac{4}{\pi}E\sum_0^\infty\frac{1}{n}\sin nv_0\left(t + \frac{x}{v}\right) \quad (1.53)$$

represents a square-wave train traveling to the left with constant velocity v, as shown in

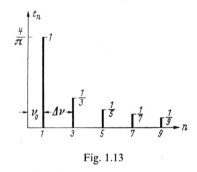

Fig. 1.13

Fig. 1.12. The amplitude of the odd harmonics decreases inversely with their harmonic order number n, as shown in the spectrum in Fig. 1.13.

As the fundamental frequency v_0 decreases, the harmonics

$$v = nv_0 \quad (1.54)$$

move closer together, since their separation (see Fig. 1.13), is

$$\Delta v = 2v_0. \quad (1.55)$$

At the same time the rectangular wavelength becomes longer according to Eq. (1.49), and in the limiting condition as v approaches 0 one obtains a single voltage jump (or step pulse), as shown in Fig. 1.14. By substitution of Eqs. (1.54) and (1.55) into Eq. (1.53) this can be represented mathematically by

$$e = \frac{2E}{\pi} \int_0^\infty \frac{\sin v(t + x/v)}{v} \, dv. \quad (1.56)$$

This shows that the amplitudes of the continuous spectrum components equivalent to such a step pulse decrease as $1/v$, as shown in Fig. 1.15. Note the difference between this continuous spectrum, associated with the

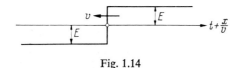

Fig. 1.14

single step, and the line spectrum, associated with the square-wave train. The presence of relatively strong high-frequency components in the pulse spectrum is of course due to the sharp rise of the pulse front or step (step function). For less sharply rising pulse fronts

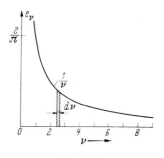

Fig. 1.15

the amplitudes of the high-frequency components of the spectrum decrease more rapidly.

Because Eq. (1.51) allows the composition of any arbitrary wave or pulse shape from a set of harmonically related functions, we shall try to see if a completely arbitrary function of the same argument,

$$e = f\left(t + \frac{x}{v}\right) \quad \text{or} \quad e = f(x + vt), \quad (1.57)$$

as shown in Fig. 1.16, will also satisfy the wave equation (1.47). If we designate the

differentiations corresponding to the argument f' and f'', we obtain

$$\frac{\partial^2 e}{\partial t^2} = f''\left(t + \frac{x}{v}\right),$$

$$\frac{\partial^2 e}{\partial x^2} = \frac{1}{v^2} f''\left(t + \frac{x}{v}\right). \quad (1.58)$$

Substitution in Eq. (1.47) again yields the condition of Eq. (1.52) for the velocity of propagation. Thus any arbitrary function of the moving coordinate $x \pm vt$ is a solution of the wave equation (1.47). This d'Alembert solution also allows a representation of the propagation of pulses along the line. The function f remains constant within the coordinate system, which moves with the velocity of light to either left or right and thus represents a pulse traveling without any distortion.

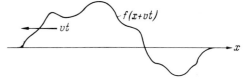

Fig. 1.16

If the line has considerable resistance, the voltage difference due to self-inductance as given by Eq. (1.1) requires the addition of a term $r\Delta x \cdot i$ due to the resistive voltage drop. This changes the differential equation (1.2) to

$$\frac{\partial e}{\partial x} = -l\frac{\partial i}{\partial t} - ri. \quad (1.59)$$

If we differentiate this equation with respect to x and substitute Eq. (1.4), we obtain a more exact differential equation for the voltage as follows:

$$\frac{\partial^2 e}{\partial x^2} = lc\frac{\partial^2 e}{\partial t^2} + rc\frac{\partial e}{\partial t}, \quad (1.60)$$

which is known as the telegrapher's equation. If we were also to consider losses in the insulation, an additional term $e \cdot r/p$ would be introduced.

This equation can be solved simply and rigorously only with a harmonic trial solution of the kind in Eq. (1.51):

$$e = E\varepsilon^{jvt}\varepsilon^{j\mu x}. \quad (1.61)$$

The result is

$$\mu = \pm \sqrt{lcv^2 - jrcv}$$

$$= \pm \sqrt{lcv} \cdot \sqrt{1 - j\frac{r}{vl}} = \frac{v}{v} + j\delta. \quad (1.62)$$

The wave number μ thus becomes complex and results after substitution in Eq. (1.61) in a spatial attenuation coefficient $-\delta x$. If the resistance r is small compared with the inductive impedance vl, the roots of Eq. (1.62) can be developed by the binomial method with fair accuracy. Of course, if this condition is met for the fundamental, it is even more closely met for the higher harmonics. With it we obtain

$$\sqrt{1 - j\frac{r}{vl}} = 1 - j\frac{r}{2vl} + \frac{1}{8}\left(\frac{r}{vl}\right)^2. \quad (1.63)$$

After substitution in Eq. (1.62) the attenuation or damping is given by

$$\delta = \pm \sqrt{lcv} \frac{r}{2vl} = \pm \frac{r}{2\sqrt{l/c}} \quad (1.64)$$

and the propagation velocity by

$$v = \pm \frac{1 - \frac{1}{8}(r/vl)^2}{\sqrt{lc}}. \quad (1.65)$$

The value for damping corresponds exactly to the form earlier developed in Eq. (1.36). With constant resistance it is independent of frequency and thus results only in attenuation of the pulse amplitude without distortion of shape. However, when the skin effect produces a higher resistance for the components of higher frequency, these components are attenuated more strongly. Thus sharp features of the pulse form are lost owing to the skin effect.

The velocity of propagation is shown in Eq. (1.65) to be dependent on frequency through the impedance term of the numerator. The lower-frequency components travel at somewhat lower speed than the higher-frequency components. For frequencies of 500 cy/sec on our open line, with an inductive impedance $vl = 5 \, \Omega/\text{km}$, the deviation from the velocity of light is

$$\Delta v = \frac{1}{8}\left(\frac{0.7}{5}\right)^2 = 0.245 \text{ percent.}$$

Thus for each kilometer of line traversed the distance covered is short by about 2.5 m. For cables, owing to their lower self-inductance, or for open lines using ground return of higher resistance, this deviation is considerably larger. Because the high-frequency components travel more nearly with the velocity of light, as seen in Eq. (1.65), this feature also introduces a distortion of pulse shape.

CHAPTER 2

THE GENERATION OF SIMPLE TRAVELING PULSES

Chapter 1 has shown that wave packets in the form of pulses traveling in both forward and rearward directions along transmission lines represent the general solution for voltage and current phenomena along the lines. Thus they can be used to examine any desired electrical processes connected with the lines. The relations between voltage and current in the pulses traveling forward (v) and rearward (r) are, respectively,

$$e_v = Zi_v, \qquad (2.1)$$

$$e_r = -Zi_r, \qquad (2.2)$$

and the total voltage and current are, respectively,

$$e = e_v + e_r, \qquad (2.3)$$

$$i = i_v + i_r. \qquad (2.4)$$

The surge impedance and velocity of propagation of the pulses are constants of the line. The form and amplitude of the pulses must be determined from the boundary conditions of a given problem.

(a) *Static charges.* We shall show that a static charge on the line can be represented by the interaction of traveling pulses, even though there are no transport phenomena involved.

Let a line open at both ends be charged to the voltage e. Because a static charge once established produces no current, for every point x along the line

$$i = 0. \qquad (2.5)$$

Thus according to Eq. (2.4),

$$i_r = -i_v \qquad (2.6)$$

everywhere. Hence, by Eqs. (2.1) and (2.2), the voltages of the two pulses are equal; hence, by Eq. (2.3),

$$e_v = e_r = \tfrac{1}{2}e \qquad (2.7)$$

for each point along the line. Thus the pulses have constant amplitude along the line.

Figure 2.1 shows this resolution of the static charge into two pulses traveling with the velocity of light. The forward and return current pulses of equal amplitude and opposite polarity cancel completely. The forward and return voltage pulses of equal amplitude and polarity add everywhere so that no actual

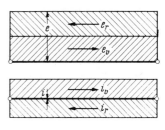

Fig. 2.1

movement can be perceived. Thus we can actually represent any static charge at a given instant of time by two pulses of constant amplitude traveling in opposite directions with the velocity of light.

(b) *Steady-state direct current.* Figure 2.2 shows a direct-current source supplying current

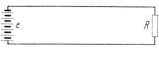

Fig. 2.2

to a resistance R via a transmission line. The current along the line is constant and given by

$$i = \frac{e}{R}. \qquad (2.8)$$

From eq. (2.4) we have

$$i_r = \frac{e}{R} - i_v. \qquad (2.9)$$

If we substitute this in Eq. (2.2) and add it to Eq. (2.1), then, with Eq. (2.3), we obtain

$$e = 2Zi_v - \frac{Z}{R}e. \qquad (2.10)$$

With Eq. (2.8) this gives for the forward-traveling current pulse

$$i_v = \left(1 + \frac{Z}{R}\right)\frac{e}{2Z} = \left(1 + \frac{R}{Z}\right)\frac{i}{2}, \qquad (2.11)$$

and from Eq. (2.9) for the rearward-traveling current pulse,

$$i_r = -\left(1 - \frac{Z}{R}\right)\frac{e}{2Z} = \left(1 - \frac{R}{Z}\right)\frac{i}{2}. \qquad (2.12)$$

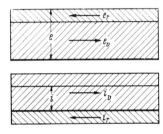

Fig. 2.3

The forward-traveling voltage pulse is from Eq. (2.1)

$$e_v = \left(1 + \frac{Z}{R}\right)\frac{e}{2} \qquad (2.13)$$

and the rearward-traveling voltage pulse, from Eq. (2.2),

$$e_r = \left(1 - \frac{Z}{R}\right)\frac{e}{2}. \qquad (2.14)$$

Figure 2.3 shows the composition of the two traveling current and voltage pulses into the constant direct current i and direct voltage e along the entire line. The pulse components all have constant amplitude along the line. The amplitudes of the components are different, however, and are determined only by the ratio Z/R.

If the load resistance R is equal to the surge impedance Z of the line, the rearward-traveling current and voltage pulses disappear completely, according to Eqs. (2.12) and (2.14), respectively. The forward-traveling pulses, given by Eqs. (2.11) and (2.13), with $Z = R$,

then simply give the total direct current i and voltage e. This shows that an ohmic load resistance of magnitude equal to the surge impedance of the line does not give rise to the generation of rearward-traveling pulses, that is, reflections into the line.

(c) *Released charge pulses.* Figure 2.4 shows a static charge along a line built up below a cloud, which is suddenly released by a lightning discharge at time $t = 0$ into the cloud. The static charge existing in the line below the cloud at the time $t = 0$ is essentially released by the lightning discharge and cannot remain static but must propagate along the

Fig. 2.4

line to allow transition to a new final steady-state solution. Thus at $t = 0$,

$$i = 0, \qquad (2.15)$$

and hence according to Eq. (2.4)

$$i_r = -i_v. \qquad (2.16)$$

Thus the voltage pulses traveling in both directions are equal in amplitude, according to Eqs. (2.1) and (2.2), and from Eq. (2.3) are

$$e_v = e_r = \tfrac{1}{2}e. \qquad (2.17)$$

They have the same spatial distribution as the released charge and half the amplitude of the voltage produced by the released charge. The current pulses are now given from Eqs. (2.1) and (2.2) as

$$i_v = -i_r = \frac{1}{2}\frac{e}{Z}. \qquad (2.18)$$

They also have the spatial distribution of the released charge and their amplitude is one-half the voltage e divided by the surge impedance of the line Z.

Figure 2.5a shows the pulses immediately after the release of the charges at $t = 0$. The suddenly produced voltage e exists only over a limited region of the line and only in this region at this instant of time do there exist the forward and rearward voltage pulses given by Eq. (2.17) and current pulses given by Eq. (2.18).

At the instant $t = 0$ the two current pulses cancel and the two voltage pulses add up to

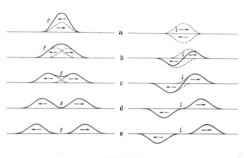

Fig. 2.5

the actual voltage e created by the charge. As time goes on, all four pulse components propagate without distortion in both directions. This is shown in Fig. 2.5b–e. Figure 2.5b shows that the forward pulses e_v and i_v have moved slightly to the right and the rearward pulses e_r and i_r have moved slightly to the left, adding up to the total voltage e and current i as shown. Figure 2.5c shows the two pulse systems still slightly overlapping, while in Fig. 2.5d,e they have completely separated and all propagate with the velocity of light away from their point of origin or generation.

Thus we see that each released charge, of any given spatial distribution, will decompose immediately after generation into two equal partial charges of half the total value. These partial charges propagate in both directions from the point of generation without distortion of form. They give rise to voltage and current pulses calculated from Eqs. (2.17) and (2.18), respectively.

This explains in detail the case shown in Chapter 1, Fig. 1.1, of high voltages along a line grounded at both ends. The pulses traveling along the line cannot be influenced by the ground terminations until such time as they have traveled from the point of generation to these terminal points of the line.

(d) *Switching between homogeneous lines.* Figure 2.6a shows two very long lines extending in opposite directions, with a switch between them. The switch is initially open and the left-hand line is at the constant voltage e. We shall develop the voltage and current phenomena that occur after the closing of the switch at time $t = 0$.

For $t < 0$ the left-hand line is charged statically and thus, in accordance with Eq. (2.7), carries the two voltage pulses

$$e_v = e_r = \tfrac{1}{2}e \qquad (2.19)$$

and the corresponding current pulses

$$i_v = -i_r = \frac{1}{2}\frac{e}{Z}. \qquad (2.20)$$

To the right of the switch all voltages and currents are zero.

At the instant $t = 0$ the closing of the switch is assumed to produce a single continuous and homogeneous line. All pulse forms present at this moment in the forward and rearward current and voltage pulse components are thus maintained for $t > 0$

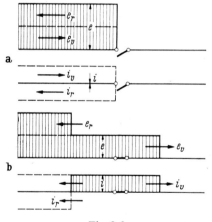

Fig. 2.6

and propagate undistorted with the velocity of light. Thus shortly after the instant of switching the front edge of the forward-traveling pulse has progressed to the right beyond the switch, while the rear edge of the rearward-traveling pulse has withdrawn to the left away from the switch. The front and rear of the pulses have maintained the sharp rectangular form that they had just prior to the closing of the switch.

Figure 2.6*b* shows the superposition of the pulses shortly after switching. A charge pulse of half the initial voltage *e* enters the right-hand line, while a discharge pulse of equal magnitude reduces the voltage of the left-hand line. At the same time current pulses flow into both the left- and right-hand lines. In both lines the current is positive. In the right-hand line it is generated by the forward-traveling voltage pulse and in the left-hand line it is generated by the retreating rear end of the rearward-traveling voltage pulse. Thus the total current pulse *i*, which is always positive, propagates with the velocity of light in both directions along the line from the switching point, as do the charge and discharge voltage pulses.

The generator at the end of the left-hand line will respond to the switching process only after the time interval taken for the discharge voltage and current pulses to travel to it, and they will then impose a demand on it. Similarly, the load at the end of the right-hand line will be influenced only after a short time interval to allow for the arrival of the forward-traveling charge voltage and current pulses. Thus the closing of the switch is announced at the terminals of the lines by the traveling pulse only after a short time interval, and only thereafter can apparatus at these terminals react on the switching process through changes in the electromagnetic conditions that they produce.

CHAPTER 3

THE GENERATION AND SHAPE OF PULSES

In power networks many electrical wave-forms are superposed. To begin with, there are the steady-state sinusoidal alternating currents, their harmonics and subharmonics, and finally there are the switching pulses and atmospherically created impulses of a transient nature and with sharp discontinuities leading to high-frequency content. All obey the basic equations for the partial waves

$$e_v = Zi_v,$$
$$e_r = -Zi_r, \tag{3.1}$$

and for the total or composite wave

$$e = e_v + e_r,$$
$$i = i_v + i_r. \tag{3.2}$$

We shall examine the generation and shape of such wave packets or pulses in several examples of practical importance.

(a) *Steady-state alternating current.* Figure 3.1 shows an alternating-current power line

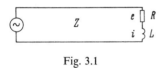

Fig. 3.1

feeding a load resistance R with inductance L in series. Across this load the voltage is given by

$$e = Ri + L\frac{di}{dt}. \tag{3.3}$$

If the alternating current is sinusoidal it is given by

$$i = I\varepsilon^{j\omega t}, \tag{3.4}$$

and thus the voltage across the line at the loaded end is

$$e = (R + j\omega L)i. \tag{3.5}$$

If we introduce into Eq. (3.5) the voltage and current waves from Eq. (3.2) and the partial voltage waves from Eq. (3.1) we obtain

$$Z(i_v - i_r) = (R + j\omega L)(i_v + i_r). \tag{3.6}$$

This gives for the ratio of rearward to forward current waves

$$\frac{i_r}{i_v} = \frac{Z - (R + j\omega L)}{Z + (R + j\omega L)}. \tag{3.7}$$

The ratio of the corresponding voltage waves is then, from Eq. (3.1).

$$\frac{e_r}{e_v} = -\frac{i_r}{i_v} = -\frac{Z - (R + j\omega L)}{Z + (R + j\omega L)}. \tag{3.8}$$

The actual partial current waves are now, from Eq. (3.7) and the second of Eqs. (3.2),

$$i_v = \frac{i}{2}\frac{Z + (R + j\omega L)}{Z},$$
$$i_r = \frac{i}{2}\frac{Z - (R + j\omega L)}{Z}, \tag{3.9}$$

and the partial voltage waves, from Eq. (3.8) and the first of Eq. (3.2),

$$e_v = \frac{e}{2}\frac{Z + (R + j\omega L)}{R + j\omega L},$$
$$e_r = \frac{e}{2}\frac{Z - (R + j\omega L)}{R + j\omega L}. \tag{3.10}$$

These equations have the same form as those developed in Chapter 2 for steady-state direct-current partial waves, namely Eqs. (2.11) to (2.14). The only difference is that the complex impedance $R + j\omega L$ takes the place of the purely direct-current resistance R. Thus we see that with a purely resistive load we can also avoid rearward (or reflected) partial traveling waves in the alternating-current case if we fulfill the condition $R = Z$. However, with an inductive load this is never possible.

Figure 3.2 exhibits the partial current waves of Eqs. (3.9) in a simple vector diagram. This shows that inductance in the load at the end of the line gives rise to two partial traveling waves of equal amplitude and opposite polarity, one traveling in the forward and the other in the rearward direction. The amplitudes of the two components are independent of the ohmic or purely resistive load and are determined solely by the ratio of the inductive

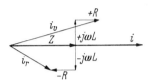

Fig. 3.2

impedance of the load to the surge impedance of the line. They cannot be reduced to zero by any impedance matching of the inductive load.

It is possible to consider any arbitrary electrical condition of the line as a superposition of a number of alternating currents of all possible frequencies ω. This then leads to the conclusion that the generation of rearward-traveling pulses or waves at the loaded end of the line can be completely avoided if the line is terminated by a pure ohmic load of magnitude equal to the surge impedance of the line.

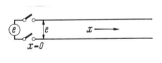

Fig. 3.3

long line which is suddenly switched to a voltage source e. We shall assume the source to have essentially zero internal impedance. Then for $x = 0$ the voltage across the line is e. Immediately after the instant of switching there can be no voltage or current pulses or waves arriving at the switching end from the line, which has been assumed to be completely free of currents and voltages prior to switching.

Thus there can be no rearward-traveling waves, so that

$$e_r = 0 \quad \text{and} \quad i_r = 0. \quad (3.11)$$

This leaves only a pulse traveling forward into the line, of voltage

$$e_v = e. \quad (3.12)$$

This produces a current

$$i_v = i = \frac{e}{Z}. \quad (3.13)$$

Thus the pulse propagates the voltage at the beginning of the line along the line with the

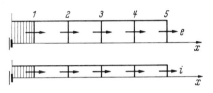

Fig. 3.4

velocity of light. In this propagation it gives rise to a current determined only by the surge impedance of the line. This pulse charges the line.

If the source is a direct voltage, then this propagates down the line with constant amplitude and velocity starting at the instant of switching. Figure 3.4 shows the voltage and

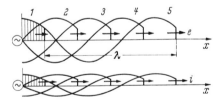

Fig. 3.5

current pulses, both of which have a sharp rectangular front and become longer and longer as time progresses and the front propagates along the line.

If an alternating-voltage source is switched onto the line, voltage and current do not propagate as a pulse of constant amplitude along the line. Instead, as shown in Fig. 3.5, the pulse front existing at the switching point is a steep rectangular one given by the instantaneous voltage at the instant of switching. This steep front propagates with the velocity of light down the line, creating behind it a sinusoidal distribution of voltage and current of the frequency ω of the voltage source. The wavelength of one complete sinusoid along the line is given by the ratio of the velocity of propagation to the frequency in cycles per second, $\omega/2\pi$,

$$\lambda = 2\pi \frac{v}{\omega}. \qquad (3.14)$$

It is of course assumed that the line is long compared with this wavelength if the waves are to develop fully. For open, that is, air-insulated lines and a frequency of 60 cy/sec,

$$\lambda = \frac{300,000}{60} = 5,000 \text{ km.}$$

The front of the forward-traveling voltage pulse is always equal to the voltage of the source at the instant of switching. As in the case of direct voltage, the front of the pulse for alternating voltage is a sharp jump in amplitude equal to the source voltage at the switching instant. Thus for alternating voltages it will have the highest value if switching occurs at the instant of peak voltage in the cycle, as may in fact happen if a spark forms between the contacts of a closing switch at the instant of peak voltage.

Immediately after switching a current is drawn from the source according to Eq. (3.13), which results in a power

$$W = ei = \frac{e^2}{Z} = i^2 Z. \qquad (3.15)$$

Thus long lines, as shown in Figs. 3.4 and 3.5, impose a load on the source of ohmic nature and of magnitude equal to the surge impedance Z because only forward-traveling components are present. The energy drawn from the source is not, however, dissipated in the line, but is guided as a wave or pulse phenomenon toward the far end of the line.

If we wish to avoid completely the generation of rearward-traveling wave trains at the far end of the line, we must terminate it in an ohmic load resistance, that is, put $\omega L = 0$, and make

$$R = Z \qquad (3.16)$$

according to Eqs. (3.9) and (3.10). All forward-traveling wave-train energy is then completely absorbed by the resistive load, though only after a time delay beyond the switching instant equal to the time required for propagation along the length of the line. Such an arrangement permits a convenient experimental

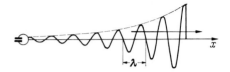

Fig. 3.6

study of traveling-wave-train phenomena in lines.

The place of the direct-current battery or alternating-current generator can also be taken by an oscillatory circuit containing inductance and capacitance. The voltages and currents in such a circuit are of course exponentially damped sinusoids. They will then also have a spatial distribution of this form along the line behind a sharp front determined as before at the switching instant and as shown in Fig. 3.6. The phenomenon does not vary in form if the oscillatory circuit is either a series or a parallel connection of

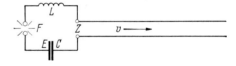

Fig. 3.7

inductance and capacitance. The wavelength along the line in this case is also given by Eq. (3.14) if ω is the frequency of self-oscillation of the circuit in cycles per 2π sec. Thus high-frequency oscillations of 10^6 cy/sec produce wavelengths along open lines of

$$\lambda = \frac{3 \cdot 10^8}{10^6} = 300 \text{ m.}$$

(c) *The generation of impulses and shock waves.* The line can be connected to a series-connected oscillatory circuit as shown in Fig. 3.7. The condenser, of capacitance C, is charged to the voltage E and the source is switched onto the line by the spark switch F. The three wave forms possible in such a circuit arrangement are shown in Fig. 3.8. The surge impedance Z of the line acts exactly like an ohmic resistance of equal value in the series circuit formed by Z, C, and the inductance L. Thus the three cases of Fig. 3.8 are respectively (a) the oscillatory or underdamped, (b) the critically damped, and (c) the overdamped circuit conditions. As shown previously, the spatial distribution along the line will have the same form as the temporal distribution in the equivalent lumped circuit.

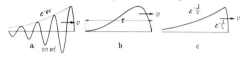

Fig. 3.8

In the oscillatory case, corresponding to Fig. 3.8a

$$e = Zi = E\frac{Z}{\sqrt{L/C}}\varepsilon^{-\rho t}\sin vt, \quad (3.17)$$

if we ignore the small voltage drop in the spark switch F and use the method developed in Ref. 1, Eqs. (20) and (21), p. 582, with

$$\rho = \frac{Z}{2L} \quad \text{and} \quad v = \frac{1}{\sqrt{LC}} \quad (3.18)$$

as the damping exponent and frequency of the oscillations, respectively.

The two quantities given by Eq. (3.18) can be used to establish values of L and C in relation to Z that give a desired frequency and damping exponent. Equation (3.17) can then be used to calculate the voltage E to which the condenser must be charged to produce a voltage e in the wave train along the line. Since the value of Z is relatively fixed, the values of L and C so determined may not be very practicable. This can be easily remedied by inserting an ohmic resistance either in series with Z, if this is too small, or in parallel with Z if it is too large, to produce a suitably larger or smaller effective resistance in the oscillatory circuit.

For the critically damped case corresponding to Fig. 3.8b, Ref. 1, Eq. (16), p. 50 gives

$$Z = 2\sqrt{\frac{L}{C}}. \quad (3.19)$$

The duration of the transient is then of the order of magnitude of

$$\tau = 2\pi\sqrt{LC}. \quad (3.20)$$

This gives for the circuit constants

$$L = \frac{\tau Z}{4\pi}, \quad C = \frac{\tau}{\pi Z}. \quad (3.21)$$

If, for example, we wish to produce in an open line of surge impedance $Z = 500\ \Omega$ such a single critically damped pulse of duration $\tau = 10^{-5}$ sec, which corresponds to a length of 3 km, we need an inductance

$$L = \frac{10^{-5} \cdot 500}{4\pi} = 0.4\ \text{mH}$$

and a capacitance

$$C = \frac{10^{-5}}{\pi \cdot 500} = 0.0064\ \mu\text{F}.$$

To produce the overdamped case shown in Fig. 3.8c, the resistance in the circuit must exceed that established by the critical condition given in Eq. (3.19). The two time constants of the overdamped circuit are

$$T_L = \frac{L}{Z} \quad \text{and} \quad T_C = CZ, \quad (3.22)$$

the first of which is very small and the second one much larger. The voltage on the line is

$$e = E(\varepsilon^{-t/T_C} - \varepsilon^{-t/T_L}). \quad (3.23)$$

The fast rise is determined by the inductive time constant and the slow decay by the capacitive time constant. For a rise-time constant $T_L = 10^{-6}$ sec and a decay-time constant $T_C = 10^{-4}$ sec for the typical open line, one must use an inductance

$$L = 10^{-6} \cdot 500 = 0.5\ \text{mH}$$

and a capacitance

$$C = \frac{10^{-4}}{500} = 0.2\ \mu\text{F}.$$

If one uses a larger capacitance, it is possible to flatten out the decay along the back of the shock wave shown in Fig. 3.8c to any desired extent. In contrast with this, a reduction in inductance will not allow a sharpening of the wavefront to any desired extent. One reason is that the circuit, owing to its physical size, particularly in the case of high voltages, has so to speak a basic minimum inductance; another is that the discharge spark F shown in Fig. 3.7 never occurs instantaneously but requires a formation time of 10^{-8} to 10^{-7} sec, which depends on many variables, cannot be accurately predicted, and is also often not very repeatable. Thus a given inductance will increase the rise time of the current above that inherent in the spark formation. At the same time this serves to give the wavefront a better-defined and repeatable shape.

(d) Collapse of an electric field in the atmosphere. During a lightning discharge the electrostatic field in the atmosphere above the ground changes and produces pulses traveling along a line, as shown in Chapter 2. If the change of electric field is not instantaneous, as assumed previously, the released charges can flow along the line with the velocity of

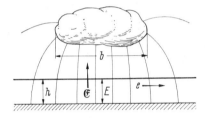

Fig. 3.9

light during the entire period of change in the field. This more gradual process reduces the peak amplitude of the voltage generated on the line.

Figure 3.9 shows a line at a height h above the ground in an atmospheric electrostatic field of field strength $\mathfrak{E}$. Thus the line is charged continuously to a voltage

$$E = h\mathfrak{E} \qquad (3.24)$$

and this produces pulses traveling away from the charging point in both directions. For each such pulse the relation between the time and space dependences of the voltage has the form of Eq. (1.11),

$$-\frac{\partial e}{\partial t} = \pm v \frac{\partial e}{\partial x}, \qquad (3.25)$$

where the $+$ sign indicates the forward- and the $-$ sign the rearward-traveling partial

pulse. Because the line is symmetric, the two pulses have the same shape and amplitude. Thus the voltage change picked up from the atmospheric field is composed of the two partial pulses:

$$\frac{dE}{dt} = \frac{\partial e_v}{\partial t} + \frac{\partial e_r}{\partial t} = 2\frac{\partial e}{\partial t} = -2v\frac{\partial e_v}{\partial x} = 2v\frac{\partial e_r}{\partial x}. \qquad (3.26)$$

The voltage rise in the outward-traveling pulse fronts is thus

$$-\frac{\partial e_v}{\partial x} = \frac{\partial e_r}{\partial x} = \frac{1}{2v}\frac{dE}{dt} = \frac{h}{2v}\frac{d\mathfrak{E}}{dt}. \qquad (3.27)$$

Figure 3.10a shows the pulse front, traveling to the right along the line, produced by an arbitrary increase in the atmospheric electric field with time. Thus as the atmospheric field below a thundercloud changes with time, it produces pulses traveling outward along the line owing to the changing voltage impressed on the line.

During the interval τ of the voltage rise the pulse front travels a distance

$$a = v\tau. \qquad (3.28)$$

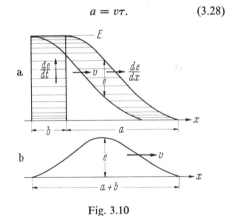

Fig. 3.10

Thus, for example, a rise time of 20 μsec produces travel of the pulse front over a distance of

$$a = 3 \cdot 10^5 \cdot 20 \cdot 10^{-6} = 6 \text{ km}.$$

The actual pulse shape along the line is produced, however, by the superposition of the two pulses propagated as shown in Fig. 3.10a from the right- and left-hand edges of the cloud of length b. All charges in the region b propagate with the same velocity as elements that compose the shaded area in Fig. 3.10a. The actual pulse shape is shown in Fig. 3.10b as the difference between the two edge pulses.

Integration of Eq. (3.27) gives us the voltage of the pulses traveling outward in both directions:

$$e = \frac{1}{2v}\int_b^{} \frac{dE}{dt}\, dx = \frac{h}{2v}\frac{d}{dt}\int_b^{} \mathfrak{E}\, dx. \quad (3.29)$$

This implies a uniform change of field with time along the length b, so that the differentiation with respect to time has been put outside the integral over the space b. Furthermore, b is assumed small compared with the distance of pulse advance a. If the change of atmospheric electric field is uniform along the length of the cloud b, and thus rectangular in shape, the field strength $\mathfrak{E}$ can also be removed from the integral, which becomes simply b, and we obtain

$$e = \frac{h}{2v}\frac{d\mathfrak{E}}{dt}\int_b^{} dx = \frac{hb}{2v}\frac{d\mathfrak{E}}{dt}. \quad (3.30)$$

However, if the field is nonuniform, we must use in Eq. (3.30) the concept of an equivalent length b of the cloud above the line. We see that the voltage pulse created is proportional to the time rate of change of the atmospheric electric field and the length of the cloud. It is limited in magnitude by the very high speed of propagation.

If the length of the cloud b is greater than the distance of pulse advance a, we see from Fig. 3.10 that the pulse originated at the left-hand edge reaches the right-hand edge only after the full voltage E has been created at this point according to Eq. (3.24). This equation then represents the highest voltage generation possible under the cloud, since it is no longer diminished by the propagated pulses.

As an example, take a typical atmospheric field change of 100 kV/m within a rise time of 20 μsec, below a cloud of equivalent length 2 km, and examine its effect on a line of average height 10 m. Pulses of voltage

$$e = \frac{10 \cdot 2}{2 \cdot 3 \cdot 10^5} \cdot \frac{100}{20 \cdot 10^{-6}} = 167 \text{ kV}$$

are produced, which travel in both directions along the line away from the cloud. The length of the pulses is, according to Fig. 3.10,

$$a + b = 6 + 2 = 8 \text{ km.}$$

If the electric field increases linearly with time, the pulses have a trapezoidal shape along the line, the head and tail occupying distances equal to the length of the cloud. Thus the change in voltage along the line will be, from Eq. (3.27),

$$\frac{de}{dx} = \frac{10}{2 \cdot 3 \cdot 10^8} \frac{100 \cdot 10^3}{20 \cdot 10^{-6}} = 83.5 \text{ V/m,}$$

which is relatively small.

In the area below the cloud the two pulses add and thus produce a total voltage of 333 kV. This is still only one-third of the voltage $h\mathfrak{E} = 1000$ kV that would be produced on a short line, open at both ends, below the cloud. Thus the pulses propagating along the line reduce the voltage of the lightning-induced effects by a factor of 3 in our example.

(e) Different pulse shapes. Traveling pulses stress the insulation of insulators, switchgear, machinery, transformers, and general electrical apparatus in a manner quite different from that produced by the normal operating voltages and currents. Essentially, we can distinguish between three typical shapes of traveling pulses, as shown in Fig. 3.11.

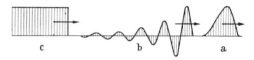

Fig. 3.11

Single pulses (Fig. 3.11a) stress the insulation primarily by their voltage amplitude, which can be very large when caused by atmospheric discharges. Because the duration of the stress is very short, often only of the order of microseconds, the insulation can withstand voltage pulses of far higher amplitude than the regular operating voltage. The increase is strongly dependent on the brevity in time of the pulse. In addition to the voltage amplitude, the change of voltage during the pulse produces stresses in the insulation between adjacent turns and layers in coil windings. In this case also the short duration of the pulses tends to increase the voltage required to produce breakdown.

Pulse trains (Fig. 3.11*b*) stress the insulation in a similar manner. They can be particularly dangerous to some types of insulators if they are of high frequency and only weakly damped. They are dangerous to the insulation in coil windings if the largest potential differences separated by a half wavelength occur between adjacent turns, as in the terminals of coil sections making up a complete transformer winding. Thus any coil winding that is constructed in sections can perhaps be stressed to the breaking point by pulse trains of a given frequency. This stress is, of course, repeated several times if the pulse train is long on account of weak damping.

Step pulses (Fig. 3.11*c*) have an extremely high voltage gradient at the front end concentrated in a short distance, of the order of meters or tenths of a meter. Because this sharp rise is followed by a slow decay of long duration, the effective stress on the insulation is higher than for short single pulses (Fig. 3.11*a*). If the amplitude decays after some given time, then of course the stress also becomes less. In coil windings, depending on the rise time, that is, the length of the pulse front, the entire voltage difference can occur within single turns or layers and can thus stress the insulation between adjacent turns or layers very heavily. Because the insulation of coil windings has to withstand only relatively low voltages in regular operation, such step pulses, even of moderate amplitude, are most dangerous to these windings.

We shall carry out our calculations primarily for step pulses because they are mathematically simple in form and thus give clear solutions. Also, as shown in Fig. 3.12, any other pulse

Fig. 3.12

shape can be made up by the superposition of a number of step pulses. In most coil windings step pulses provide the highest stress for which the insulation must be designed. However, we shall also treat the cases in which single pulses and pulse trains produce effects of special significance.

CHAPTER 4

THE INFLUENCE OF LINE TERMINATIONS

When a pulse of arbitrary shape reaches the end of a line, it cannot propagate further without undergoing a change. To examine these changes we can use again the simple Eqs. (1.16)–(1.19) for the currents and voltages of the pulses traveling in the two directions.

For the partial pulses,

$$e_v = Zi_v, \qquad (4.1)$$

$$e_r = -Zi_r, \qquad (4.2)$$

and for the total voltages and currents,

$$e = e_v + e_r, \qquad (4.3)$$

$$i = i_v + i_r. \qquad (4.4)$$

First let us examine two typical cases, namely, the line terminated in either an open or a short circuit. Let us treat initially any arbitrary

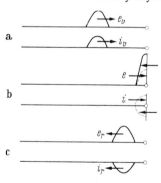

Fig. 4.1

the current is reduced to zero. This boundary condition,

$$i = 0, \qquad (4.5)$$

can be met according to Eq. (4.4) only if to the forward-current-pulse component i_v there is added a backward-current-pulse component i_r of magnitude

$$i_r = -i_v. \qquad (4.6)$$

This component must have an associated voltage, by Eqs. (4.2), (4.6), and (4.1),

$$e_r = -Zi_r = Zi_v = e_v. \qquad (4.7)$$

The total voltage at the line terminal is, by Eq. (4.3), the sum of the forward and rearward voltage pulses,

$$e = e_v + e_r = 2e_v. \qquad (4.8)$$

Thus at the open end of the line there is an instantaneous voltage of twice the magnitude of the pulse arriving at this point. This voltage

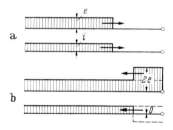

Fig. 4.2

pulse shape and portray graphically short single pulses as well as long step pulses.

(a) *Open-circuit line termination.* A single pulse (Fig. 4.1a) that travels in the forward direction along a homogeneous line is a complete solution and requires no pulses traveling in the opposite direction. As it reaches the open circuit at the end of the line, it can develop an arbitrary voltage; however,

doubling is caused by the boundary conditions (4.5) and (4.6), which give rise to a reflected pulse in which the voltage is the same in magnitude and polarity while the current is the same in magnitude but opposite in polarity. Thus the reflected pulse at the open end is the mirror image of the arriving pulse and is propagated backward from the open end just as the arriving pulse is absorbed at this point. This, then, is a case of total reflection with a change in current polarity and no change in voltage polarity.

Figure 4.1*b* shows the instant at which the center of the symmetric single pulse has reached the open end. Figure 4.1*c* shows a time just after all of the arriving pulse has

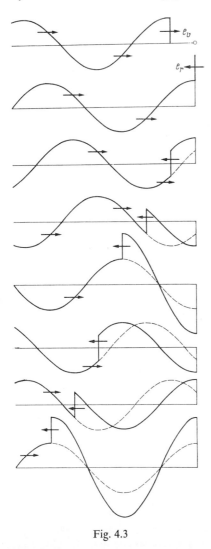

Fig. 4.3

passed the open end. In this case only the reflected pulse is present, traveling to the left with the mirror-image shape of the originally incident pulse along the homogeneous line.

Equations (4.5)–(4.8) are valid for any pulse shape, as shown for single pulses in Fig. 4.1 and for step pulses in Fig. 4.2. Figure 4.2*a*

shows a step pulse shortly before arrival at the open end of the line. It may have been caused by switching at the far end of the line, as described in Chapter 3. At the instant of reaching the open end the mirror-image reflection produces a rearward-traveling pulse of rectangular front shape, equal voltage, and equal and opposite current, again given by Eqs. (4.7) and (4.6), respectively. This reflected pulse is now superimposed on the incident one and results essentially in a propagation backward from the open end of the line of the two boundary conditions, namely, a doubling of the voltage and reduction of the current to zero. This is shown in Fig. 4.2*b*. Thus the arrival of a very long step pulse at the open end of a line produces by reflection a doubling of the voltage on the line that not only appears at the end but is propagated backward along the entire length of the line.

This doubling of voltage is always found at the open end of a line. How far this doubling is propagated back along the homogeneous line depends entirely on the shape of the incident pulse. In every case the incident voltage pulse is totally reflected with the same polarity, and the incident current pulse with opposite polarity, without any distortion in shape.

If a stepped sinusoid, as shown in Fig. 4.3, arrives at the open end of the line, it must obey the same laws of reflection. Figure 4.3 shows the sequence of superposed voltage pulses before, at, and after arrival of the original pulse at the open end of the line. We see that a sinusoidal standing voltage wave is set up on the line behind the reflected rectangular front. This standing wave has nodal zero points every half-wavelength, with the first one a quarter-wavelength to the left of (that is, behind) the open end. Such standing waves can be generated by pulse trains of high frequency in lines, and in transformers and other windings of great length.

(*b*) *Short-circuit line termination.* If the homogeneous line terminates in a short circuit between the two conductors, there can be no voltage at this point. This establishes the boundary condition

$$e = 0, \tag{4.9}$$

and thus, by Eq. (4.3)

$$e_r = -e_v. \qquad (4.10)$$

Thus here too we have a reflected rearward-traveling voltage pulse of magnitude equal to the incident one but of opposite polarity. This gives rise, according to Eqs. (4.2), (4.10), and (4.1), to a reflected current pulse

$$i_r = -\frac{e_r}{Z} = \frac{e_v}{Z} = i_v. \qquad (4.11)$$

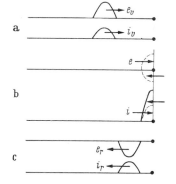

a

b

c

Fig. 4.4

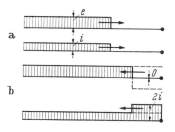

a

b

Fig. 4.5

Thus it is equal in magnitude and polarity to the incident current pulse. The total current at the short-circuited end is then

$$i = i_v + i_r = 2i_v, \qquad (4.12)$$

that is, twice the originally incident pulse.

Figure 4.4 shows this reflection for a single pulse and Fig. 4.5 for a step pulse. We see that in the case of reflection by a short circuit the voltage and current pulses have interchanged roles. Both are also reflected without distortion of shape, but now the current pulse maintains polarity as well as amplitude, whereas the voltage pulse reverses polarity

while maintaining amplitude. Thus at the short-circuited end the voltage is reduced to zero and the current is doubled. Some time after arrival of the incident pulse (Fig. 4.4c) we have an equal rearward-traveling pulse of merely opposite voltage polarity. The charge in the pulse is simply transferred to the opposing conductor at the short circuit. Figure 4.5b shows that step pulses lead in this case also to a change of current and voltage along the entire line propagated backward from the short-circuited end as a doubling of the current and a reduction of the voltage to zero.

(*c*) *Free oscillations in lines of finite length.* Figure 4.6 shows a single pulse traveling along

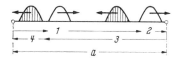

Fig. 4.6

a line open at both ends. We restrict our examination to the voltage pulse, which is completely reflected, accompanied by a temporary doubling, at each end. Let the sequence of events start with the pulse in position 1; it then propagates with the velocity of light to position 2, after reflection at the

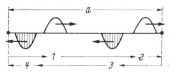

Fig. 4.7

end to 3, to 4, and, after reflection at the opposite end, back again to 1. Thus the pulse travels to and fro continuously along the line with the velocity of light.

If we take into account the resistance losses in the line conductors and insulation, we see that the pulse decays exponentially with time and distance traveled, according to the equations developed in Chapter 1. In this process, both the charge and the heat due to resistance losses are uniformly distributed over the length of the line.

If both ends of the line are short-circuited, as shown in Fig. 4.7, the polarity of the voltage pulse is reversed at each reflection. Thus after one complete cycle, including two reflections, the pulses reach the same position with the same polarity and thus the process is as repetitive as for the open-circuit case.

If we designate the period τ as the time taken to travel a complete cycle of two line lengths a with the velocity v, then

$$v\tau = 2a, \qquad (4.13)$$

and thus the period of oscillation of either an open or a short-circuited line is

$$\tau = \frac{2a}{v}. \qquad (4.14)$$

The number of oscillations in 2π sec is

$$\nu = \frac{2\pi}{\tau} = \pi\frac{v}{a}, \qquad (4.15)$$

which is called the natural, free, fundamental, or "eigen" frequency of the line. Equation (4.15) is in complete agreement with the first solution of Eq. (1.50).

Figure 4.8 shows a line that is open at one end and short-circuited at the other. If the

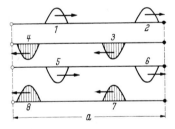

Fig. 4.8

sequence of events for a voltage pulse starts in position 1, the pulse propagates with the velocity of light to position 2, after reflection and polarity reversal to 3 and 4, after reflection to 5 and 6, after reflection and polarity reversal to 7 and 8, and finally back to 1 again, after a fourth reflection and two complete round trips along the line. Thus here too we have a cyclic or periodic process. However, because the line length is traversed four times per period,

$$\tau = \frac{4a}{v}. \qquad (4.16)$$

for a line of length a open at one end and short-circuited at the other. In this case resistance losses will also produce a gradual decay of the pulses as they travel along the line.

The last case (Fig. 4.9) is similar to an

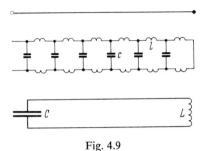

Fig. 4.9

oscillatory-circuit, but with the inductance and capacitance, distributed uniformly along the length of the line. If we represent these distributed values as concentrated in an inductance at the short-circuited end and a capacitance at the open end, the line will constitute an oscillatory circuit with a natural frequency (see Ref. 1, Eq. (15), p. 49)

$$\nu = \frac{1}{\sqrt{LC}} \qquad (4.17)$$

However for the line with distributed l and c, Eq. (4.16) gives

$$\nu = \frac{2\pi}{\tau} = \frac{\pi v}{2a}. \qquad (4.18)$$

If we now insert the velocity of propagation according to Eq. (1.14) and lump the distributed inductance l and capacitance c into the total values L and C for the entire length of the line, we obtain

$$\nu = \frac{\pi}{2}\frac{1}{\sqrt{al \cdot ac}} = \frac{\pi}{2}\frac{1}{\sqrt{LC}}. \qquad (4.19)$$

This shows that the natural frequency of a line open at one end and short-circuited at the other with distributed inductance and capacitance is greater by a factor $\pi/2$ than that of a similar oscillatory circuit having lumped inductance and capacitance of the same values. Equation (4.18) gives the natural frequency of such a line as a function only of its length and the velocity of propagation of a pulse.

A line of length $a = 10$ km open at one end and short-circuited at the other thus has a natural frequency

$$\nu = \frac{\pi \cdot 3 \cdot 10^5}{2 \cdot 10} = 47{,}000 \text{ cy}/2\pi \text{ sec}$$

or $\quad f = 7{,}500$ cy/sec.

A cable of length 1 km has a natural frequency

$$\nu = \frac{\pi \cdot 1.5 \cdot 10^5}{2 \cdot 1} = 235{,}000 \text{ cy}/2\pi \text{ sec}$$

or $\quad f = 37{,}500$ cy/sec.

Thus the natural frequency of lines of moderate length is of the order of magnitude of thousands of cycles (kilocycles) per second. Only for open lines longer than 1000 km and cables longer than 500 km does the natural frequency begin to approach the fundamental operating frequency of 60 cy/sec.

(d) Switching on of short lines. If a given line is suddenly energized by switching onto a voltage source, voltage at the switching point is propagated down the line in the form of a step pulse with a steep front. This is not influenced initially by the load or other conditions at the far end of the line. The pulse travels with the velocity of light toward the far end, where it is reflected, undergoing a change, and then travels back to the switching point. Only upon arrival of this reflection at the switching point are the processes here influenced by the open circuit, short circuit, or load at the far end of the line. The effect of the far-end termination thus enters into the calculation of voltage and current relations at the switching point through the amplitude and shape of the reflected pulses. Thus the steady-state currents and voltages in the line are not established immediately after switching but are formed in transient fashion by the to-and-fro travel of the pulses. The transient persists until the traveling pulses have decayed, that is, been damped out.

Let us examine the development of current and voltage in short-circuited and open lines shortly after switching onto a constant voltage source of negligible internal impedance. Let the line be so short that the damping of the pulses is negligible during the first few periods of traversal of the line. As was just shown

above, it is then also very short in comparison with the wavelength associated with the operating frequency of 60 cy/sec. Thus the time interval to be examined is only a small fraction of the steady-state alternating-current period, and we can disregard the operating low-frequency changes in the voltage source; for major changes, say during about a tenth of the operating period, the traveling pulses, as we have previously shown, are completely damped out. Thus for all such cases of switching for both direct and alternating voltages we can assume a constant voltage source of negligible internal impedance. In practice this can be represented by a very large capacitor or a battery of high load capacity.

The source then always imposes a constant voltage E at the switching point. Thus each rearward-traveling voltage pulse e_r impinging from the line onto the source (essentially a short circuit for pulses) must, by Eq. (4.3), produce a reflected pulse traveling into the line

$$e_v = -e_r, \qquad (4.20)$$

since otherwise the voltage e at the switching point would not remain equal to E. This corresponds to a current pulse advancing into the line that is of the same polarity as the voltage, in accordance with Eq. (4.1), and, by Eq. (4.2), is related to the incident current pulse so that

$$i_v = i_r. \qquad (4.21)$$

Thus the laws of reflection at a constant voltage source of negligible internal impedance are seen to be entirely analogous to those for a short-circuited line termination given in Eqs. (4.10) and (4.11).

Figure 4.10 shows a sequence of instants during the charging of a line open at the far end. Figure 4.10a shows the initial pulses traveling into the line. Figure 4.10b, like Fig. 4.2, shows the first reflection at the open end, leading to doubling of voltage and reduction of current to zero. The rearward-traveling positive voltage pulse then impinges on the source and is reflected with reversed polarity according to Eq. (4.20). In Fig. 4.10c this is shown to lead by superposition to a reduction of the doubled voltage $2E$ on the line back to the source voltage E. This negative voltage pulse corresponds to a forward-traveling charging-current pulse, also of negative polarity, and a reflection according to Eq. (4.21) of the rearward-arriving discharge-current pulse. On the second arrival at the open end of the line the negative voltage pulse

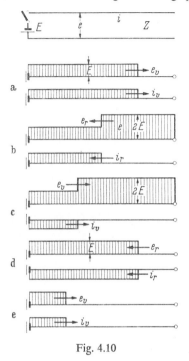

Fig. 4.10

is reflected without change in polarity and thus discharges the line entirely to zero voltage. The current pulse is reflected with a change of polarity and thus also completely discharges the line. Thus, as shown in Fig. 4.10*d*, the line is again free of voltage and current after a twofold traversal in both directions. The next reflection at the source, which occurred between Fig. 4.10*d* and 4.10*e*, has produced conditions exactly the same as those of Fig. 4.10*a* and thus the cycle can be repeated.

This representation shows that after switching the open line is alternately charged to twice the impressed voltage and completely discharged, and this is accompanied by a current alternating between positive and negative polarity. Voltage and current thus oscillate about their final values with a period or frequency that can be calculated from Eq. (4.16) or (4.18), respectively. In essence, then, the line behaves like a lumped *LC*

circuit when it is switched. There are, however, finer details, particularly in the form of spatial distribution, that are subject to more complicated processes. Above all, the oscillation of voltage and current is not sinusoidal but has the form of quite sharply rectangular pulses.

In a real line there will be added ohmic resistance not shown in Fig. 4.10. This will gradually damp out the step pulses. Thus the voltage will gradually approach the mean value between 2*E* and 0 and thus become equal to that of the source *E*. The current pulses are also damped out during this time to their mean value of 0.

If one were to measure the current at the switching point and the voltage at the open

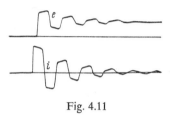

Fig. 4.11

end of the line, one would obtain, in accordance with Fig. 4.10, a rectangular pulse train between the limits of opposite current values and the voltage values of 2*E* and 0. Figure 4.11 shows an oscillogram taken by K. W. Wagner on an artificial line in which distributed inductance and capacitance per unit length were made sufficiently large that the traveling-pulse oscillations were not distorted by the time constants of the coil-type mirror oscillograph used. The gradual decay of the switching transient due to losses in the line is very clear. It is evident that the resulting pulse train is composed of half-wave rectangular single pulses.

Figure 4.12 shows the switching on of a line short-circuited at the far end. The first forward pulse in Fig. 4.12*a* is exactly like that of Fig. 4.10*a*. However, upon arrival at the shorted end, as shown in Fig. 4.5, the reflected pulses reduce the voltage to zero and increase the current to twice the value of the original incident pulse. Thus Fig. 4.12*b* shows a backward-traveling pulse of negative voltage and positive current. Upon arrival at the source, the voltage pulse is reflected according to Eq. (4.20) with a reversal of polarity, and thus changes the line anew. The current pulse is reflected according to Eq. (4.21) without a

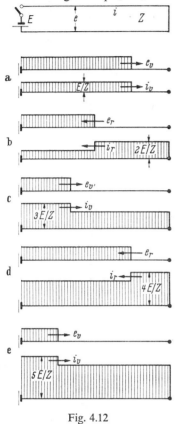

Fig. 4.12

change of polarity. Thus the superposition produces (Fig. 4.12*c*) a current pulse of three times the amplitude of the original one shown in Fig. 4.12*a*. The second reflection at the shorted end (Fig. 4.12*d*) again produces a voltage reduction to zero and an increase of current to four times the original first incident

pulse value. Figure 4.12*e* shows that after reflection at the source end we have again a forward-traveling voltage pulse of source-voltage amplitude and a current pulse creating five times the amplitude of current originally flowing into the line. The same processes repeat during successive reflections of the pulses; the voltage alternates between E and 0 and the current increases continuously.

This behavior of the pulses shows that the source repeatedly tries to charge the line to its voltage E but cannot succeed because the

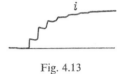

Fig. 4.13

short-circuited end always reflects a discharge pulse, thus negating the charging of the line. The source reacts with a continuous supply of current pulses of the same polarity, which results in a stepwise or staircase increase of current, which would continue indefinitely in lossfree lines. Thus the short-circuit current of a line develops in a finite series of steps in staircase fashion, in contrast to the sudden full-stop increase for an ideal resistance or the gradual exponential increase for an ideal inductance.

In this case the losses in the line also gradually reduce the amplitude of the oscillating pulses. The steps of the continually increasing current staircase become smaller and smaller, until the total value reaches the steady-state value of the short-circuit current given by the resistance of the line. Figure 4.13 is the oscillogram of the current increase in an artificial line terminated in a short circuit taken by K. W. Wagner.

CHAPTER 5

DISTORTION OF PULSE SHAPE

We saw in Chapter 1 that the shape of traveling pulses in homogeneous lines will not be distorted if the effects of the resistances of the line conductor and the insulation are relatively small. This is not always the case. The conductor resistance is considerably increased by skin effects owing to high-frequency components associated with the steeper pulse-front features. The insulation resistance is reduced by corona and glow-discharge-current effects on the surfaces of the conductors during periods of high voltage. Both effects produce distortions of the initial pulse shape.

(a) *Damping by corona effects.* When the amplitude of the voltage pulse e exceeds the corona voltage limit e_0, a glow-discharge current i_g flows from the surface of the conductor into the air insulation. This causes a very heavy energy loss in the pulses. As this current is approximately proportional to the excess voltage $e - e_0$, we can set for the loss per unit length of line

$$W_g = ei_g = \frac{e(e - e_0)}{p}, \qquad (5.1)$$

where p is an insulation resistance corresponding to the glow-discharge losses in the line. It is measured in ohm kilometers and is considerably lower than the usual insulation resistance for normal voltage on the line.

By Eq. (1.27) the energy of the pulse as a function of its voltage is

$$W = \frac{e^2}{Z}, \qquad (5.2)$$

and its change in each line element is

$$dW = \frac{2e\,de}{Z}. \qquad (5.3)$$

Because this energy decrease is caused by the glow-discharge loss in the element of length dx, the energy balance gives

$$-\frac{2e\,de}{Z} = \frac{e(e - e_0)}{p}\,dx, \qquad (5.4)$$

from which the differential equation for voltage is

$$\frac{de}{e - e_0} = -\frac{Z}{2p}\,dx. \qquad (5.5)$$

The solution is

$$e - e_0 = (E - e_0)\varepsilon^{-Zx/2p}. \qquad (5.6)$$

Here E is the initial value of the voltage pulse and we see that in its travel along the length x

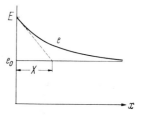

Fig. 5.1

of the line it is reduced exponentially to the corona-voltage limit, as shown in Fig. 5.1.

The damping-distance constant

$$X = \frac{2p}{Z} \qquad (5.7)$$

depends on the kind of conductor as well as on the shape and polarity of the pulses. A typical mean value of the glow-discharge resistance of 100-kV lines to earth is $p = 2500\ \Omega$ km, which gives

$$X = \frac{2 \cdot 2500}{500} = 10 \text{ km}.$$

For all excess-voltage pulses, then, those portions for which the voltage exceeds the corona limit, that is, the limit for onset of

36

glow-discharge current, are reduced exponentially with time and hence progress as shown in Fig. 5.2. During this process the relative shape of these portions is not distorted. The portions for which the voltage lies below the limit initially are not affected at all. Because the onset of glow discharge occurs in most high-voltage lines at a voltage only about 30 percent above the peak operating voltage,

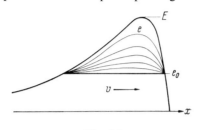

Fig. 5.2

we see that any traveling pulse, however high in initial voltage, will have become of moderate amplitude after the traversal of a line length equal to a few damping distances, that is, 20 to 30 km.

(b) *Smoothing of the pulses by skin effects.* Traveling pulses in power lines can develop either between two conductors of copper or aluminum or between one such conductor and the ground return. They can further develop between a grounding conductor and the actual ground or line conductor. The ohmic resistance for pulse currents is quite different for these different modes of propagation.

The resistance of nonmagnetic conductors per unit length for high-frequency components (Ref. 1, Eq. (41), p. 404) is

$$r = \frac{2}{d}\sqrt{fs} = \frac{1}{d}\sqrt{\frac{2}{\pi}s\omega}; \qquad (5.8)$$

here ω, the angular frequency, is introduced; s is the specific resistance and d the diameter of the conductor. Steel conductors with effective permeability μ have far higher resistance, namely,

$$r_\mu = \frac{1}{d}\sqrt{\frac{2}{\pi}s\mu\omega}. \qquad (5.9)$$

Finally, the resistance of ground-surface

layers, in which the return currents of pulses often travel, is (Ref. 1, Eq. (38), p. 403)

$$r_e = \frac{1}{h}\sqrt{\frac{s\omega}{\pi}}, \qquad (5.10)$$

where h is the height above ground of the forward conductor. All these resistances are

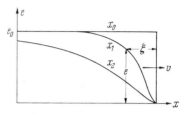

Fig. 5.3

proportional to the square root of the frequency and thus we can express them as

$$r = \rho\sqrt{\omega}, \qquad (5.11)$$

where ρ is a frequency-invariant basic constant which can be determined from Eqs. (5.8)–(5.10). For the complete circuit pertaining to pulse currents we can then simply add the two ρ's for the forward and return conductors.

The resistance damping of high-frequency pulse components along the distance x is, by Eq. (1.36),

$$\frac{e}{e_0} = \varepsilon^{-rx/2Z} = \varepsilon^{-\rho\sqrt{\omega}x/2Z}. \qquad (5.12)$$

Thus the damping depends not only on the distance traveled but also on the frequency. Hence the distortion of the pulse shape increases with increasing ω, that is, with increasing high-frequency content caused by the abrupt changes with time in the pulse envelope. For step pulses, the steep jump-function front shown in Fig. 5.3 implies an infinite time derivative at the front or zero rise time. Owing to the high-frequency content, we must then expect that after a short time this sharp rise will be broadened over a longer interval to reduce its steepness.

According to a rule due to Heaviside, we may introduce the time derivative directly to replace the continuous frequency spectrum of such step pulses. If we count time from the initiation of the pulse, we can then put

$$\omega = \frac{1}{t}. \qquad (5.13)$$

The voltage of the pulse as a function of distance traveled and time taken is then

$$\frac{e}{e_0} = \varepsilon^{-\rho x/2Z\sqrt{t}}. \qquad (5.14)$$

We can interpret this equation to state that a given voltage amplitude e falls behind the pulse front by a time interval t during its travel over a distance x along the line. This corresponds to a space interval

$$\xi = vt, \qquad (5.15)$$

which is shown in Fig. 5.3. If we substitute Eq. (5.15) in Eq. (5.14), we obtain the spatial behavior of the pulse front in the form

$$\frac{e}{e_0} = \varepsilon^{-\rho x\sqrt{v}/2Z\sqrt{\xi}}. \qquad (5.16)$$

It is shown in Fig. 5.3 for several distances of travel x_0, x_1, x_2; however, the starting instant of the pulse is always pushed back in this illustration to the original starting instant for x_0. Chart II shows this function in more detail.

Thus the front of the rectangular step pulse is flattened out as the pulse travels along the line. Since, by Eq. (5.16), the square root of the space measure ξ is proportional to the distance x covered, ξ increases as the square of the distance traversed, that is, x^2. Thus ξ increases slowly at first but more rapidly after a time. From Fig. 5.3 we see that the voltage rise in the pulse front is small at first but then becomes more rapid and finally slows down again, to approach its ultimate value gradually. Because the exponent in Eq. (5.14) depends on x and t quite differently, we must recognize that the pulse variation with time at a given point of the line is no longer equal to the pulse variation with distance along the line at a given time.

In analogy with our earlier considerations, let us designate as the front-rise constant K of the original step pulse the value of ξ for which the voltage has approached its final

value to within $1/\varepsilon$ or 36.8 percent. From Eq. (5.16) we then obtain

$$\ln(1 - 1/\varepsilon) = -\frac{\rho\sqrt{v}}{2Z}\frac{x}{\sqrt{K}}, \qquad (5.17)$$

and this gives for the front-rise constant

$$K = \frac{\rho^2 v x^2/4Z^2}{\ln^2[\varepsilon/(\varepsilon - 1)]}. \qquad (5.18)$$

The denominator is simply

$$\ln^2\left(\frac{\varepsilon}{\varepsilon - 1}\right) = \frac{1}{4.8},$$

and we see that the front-rise constant depends only on the properties of the line and increases with the square of the distance traveled x.

For pulses traveling between two open conductors, ρ is given by Eq. (5.11) as twice the value in Eq. (5.8). Thus the front-rise constant becomes

$$K = \frac{9.6}{\pi}\frac{sv}{Z^2 d^2}x^2. \qquad (5.19)$$

For pulses traveling between a line conductor and a steel grounding conductor, the latter has an overwhelming influence. Thus, according to Eq. (5.9),

$$K_\mu = \frac{2.4\mu}{\pi}\frac{sv}{Z^2 d^2}x^2. \qquad (5.20)$$

Finally, for pulses traveling between a line conductor and ground and also approximately for pulses between a steel grounding conductor and ground, the overwhelming effect of the ground resistance gives, according to Eq. (5.10),

$$K_e = \frac{1.2}{\pi}\frac{sv}{Z^2 h^2}x^2. \qquad (5.21)$$

Table 5.1 gives the front-rise lengths of pulses for different distances traveled x. The Table is based on a 100-kV line of diameter $d = 1.2$ cm, specific resistance $s = 1.8 \cdot 10^3$ cm^2/sec, and $h = 12$ m. The steel grounding conductor has the same diameter, resistance $s = 13 \cdot 10^3$ cm^2/sec, and for the high pulse currents a permeability $\mu = 100$. The ground resistance is assumed to be for a moist soil, $s = 10^{13}$ cm^2/sec, and the surge impedance to be uniformly 500 Ω.

In Table 5.1 we see that the reduction in

TABLE 5.1. Front-rise length of step pulses.

Distance traveled (km)	Two copper lines (m)	One line and steel ground conductor (km)	One line and ground return (km)
1	0.046	0.0083	0.033
10	4.6	0.83	3.3
100	460	83	330

steepness of the pulse front is small between two copper conductors, but that it becomes larger for return along a steel ground conductor and larger still for a ground return. In the last two cases the danger due to the steep front rise of the pulses is already reduced significantly after travel over a few kilometers.

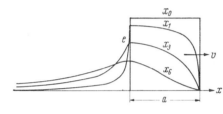

Fig. 5.4

If the pulse current returns through both the ground and a steel ground conductor, we obtain a reduction depending on the division of current between the two return paths and thus given by the parallel resistance of the two paths, which is always less than the smaller of them.

Single pulses of initially rectangular shape, as shown in Fig. 5.4, can be treated as the superposition of two step pulses of opposite sign separated by a short time interval. This superposition results in a loss of peak-voltage amplitude as the pulse progresses along the line. Because the highest voltage occurs almost at the end of the original square pulse, that is, for $\xi = a$, we obtain for the reduction of peak voltage, from Eq. (5.16),

$$\frac{e}{e_0} = \varepsilon^{-(\rho/2Z)\sqrt{v/ax}}. \qquad (5.22)$$

From this we see that the reduction is greater for initially shorter pulses and that apart

from this it increases in general exponentially with increasing distance traveled x. Thus the damping constant for the voltage reduction of the single square pulse is

$$X = \frac{2Z}{\rho}\sqrt{\frac{a}{v}}, \qquad (5.23)$$

and its time constant is

$$T = \frac{X}{v} = \frac{2Z}{\rho v}\sqrt{\frac{a}{v}}. \qquad (5.24)$$

By insertion of the different values of ρ from Eqs. (5.8), (5.9), and (5.10), we obtain for pulses traveling between two conductors,

$$X = \sqrt{\frac{\pi}{2}\frac{a}{vs}}\,Zd, \qquad (5.25)$$

for pulses between one conductor and one steel ground conductor,

$$X_\mu = \sqrt{\frac{2\pi a}{vs\mu}}\,Zd, \qquad (5.26)$$

and for pulses between one conductor and ground,

$$X_e = 2\sqrt{\frac{\pi a}{vs}}\,Zh. \qquad (5.27)$$

For a square pulse of length $a = 1$ km, corresponding to a time interval of 3.3 μsec, the spatial and temporal damping constants are given in Table 5.2 for lines of the dimensions used earlier. We see from this table

TABLE 5.2. Damping constants due to skin effect.

Damping constant	Traveling pulse on—		
	Two copper lines	One line and steel ground conductor	One line and ground return
Spatial, X (km)	325	24	12
Temporal, T (μs)	1080	80	40

that the voltage reduction is very small for short distances along two good conductors made of copper. It increases to quite considerable values if one of the conductors is made of steel, and is largest if the ground is used as a return conductor. In this process the original sharp rectangular pulse is broadened into a rounded hill with a long drawn-out tail, as shown in Fig. 5.4. In this process the rise distance of the hill remains constant at a, while the tail becomes longer and longer. For longer pulses Eqs. (5.25)–(5.27) show that the damping reduction decreases with the square root of the original pulse length a, but increases with decreasing height of the

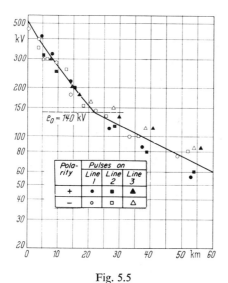

Pola-rity	Pulses on		
	Line 1	Line 2	Line 3
+	●	■	▲
−	○	□	△

Fig. 5.5

line h and decreasing surge impedance Z. If the return current uses both a steel ground conductor and the ground, it will produce less damping than one alone. For equal currents on both, the spatial damping constant, according to Eq. (5.23), would become the sum of the two separate ones and thus in the example in Table 5.2 X would become 36 km.

The most dangerous traveling pulses are generally produced by lightning and travel between a line conductor or a steel grounding conductor and the ground. It must be noted that the reduction of steepness during their rise time as given in Eqs. (5.19)–(5.21) and their peak-voltage amplitude as given by Eqs. (5.22)–(5.27) depend strongly on the specific resistance of the ground. Under

comparable conditions the rise period is proportional to s and the voltage reduction to the square root of s. Thus the condition of the intervening ground strongly affects the danger of pulses from distant lightning storms. Wet soil increases these dangers by good transmission, whereas dry soil decreases them.

If the skin and corona effects occur simultaneously for high overvoltage pulses, the highest voltage pulse is, by Eqs. (5.6) and (5.22),

$$e = [(E - e_0)\varepsilon^{-x/X_k} + e_0]\varepsilon^{-x/X_s}, \quad (5.28)$$

where X_k is the space constant of the corona effect and X_s is the space constant of the skin effect. In Fig. 5.5 are presented some measurements that were taken on a 110-kV line on which artificial traveling pulses of 300 to 500 kV were impressed. Thus the corona voltage limit e_0 [= 140 kV] was exceeded by a factor of more than 2. Voltage is plotted on a logarithmic scale, and from the line plotted to fit the measurements we see that above the corona voltage limit e_0 Eq. (5.28) and below the limit Eq. (5.22) give a good fit with the measurements. At e_0 there is a definite kink in the plotted line. As mean values for the different experimental arrangements we obtain for the damping constants $X_k = 15$ km and $X_s = 45$ km, which are in good agreement with our earlier calculations.

If the glow-discharge current on the line is very large, its capacitance will change owing to the increase in effective diameter of the conductors. The inductance, however, is unaltered because the axial currents continue to flow only in the conductors themselves. Thus the surge impedance of the line with glow-discharge currents is somewhat reduced. According to Eq. (5.7), this reduces damping by corona effects, but it increases damping due to skin effects in accordance with Eq. (5.23). The difference in surge impedance is not great because, as shown in Eq. (1.41), the conductor diameter enters only logarithmically. For an arrangement of multiple conductors, Eq. (1.45) shows that the influence is negligible owing to their equivalent diameter.

PART II

COMPOSITE LINES

CHAPTER 6

TRANSMISSION AND REFLECTION

The ratio of voltage to current of a traveling pulse is given by the surge impedance of the line. Thus a traveling pulse that arrives at a junction between two lines of different surge impedances must undergo a change as it travels across this junction. We shall investigate the change in pulses for various junctions.

(*a*) *Junction of two lines.* Figure 6.1 shows a simple junction between two lines. To distinguish between the two lines we shall use

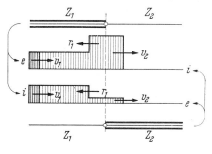

Fig. 6.1

the subscripts 1 and 2 for all symbols connected with the lines. For example, consider the junction between an underground cable of surge impedance Z_1 and an overhead line of surge impedance Z_2. At the junction the voltages and currents in the two lines must be equal; hence

$$e_1 = e_2 \qquad (6.1)$$

and

$$i_1 = i_2. \qquad (6.2)$$

Let a forward-traveling voltage pulse e_{v1} be given in line 1, the underground cable. If the line to the right is free of voltage, then after arrival of this pulse at the junction, line 2, the overhead line, can only contain a forward-traveling pulse e_{v2}. Line 1, however, has a reflected pulse e_{r1} impressed upon it which travels in a rearward direction away from the

junction. Thus resolution of the total voltages in Eq. (6.1) into partial pulses traveling in both directions gives

$$e_{v1} + e_{r1} = e_{v2}. \qquad (6.3)$$

Similarly, the resolution of currents in Eq. (6.2) into partial pulses i_{v1}, i_{v2}, and i_{r1}, together with substitution of the surge impedances, gives

$$\frac{e_{v1}}{Z_1} - \frac{e_{r1}}{Z_1} = \frac{e_{v2}}{Z_2}. \qquad (6.4)$$

From these two equations we can derive the reflected and transmitted pulses as follows. Multiply Eq. (6.4) by Z_1 and add Eq. (6.3), to obtain

$$2e_{v1} = e_{v2}\left(1 + \frac{Z_1}{Z_2}\right). \qquad (6.5)$$

This gives e_{v2} directly as a function of e_{v1}:

$$e_{v2} = \frac{2Z_2}{Z_1 + Z_2} e_{v1}. \qquad (6.6)$$

The corresponding current pulses can be simply derived by dividing both members by the surge impedance:

$$i_{v2} = \frac{2Z_1}{Z_1 + Z_2} i_{v1}. \qquad (6.7)$$

The reflected voltage pulse e_{r1} generated at the junction is derived from Eq. (6.3) by substituting in it Eq. (6.6) to obtain

$$e_{r1} = \frac{Z_2 - Z_1}{Z_1 + Z_2} e_{v1}. \qquad (6.8)$$

The corresponding reflected current pulse is again obtained by dividing by the surge impedance:

$$i_{r1} = \frac{Z_1 - Z_2}{Z_1 + Z_2} i_{v1}. \qquad (6.9)$$

The change of sign between Eqs. (6.8) and (6.9) is due to the fact that, in accordance with Eq. (4.2), $e_{r1} = -Z_1 i_{r1}$.

43

Equations (6.6)–(6.9) give the laws of transmission and reflection for traveling pulses at a junction between two lines. We see that the process of transmission and reflection is governed solely by the magnitude of the surge impedances of the two lines. The transmitted pulse always has the same polarity as the incident pulse, but the reflected pulse undergoes a change of polarity in either its voltage or its current.

If a pulse travels from a line of lower impedance into a line of higher impedance, so that Z_2 is greater than Z_1, then according to Eq. (6.6) the amplitude of the voltage pulse is increased and according to Eq. (6.7) the amplitude of the current pulse is decreased

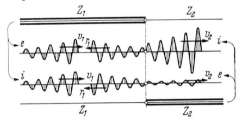

Fig. 6.2

after arrival at the junction. This is shown in Fig. 6.1 for the case of an underground cable joined to an overhead line. In the limiting case when Z_2 becomes very much larger than Z_1 the junction acts almost as an open-circuit termination. Thus it produces a severe reduction in current amplitude and in the limit a twofold increase in voltage amplitude. However, in this case the current and voltage pulses are transmitted in a forward direction into the second line beyond the junction as well as reflected back along line 1.

The lower portion of Fig. 6.1 shows an overhead line of higher impedance Z_1 joined to an underground cable of lower impedance Z_2. In this case the junction produces, in accordance with Eq. (6.6), a reduction in voltage amplitude and, in accordance with Eq. (6.7), an increase in current amplitude. In the limiting case when Z_2 becomes very much smaller than Z_1 the junction acts almost as a short-circuit termination. The voltage is reduced severely and the current is increased to almost twice the amplitude of the incident value.

Both changes of surge impedance are shown in Fig. 6.1 for the passage of a rectangular step pulse at a junction. Figure 6.2 shows the transmission and reflection at a junction of a multiple pulse train and a single pulse will give similar results. In both figures the pulse shapes of voltage and current given for the case $Z_2 > Z_1$ also give, owing to the reciprocity between Eqs. (6.6) and (6.7) on the one hand and between Eqs. (6.8) and (6.9) on the other, the pulse shapes of current and voltage respectively for the case $Z_2 < Z_1$.

The power of the forward-traveling pulse is reduced at the junction, since a part of the arriving pulse is reflected.

The power in the pulse arriving at the junction is, by Eq. (1.27),

$$W_1 = \frac{e_{v1}^2}{Z_1}, \tag{6.10}$$

and the transmitted pulse power is

$$W_2 = \frac{e_{v2}^2}{Z_2}. \tag{6.11}$$

Thus the ratio is, in accordance with Eq. (6.6),

$$\frac{W_2}{W_1} = \frac{Z_1}{Z_2}\left(\frac{2Z_2}{Z_1 + Z_2}\right)^2 = \left(\frac{2}{\sqrt{Z_1/Z_2} + \sqrt{Z_2/Z_1}}\right)^2. \tag{6.12}$$

Since this ratio remains constant if the values of Z_1 and Z_2 are interchanged, the loss of power at a junction is independent of the direction in which a pulse is transmitted through it.

The reflected pulse power is

$$W_r = \frac{e_{r1}^2}{Z_1}. \tag{6.13}$$

Its ratio to the incident pulse power is, from Eq. (6.8),

$$\frac{W_r}{W_1} = \left(\frac{Z_1 - Z_2}{Z_1 + Z_2}\right)^2. \tag{6.14}$$

Thus it is always positive and depends on the square of the difference between the surge impedances of the two lines. For lines of only slightly different surge impedances, the reflected pulse power becomes negligibly small. Also, as seen in Eq. (6.12), the power content of the transmitted pulse is about equal to that of the incident pulse.

By use of the laws of pulse transmission and reflection in Eqs. (6.6)–(6.9) we can follow the propagation of traveling pulses in any electrical network. Figure 6.3 shows a bifurcation in a line of uniform structure in all three sections. A pulse arriving at the junction is divided into several parts. The transmitted pulses travel in parallel on their lines. At the junction, then, the impedance seen is that due to the two equal surge impedances in parallel and thus equal to half the surge impedance of the lines involved. The junction also gives rise to a reflected pulse that reduces the voltage on the line on which the incident pulse reaches the junction.

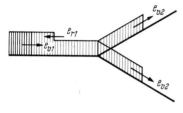

Fig. 6.3

In the case of simple bifurcation, as in Fig. 6.3, the voltage is reduced, in accordance with Eq. (6.6), to

$$e_{v2} = \frac{2 \cdot \frac{1}{2}}{1 + \frac{1}{2}} e_{v1} = \frac{2}{3} e_{v1}.$$

Thus three-phase windings experience upon the arrival of pulses a lower voltage when delta connected than when star connected.

If three, four, or more lines are joined at one point and a traveling pulse arrives along one of them at this junction, it is propagated forward as a transmitted pulse into all the other lines and only reflected as a backward-traveling pulse along the line of arrival. If, as is usually the case, the combined surge impedance of all the other lines in parallel is smaller than the surge impedance of the line of arrival, then, in accordance with Eq. (6.6), a reduced voltage pulse is transmitted beyond the junction.

For the case of n equal lines, joined as shown in Fig. 6.4 for $n = 4$, the combined surge impedance of all the outgoing lines is

$$Z_2 = \frac{Z}{n - 1}, \qquad (6.15)$$

where all the lines have surge impedances Z, and $Z_1 = Z$ for the single line along which the pulse arrives at the junction. The transmitted voltage pulse, from Eq. (6.5) or (6.6), is

$$e_{v2} = \frac{2}{1 + Z_1/Z_2} e_{v1} = \frac{2}{n} e_{v1}, \qquad (6.16)$$

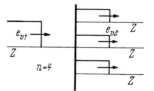

Fig. 6.4

and the reflected voltage pulse, from Eq. (6.8), is

$$e_{r1} = \frac{1 - Z_1/Z_2}{1 + Z_1/Z_2} e_{v1} = \left(\frac{2}{n} - 1\right) e_{v1}. \qquad (6.17)$$

The sum of the transmitted current pulses is, by Eq. (6.7),

$$i_{v2} = 2\left(1 - \frac{1}{n}\right) i_{v1} \qquad (6.18)$$

and the reflected current pulse, by Eq. (6.9), is

$$i_{r1} = \left(1 - \frac{2}{n}\right) i_{v1}. \qquad (6.19)$$

Most important in the operation of networks is the voltage that arises at a junction or a bus bar upon the arrival of a traveling

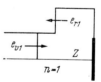

Fig. 6.5

pulse. We see from Eq. (6.16) that its amplitude is twice that of the incident pulse divided by n, the number of equal lines connected at the junction. Such a junction, then, provides an excellent way of reducing damaging traveling pulses. For $n = 1$, Eq. (6.16) gives a doubling of the voltage, as was shown in Chapter 4 to occur for any open-circuited line termination and as shown in Fig. 6.5 for such a terminal station. For $n = 2$ we obtain a switching station as shown in Fig. 6.6 through which, according to Eq. (6.16), the pulses travel without change. For $n = 3$ we have the simple bifurcation shown in Fig. 6.3, which, as we saw, reduces the voltage amplitude to two-thirds of the incident value. For $n = 4$ the

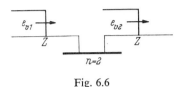

Fig. 6.6

voltage amplitude is reduced by a factor of 2, as shown in Fig. 6.4. For $n = 5$, as shown in Fig. 6.7, one approaches the condition prevailing in major switching stations and the voltage amplitude is reduced to 40 percent of the incident value.

More dangerous than the parallel connection at junction points is the series connection of dissimilar lines. Figure 6.8 shows a cable charged from a constant voltage source E by a rectangular step pulse propagating along it. At the end of the cable there is a junction to

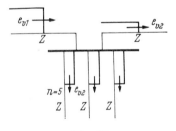

Fig. 6.7

an overhead line whose surge impedance is several times that of the cable, so that the amplitude of the voltage step pulse is amplified in transmission to almost $2E$. At the far end the overhead line is joined to a coil of high inductance, such as a transformer winding, whose surge impedance is high compared to that of the overhead line. Thus in transmission past this junction into the coil winding the voltage pulse is again amplified, to almost $4E$. If, finally, the pulse has traveled to the open end of the coil winding, it is totally reflected here and attains an amplitude of up to $8E$,

if all transmission and reflection losses can be neglected. Such amplification of traveling voltage pulses will always occur if they travel in sequence into lines of monotonically increasing surge impedance. Conversely, decreasing surge impedance produces a decay of voltage pulses.

Figure 6.9 shows a three-phase line connected to a transformer with voltages e between the individual lines and E between all

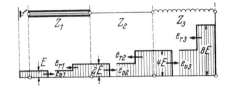

Fig. 6.8

the lines and ground. The voltage E arises from traveling pulses produced in the system by atmospheric discharges and it is seen that these pulses enter the three windings at the same time and arrive simultaneously at the star junction, which thus constitutes an open-circuit termination. Thus such pulses produce high voltages not only at the beginning of the windings but also in particular at the floating

Fig. 6.9

star point. In contrast, e represents traveling pulses generated by simultaneous switching in the line, and the star point acts as a short circuit for them, so that high voltages can be produced only at the beginning of the windings.

(b) Series resistance inserted at a junction. Figure 6.10 shows an ohmic resistance R connected between an overhead line and an underground cable. The purpose of this series-connected resistance is to avoid the occurrence of high reflected voltage pulses. The resistance leaves the current summation at the junction unaltered, that is,

$$i_1 = i_2. \qquad (6.20)$$

However, the voltages in the two lines differ by the voltage drop in the resistance; thus,

$$e_1 = e_2 + Ri_2. \qquad (6.21)$$

Here the resistance is assumed to be physically concentrated in a space short in comparison with that over which the pulse extends.

If we again decompose the pulse in line 1 into forward- and backward-traveling components, while line 2 contains only a forward transmitted pulse, then from Eq. (6.21) the voltage pulses are

$$(6.22)$$

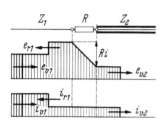

Fig. 6.10

and from Eq. (6.20) the current pulses are

$$\frac{e_{v1}}{Z_1} - \frac{e_{r1}}{Z_1} = \frac{e_{v2}}{Z_2}. \qquad (6.23)$$

If we multiply Eq. (6.23) by Z_1 and add Eq. (6.22), we get for the forward-traveling transmitted pulse in line 2

$$e_{v2} = \frac{2Z_2}{Z_1 + Z_2 + R} e_{v1}. \qquad (6.24)$$

We note that the transmitted pulse can be reduced to any desired value by a suitable series resistance even for high values of Z_2. But R must then be substantially greater than the sum of Z_1 and Z_2, which leads in practice to values of several hundred ohms; this would produce serious losses owing to the normal steady-state operating currents. However, a large inductance or choke of low resistance can be provided as a parallel shunt for the operating currents.

The reflected voltage pulse can now be obtained from Eq. (6.22):

$$e_{r1} = \frac{Z_2 - Z_1 + R}{Z_2 + Z_1 + R} e_{v1}. \qquad (6.25)$$

Even very large values of R will only produce an amplitude equal to that of the incident pulse.

Equation (6.10) gave the power contained in the incident rectangular step pulse. The power loss in the resistance is

$$W_R = Ri_2^2 = R \frac{e_{v2}^2}{Z_2^2}. \qquad (6.26)$$

The ratio of the two powers gives the efficiency of the protective resistance. It is, by Eq. (6.24),

$$\frac{W_R}{W} = \eta = \frac{4RZ_1}{(Z_1 + Z_2 + R)^2}. \qquad (6.27)$$

This efficiency has a maximum when the protective resistance is taken as

$$R = Z_1 + Z_2, \qquad (6.28)$$

and this gives

$$\eta_{\max} = \frac{Z_1}{Z_1 + Z_2}. \qquad (6.29)$$

To extract as much energy as possible from a traveling pulse at a junction this arrangement is useful for the transmission from lines of high Z_1, such as overhead lines, into lines of low Z_2, such as underground cables, as shown in Fig. 6.10.

For surge impedances of $Z_1 = 500 \; \Omega$ and $Z_2 = 50 \; \Omega$ it is possible to absorb up to

$$\eta_{\max} = \frac{500}{500 + 50} = 91 \text{ percent}$$

of the energy of a traveling pulse. If a resistance is inserted in a homogeneous line of equal Z on both sides of the junction, only a maximum of 50 percent of the energy can be absorbed. For the opposite direction of pulse transmission and hence interchanged values of Z_1 and Z_2 a maximum of

$$\eta_{\max} = \frac{50}{50 + 500} = 9.1 \text{ percent}$$

of the energy could be absorbed, which is negligible in practice.

When the protective resistance for maximum energy absorption is used as given in Eq. (6.28), the transmitted voltage pulse becomes

$$e_{v2} = \frac{Z_2}{Z_1 + Z_2} e_{v1}, \qquad (6.30)$$

which is just half the value without protective resistance, as given in Eq. (6.6).

(c) *Parallel resistance inserted at a junction.* As we have just seen, there is a direction of pulse transmission across a junction in which only a small fraction of the energy in the pulse is dissipated in the series resistance. For this direction of travel a parallel resistance connected across the line at the junction, as shown in Fig. 6.11, is more effective. Now the voltage in both lines is the same,

$$e_1 = e_2, \tag{6.31}$$

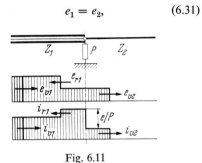

Fig. 6.11

but the currents in the two lines differ by the amount flowing in the parallel resistance P,

$$i_1 = i_2 + \frac{e_2}{P}. \tag{6.32}$$

Thus the voltage equality of the partial pulses is, by Eq. (6.31),

$$e_{v1} + e_{r1} = e_{v2} \tag{6.33}$$

and the current equality, by Eq. (6.32),

$$\frac{e_{v1}}{Z_1} - \frac{e_{r1}}{Z_1} = \frac{e_{v2}}{Z_2} + \frac{e_{v2}}{P}. \tag{6.34}$$

If we multiply Eq. (6.34) by Z_1 and add Eq. (6.33), we obtain the forward-traveling pulse in line 2,

$$e_{v2} = \frac{2\dfrac{1}{Z_1}}{\dfrac{1}{Z_1} + \dfrac{1}{Z_2} + \dfrac{1}{P}} e_{v1}, \tag{6.35}$$

and thus the reflected pulse in line 1 becomes, by Eq. (6.33),

$$e_{r1} = \frac{\dfrac{1}{Z_1} - \dfrac{1}{Z_2} - \dfrac{1}{P}}{\dfrac{1}{Z_1} + \dfrac{1}{Z_2} + \dfrac{1}{P}} e_{v1}. \tag{6.36}$$

Equation (6.35) shows that by making the parallel resistance small enough we can reduce the amplitude of the transmitted pulse to any arbitrarily small value. However, in such a case we must connect a condenser in series with the resistance to prevent the flow of the operating steady-state current through this bypass instead of through the working load.

The loss of power in the parallel resistance is

$$W_R = \frac{e^2}{P} = \frac{e_{v2}^2}{P}. \tag{6.37}$$

The ratio of the power loss to the power in the arriving pulse, from Eqs. (6.10) and (6.35), is

$$\frac{W_R}{W} = \eta = \frac{4\dfrac{1}{Z_1 P}}{\left(\dfrac{1}{Z_1} + \dfrac{1}{Z_2} + \dfrac{1}{P}\right)^2}. \tag{6.38}$$

Thus for a value of parallel resistance given by

$$\frac{1}{P} = \frac{1}{Z_1} + \frac{1}{Z_2} \tag{6.39}$$

the efficiency becomes a maximum,

$$\eta_{\max} = \frac{Z_2}{Z_1 + Z_2}. \tag{6.40}$$

Thus the arrangement becomes highly effective when the traveling pulses enter a line of higher surge impedance Z_2, as shown in Fig. 6.11.

For the previously used values of Z for underground cables and overhead lines the pulse will now lose 91 percent of its energy at the junction when it arrives from the direction shown, but only 9.1 percent when it arrives from the opposite direction.

For the optimum parallel resistance given by Eq. (6.39), the transmitted voltage pulse is, by Eq. (6.35),

$$e_{v2} = \frac{Z_2}{Z_1 + Z_2} e_{v1}, \tag{6.41}$$

which is again one-half the amplitude that it would have without the protective resistance.

Since without such a resistance a pulse traveling from a line of low surge impedance into one of high surge impedance produces a large increase in voltage amplitude, it is possible to use a spark gap connected in series with it to switch the parallel resistance *P* into operation only upon the actual occurrence of an overvoltage for which the spark gap is set. Such a spark gap then will switch the protective resistance in for overvoltages that may occur at the transition between underground cables and overhead lines or overhead lines and transformer windings. The protective resistance will then rapidly dissipate the pulse energy so as to reduce both the amplitude and the duration of overvoltage.

The energy of the arriving pulse cannot be wholly dissipated in either a series or a parallel resistance. If the surge impedances of the lines joined are substantially different, the examples given indicate that a proper choice of values and connection can produce almost 100-percent loss of energy. However, for two homogeneous lines with $Z_1 = Z_2$ protective resistances can at best dissipate 50 percent of the energy, as seen from Eqs. (6.29) and (6.40), if the series resistance is twice and the parallel resistance is half the surge impedance. Complete dissipation of the pulse energy could result only if in Eq. (6.29) for the case of the series resistance we made $Z_2 = 0$, that is, if we short-circuited the line just beyond the resistance. Similarly, for the case of the parallel resistance we would have to put $Z_2 = \infty$ in Eq. (6.40), thereby opening the line just beyond the resistance. Both cases now represent the simple termination of a line of surge impedance Z in a pure resistance load R treated in Chapter 2.

In that chapter we saw that, if the load resistance is equal to the surge impedance of the line, traveling pulses are not reflected and thus dissipate all their energy in the load resistance. We now must recognize that this is the only way in which the energy of traveling pulses can be completely dissipated. It is necessary to let them travel to the end of a line and there into a resistance of the right magnitude. We shall see in Chapter 16 how in practice arrangements can be made to realize such effects, even for a continuous line.

In the last few examples and figures we have shown the traveling pulses as rectangular step pulses, to make the treatment simple. However, all the equations were developed for the general case and thus apply equally for single pulses or pulse trains. These also

will be reduced in amplitude by series- or parallel-connected resistances and part of their energy will be dissipated in heat.

(*d*) *Nonlinear arresters.* For arresters to carry only a very small fraction of the normal steady-state operating currents and still be effective in almost completely reducing overvoltages such as those due to lightning, they

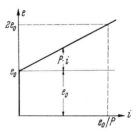

Fig. 6.12

must have a nonlinear voltage-to-current relation, as shown idealized in Fig. 6.12. In practice such a nonlinear characteristic can be produced by vacuum tubes, thyristors, diodes, spark gaps, or any other suitable

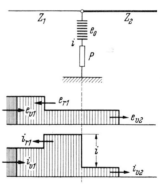

Fig. 6.13

nonlinear device used singly or in combination. For very small currents the voltage rises very rapidly to a limiting value e_0 and for larger currents it increases linearly beyond e_0 as if an equivalent ohmic resistance P. In practice, e_0 is arranged to be greater than the steady-state operating voltage on the line, so that the arrester is called into action only during the occurrence of overvoltages.

As shown in Fig. 6.13, the voltage at the junction is

$$e_1 = e_2 = e_0 + Pi, \qquad (6.42)$$

while the three components of current are in balance,

$$i_1 = i_2 + i = i_2 + \frac{e_2}{P} - \frac{e_0}{P}. \qquad (6.43)$$

Here i is obtained from Eq. (6.42). The last term, e_0/P, represents, as is shown in Fig. 6.12, the current drawn by the arrester for an overvoltage 100 percent greater than e_0, and this should be as large as possible for a good arrester.

For the partial voltage pulses traveling in both directions we obtain, by Eq. (6.42),

$$e_{v1} + e_{r1} = e_{v2} \qquad (6.44)$$

and for the partial current pulses, by Eq. (6.43),

$$\frac{e_{v1}}{Z_1} - \frac{e_{r1}}{Z_1} = \frac{e_{v2}}{Z_2} + \frac{e_{v2}}{P} - \frac{e_0}{P}. \qquad (6.45)$$

If we eliminate the rearward-traveling partial pulse by addition of these two equations, the transmitted forward-traveling voltage pulse in line 2, which also gives the voltage at the junction, is

$$e_{v2} = \frac{e_0 + 2P\,e_{v1}/Z_1}{1 + P(1/Z_1 + 1/Z_2)}. \qquad (6.46)$$

While this voltage is larger than that of a resistance arrester of equal value, given by Eq. (6.35), it is possible here for the nonlinear arrester to use a very much smaller value of P without causing any appreciable effects on the steady-state operating currents.

If the arrester is connected to a bus bar, at which, as shown in Fig. 6.4, n lines of equal surge impedance are joined, the combined surge impedance Z_2 for the transmitted pulse

is given by Eq. (6.15) and the voltage pulse at the bus bar and transmitted forward is reduced to

$$e_{v2} = \frac{e_0 + 2P\,e_{v1}/Z}{1 + nP/Z}. \qquad (6.47)$$

From these equations we see that the ratio of P, the ohmic resistance, to Z, the surge impedance of the lines, determines the effectiveness of the arrester in reducing overvoltages. Table 6.1 shows the voltage reduction for a simple continuous line ($n = 2$) as a function

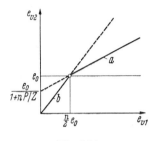

Fig. 6.14

of P/Z in the form of values e_{v2}/e_0 for different values of e_{v1}/e_0.

We see that very high imposed overvoltages can be appreciably reduced only by low values of resistance in the arrester.

Equation (6.47) shows that with the arrester in action the transmitted voltage pulse e_{v2} is a linear function of the incident voltage pulse e_{v1}. Thus in Fig. 6.14 it is represented by line a, which is shown broken for values of e_{v2} less than e_0. For these lower values the arrester is not in action and thus the simpler equations (6.6) and (6.16) apply; these also give a simple proportionality between the incident and transmitted voltage pulses. This is represented by the line b, which is shown broken for values of e_{v2} greater than e_0. The solid line

TABLE 6.1. Voltage pulse amplitude e_{v2}/e_0 transmitted in a homogeneous line.

Incident pulse, e_{v1}/e_0	P/Z						
	0	0.1	0.2	0.3	0.5	0.7	1
2	1	1.17	1.29	1.38	1.5	1.58	1.67
3	1	1.33	1.57	1.75	2.0	2.17	2.33
5	1	1.67	2.14	2.5	3.0	3.33	3.67
7	1	2.0	2.72	3.25	4.0	4.50	5.00
10	1	2.5	3.57	4.38	5.5	6.25	7.00

composed of portions a and b now represents the voltage-pulse relation for the complete junction, including the arrester. Figure 6.15 expresses graphically the values given in Table 6.1, and shows quite generally the voltage reduction obtainable by the use of an arrester along a homogeneous line for $n = 2$.

The reflected voltage pulse on line 1 is, by Eq. (6.44),

$$e_{r1} = \frac{e_0 - e_{v1}\left[1 + P\left(\frac{1}{Z_2} - \frac{1}{Z_1}\right)\right]}{1 + P\left(\frac{1}{Z_1} + \frac{1}{Z_2}\right)}$$

$$= \frac{e_0 - e_{v1}\left[1 + (n-2)\dfrac{P}{Z}\right]}{1 + n\dfrac{P}{Z}}. \qquad (6.48)$$

Here the last expression on the right is for the case of a bus bar joining n equal lines. Thus again there is a linear relation between the incident and reflected voltage pulses. The current in the arrester is, by Eq. (6.42),

$$i = \frac{\dfrac{2}{Z_1}e_{v1} - e_0\left(\dfrac{1}{Z_1} + \dfrac{1}{Z_2}\right)}{1 + P\left(\dfrac{1}{Z_1} + \dfrac{1}{Z_2}\right)} = \frac{2e_{v1} - ne_0}{Z\left(1 + n\dfrac{P}{Z}\right)}.$$

$$(6.49)$$

Again the last expression on the right is for an arrester at a bus bar joining n equal lines.

For a continuous line with $n = 2$ and $Z = 500\ \Omega$, an incident voltage pulse $e_{v1} = 60$ kV, a limiting voltage $e_0 = 20$ kV, and an arrester resistance $P = 100\ \Omega$, the transmitted voltage pulse is reduced to

$$e_{v2} = \frac{20 + 2\dfrac{100}{500}60}{1 + 2\dfrac{100}{500}} = 31.5 \text{ kV}$$

and the current in the arrester is

$$i = \frac{2 \cdot 60 - 2 \cdot 20}{500\left(1 + 2\dfrac{100}{500}\right)}10^3 = 114 \text{ A},$$

while the current in the incident pulse was $60,000/500 = 120$ A. Thus the arrester draws a very substantial fraction of the pulse current away from the line and thereby reduces the voltage considerably.

Because these last derivations made no assumptions as to pulse shape, they are valid

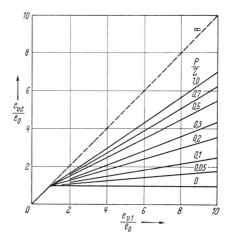

Fig. 6.15

not only for rectangular step pulses but also for pulses of any other shape. For example, we may represent the traveling-pulse front by a linearly rising voltage, as shown in Fig. 6.16. Then the incident voltage at the junction increases linearly as a function of time. Thus

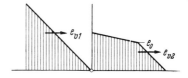

Fig. 6.16

in the voltage diagram of Fig. 6.14 we move linearly with time along the e_{v1} abscissa. We obtain for e_{v2} the voltage as a function of time and the line segments b and a and thus a pulse front as shown in Fig. 6.16. Thus the nonlinear arrester flattens the rapidly rising front of traveling pulses as soon as it is brought into action by an excess of voltage above e_0.

CHAPTER 7

SPARK DISCHARGES ON LINES DURING SWITCHING

Many switching processes in electrical systems are initiated by spark or arc discharges. These arise with very great speed, so that we can count on almost instantaneous switching on. However, the extinction of these discharges is far less rapid, owing in large part to the heating of the electrodes, and this leads to an appreciable time duration for switching-off processes. Both features are important for the generation of transient pulses during switching by such processes.

(a) *Sudden ground fault or short circuit.* If the insulation of a line under voltage suddenly

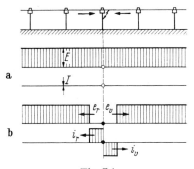

Fig. 7.1

breaks down, for example owing to insulation damage or overvoltage, the breakdown spark, because its voltage drop is low in comparison with the operating voltage, represents essentially a complete short circuit. Figure 7.1 shows as an example the flashover of an overhead-line insulator, which occurs to the conducting pole and thus to the ground.

Immediately before the flashover, for times $t < 0$, the voltage of the line with respect to ground is E, and its operating current I is set equal to zero for the sake of simplicity (Fig. 7.1a). The magnitude of E is given by the flashover or breakdown voltage of the insulator. After the flashover, that is for times

$t > 0$, the voltage at the flashover point is approximately reduced to

$$e = 0. \tag{7.1}$$

Thus the line voltage is suddenly reduced and a discharge pulse propagates with the velocity of light in both directions away from the flashover point, of magnitude

$$e_r = e_v = -E. \tag{7.2}$$

This pulse is superimposed on the original voltage (Fig. 7.1b). Because the failure of the insulation usually occurs almost instantaneously, these discharge pulses have a very steep rise, as shown in Fig. 7.1b.

These discharge-voltage pulses give rise to corresponding current pulses, whose amplitude is determined by the surge impedance of the line; from Eqs. (1.16) and (1.17), they are

$$i_r = -i_v = \frac{E}{Z}. \tag{7.3}$$

The current in the discharge breakdown is the sum of the two current pulses traveling in opposite directions; thus

$$i_f = 2\frac{E}{Z}. \tag{7.4}$$

For high network voltages this current can assume very high values. For example, in a 100-kV network with an insulator breakdown voltage to ground of $E = 600\ \text{kV}$ and a line surge impedance of $Z = 500\ \Omega$, it will be

$$i_f = 2\frac{600,000}{500} = 2400\ \text{A}.$$

This current does not, of course, persist indefinitely but changes after completion of the traveling-pulse phenomena to the ground-fault or short-circuit current fed by the generators of the network into the fault or breakdown point.

The rectangular voltage step pulses generated by such spark or arc discharges are, however, far more dangerous than these currents. If the breakdown occurs under normal operating voltage, as for example from a faulty insulator, these step pulses are of equal magnitude to the operating voltage. Frequently, however, flashover of insulators or rupture of cable insulation is caused by excess voltages produced by atmospheric charges or network transients and these may be several times the normal operating voltage. The discharge pulses generated by such

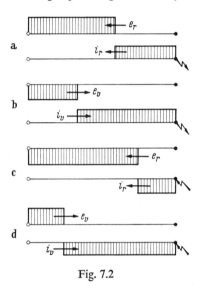

a

b

c

d

Fig. 7.2

breakdowns are not modified by all overvoltage protective devices, because they actually represent a discharge and hence voltage reduction on the line, while such devices usually respond to voltage increases. Thus such pulses can frequently enter protected stations and cause considerable damage.

Exactly analogous processes appear when the breakdown occurs between two conductors instead of between one conductor and ground. In this case the values of voltage and surge impedance are those applicable to the two conductors between which breakdown occurs.

(*b*) *Oscillatory discharges.* Figure 7.2 shows a discharge pulse generated by a ground fault or short circuit and reflected at the open end of a line. This reflection was shown in Chapter

4 to reduce the current to zero and reverse the polarity of the voltage (Fig. 7.2*b*). When this reflected pulse arrives back at the fault, another reflection occurs, accompanied by a polarity reversal of the voltage charging pulse and a current discharge pulse of corresponding polarity. Thus the current in the fault reverses polarity at the instant when the pulse reflected from the open end of the line has arrived back

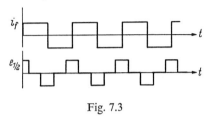

Fig. 7.3

at the fault and is again reflected there. Figure 7.2 shows further stages in the process.

The fault current i_f is shown as a function of time in Fig. 7.3; it is seen to be a square wave. By inspection of Fig. 7.2 we can see that the voltage at the open end of the line has the same square-wave relation as a function of time. However, voltages and currents

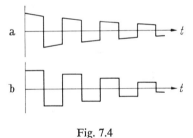

Fig. 7.4

at other points along the line, while they also have a rectangular shape, undergo more frequent changes. Thus, for example, in the middle of the line the voltage as a function of time is shown as $e_{1/2}$ in Fig. 7.3. Every short-circuited discharge of lines initiated by such instantaneous discharges has such rectangular-shaped oscillatory time functions. The frequency is given by Eq. (4.18).

If we consider the resistance of the line conductors, we obtain an exponential attenuation of the rectangular pulses in a geometric series, as shown in Fig. 7.4*a*. If, however, the voltage of the discharge predominates in the attenuation process, then, as shown in Chapter 43 of Ref. 1, the rectangular pulses are maintained by the constant discharge voltage and an arithmetic series results (Fig. 7.4*b*). In both cases pulse trains with extremely short rise and decay times are generated.

Of the rectangular step pulses generated by a discharge (Fig. 7.1), only one at most can reach an open line termination. The other one must ultimately go back to the current source that feeds the network. If this source is of sufficiently low internal impedance, it produces a reflection, as shown in Fig. 4.12. By repeated addition of the slightly attenuated current pulses the short-circuit or ground-fault current is gradually built up in staircase fashion to the final value that the current source is capable of feeding into the fault. Because the traveling pulses occur only during the first few milliseconds, this staircase form of growth exists only at the very beginning of the direct- or alternating-current curve. Thus it is usually not recorded by mechanical oscillographs. The more slowly developing transients discussed in Ref. 1 are thus also not affected by it.

(*c*) *Discharge through a resistance.* If the discharge of the line under voltage E takes place through a resistance R at one end, as shown in Fig. 7.5, then the voltage at R drops

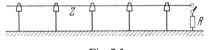

Fig. 7.5

to Ri_r and not to zero, where i_r is the discharge current in the resistance. Because the voltage of the line is composed of the original voltage E and the rearward-traveling switching pulse, we obtain

$$E + e_r = Ri_r. \qquad (7.5)$$

If in Eq. (7.5) we substitute

$$e_r = -Zi_r, \qquad (7.6)$$

we obtain for the current

$$i_r = \frac{E}{Z + R}. \qquad (7.7)$$

Thus the discharge current is reduced by the resistance, which is simply added to the surge impedance of the line. The voltage of the discharge pulse is, by Eq. (7.6),

$$e_r = -\frac{Z}{Z + R}E. \qquad (7.8)$$

It also becomes smaller with increasing resistance. For $R = Z$ the line discharges to half the original voltage.

If, as in Fig. 7.6, the line is open circuited at the opposite end, the discharge pulse is reflected there with voltage of equal magnitude

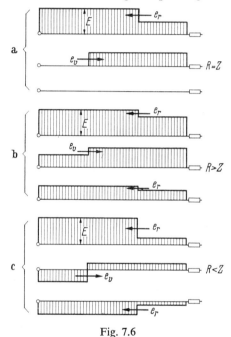

Fig. 7.6

and polarity and current of equal magnitude and reversed polarity. For $R = Z$ (Fig. 7.6*a*) there is complete discharge to zero voltage and current; for $R > Z$ (Fig. 7.6*b*) there is only partial discharge; and for $R < Z$ (Fig. 7.6*c*) there is a reversing discharge to a voltage of opposite polarity but smaller amplitude.

When the reflected pulses return to R they are reflected, in accordance with the condition at the end of the line,

$$e = Ri. \qquad (7.9)$$

If in Eq. (7.9) we combine the partial pulses to form the total pulses and substitute the voltage pulses for the current pulses in accordance with Eq. (7.6), we obtain

$$Z(i_v - i_r) = R(i_v + i_r). \qquad (7.10)$$

This gives the reflected current pulse,

$$i_r = \frac{Z - R}{Z + R} i_v, \qquad (7.11)$$

and with Eq. (7.6) the reflected voltage pulse,

$$e_r = \frac{R - Z}{R + Z} e_v. \qquad (7.12)$$

For $R = \infty$ or 0 this gives directly the laws of reflection for open or short-circuited line terminations developed in Chapter 4, in which the pulses are reflected without change in amplitude. Because in Eqs. (7.11) and (7.12) R increases the denominator and decreases the numerator by equal amounts, it reduces the reflected pulse amplitude in both cases. For $R > Z$ the polarity of the voltage pulse is not changed, while for $R < Z$ it is reversed. The polarity of current pulse remains unchanged for $R < Z$ but is reversed for $R > Z$. If the resistance is made exactly equal to the surge impedance,

$$R = Z = \sqrt{\frac{\bar{l}}{\bar{c}}}, \qquad (7.13)$$

the current and voltage pulses are completely dissipated into heat by the resistance.

Figure 7.6 shows the reflection at R of the pulse returning from the open end of the line for three different resistances. In Fig. 7.6a, $R = Z$ and the discharge pulse is completely dissipated upon its first return to R. During this single round trip the discharge-voltage pulse is one-half the original voltage on the line. The current in R is a single rectangular pulse whose amplitude is given by Eq. (7.7).

In Fig. 7.6b, $R > Z$ and the initially smaller discharge pulse is further reduced by the resistance. Thus, by a sequence of round trips of the pulse, the line is discharged in small steps that decrease according to a geometric series.

In Fig. 7.6c, $R < Z$ and the arriving pulse is reduced but reversed in polarity during reflection. This leads to a process similar to that in the short-circuited line shown in Fig. 7.2. Thus the discharge is oscillatory but

also damped according to a geometric series owing to dissipation of the energy in the load resistance.

The limiting condition for oscillatory discharge is given by Eq. (7.13), whereas in Ref. 1, p. 50, Eq. (16), it is shown for quasi-stationary circuits to have twice this value. Note, however, that for the quasi-stationary oscillatory circuit the limiting value of R produced an infinitely long transient while in the case of the distributed line the pulse is abruptly ended after one round trip.

In a similar way, pulses are generated when a line that has a resistance load connected

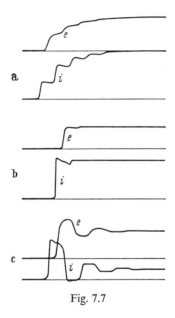

Fig. 7.7

across it at one end is charged by switching at the other end. Figure 7.7 shows oscillograms taken by K. W. Wagner which give the increase in current at the beginning and the voltage at the center of an artificial line loaded in this manner In Fig. 7.7a $R = Z/9$, in Fig. 7.7b $R = Z$, and in Fig. 7.7c $R = 7Z$. Only for $R = Z$ is there substantially no reflection from the resistively loaded end, in accordance with Eq. (7.13). For $R < Z$ there is the staircase sequence of increases and for $R > Z$ the oscillatory damped sequence, the final value being approached in both cases as given by Eqs. (7.11) and (7.12).

If one wishes to carry out experiments on lines without distortion due to reflection at either end, it is necessary to insert a resistance $R = Z$ at each end, as shown in Fig. 7.8. By this means any traveling pulse generated anywhere along the line is completely dissipated at either end. Of course the resistance at the

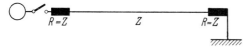

Fig. 7.8

source now introduces a voltage drop which must be compensated by an increase of the source voltage to twice its former value.

(d) *Sudden opening of a line.* Figure 7.9 shows a line that is suddenly opened at a point. The voltage E and current I existing at time $t < 0$ cannot continue beyond the switching instant at $t = 0$. Instead, for $t > 0$ the current at the point of interruption must become

$$i = 0. \tag{7.14}$$

Thus the prior steady-state current I must be canceled by current discharge pulses propagated in both directions from the switch and of magnitude

$$i_r = i_v = -I. \tag{7.15}$$

The current pulses give rise to voltage pulses that are superimposed on the steady-state

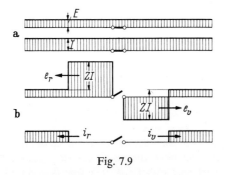

Fig. 7.9

operating voltage. Their magnitude is determined by the surge impedance; by Eqs. (1.16) and (1.17), it is

$$e_r = -e_v = ZI = \sqrt{\frac{l}{c}}\, I. \tag{7.16}$$

The pulses are of opposite polarity on opposite sides of the switch, so that the voltage drop across the switch has twice this value, namely

$$e_s = 2ZI. \tag{7.17}$$

Figure 7.9b shows these relations.

In these equations and in Figs. 7.1 and 7.9 we notice a strong similarity between the short- and open-circuiting of lines. Current and voltage have merely interchanged roles. Equation (7.16) is very similar to Eq. (4), Ref. 1, p. 72, for the overvoltage in the case of a lumped circuit containing capacitance and inductance. This is due to the similarity of the physical phenomena involved.

A surge impedance $Z = 500\ \Omega$ would give rise, according to Eq. (7.17), to an overvoltage pulse across the switch of

$$e_s = 2 \cdot 500 \cdot 100 = 100{,}000\ \text{V}$$

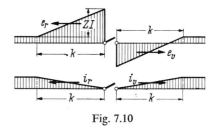

Fig. 7.10

upon interruption of a current $I = 100$ A. In practice, such high overvoltages usually do not occur, because an arc is formed during interruption of the line which prevents a sudden interruption of the current.

If the arc is extinguished in a time interval of the order of magnitude of 0.01 sec, which corresponds to about half a period of the usual 50- or 60-cy/sec alternating current, sharp step pulses are not generated. Instead the current decay and thus also the voltage rise are distributed in space, as shown in Fig. 7.10, over a distance k given by the product of τ, the time taken to interrupt the current, and v, the velocity of light. Thus the front-rise length of the pulse is

$$k = v\tau, \tag{7.18}$$

because the initial portion of the pulse has already traveled along the line through this distance when the switching process is completed. For $\tau = 0.01$ sec, and in an overhead line, the front rise of the pulse is distributed over 3000 km. This gives a very gradual rise of voltage along the line. Even if the current is interrupted more rapidly, for example by forced cooling of the arc, so that an interruption time $\tau = 0.0001$ sec is obtained, the rise is still distributed over a distance of 30 km, which again represents a gradual voltage rise.

Only if the lines are of length comparable to the front-rise length of the pulses, as shown in Fig. 7.10, can the voltage pulses be fully developed as given by Eq. (7.17). For shorter lengths of line the voltage pulses are altered by multiple reflections between the far end of the line and the switching point. This shows that in general neither very large nor steeply rising voltage pulses are generated in short lines by the interruption or switching off of current. Such damaging pulses are generated usually by disturbances in or phenomena secondary to the switching-off process.

If an alternating current is interrupted at the very instant at which it passes through zero, the reason for the generation of traveling pulses is completely removed.

(e) Interruption of a short-circuit current. Figure 7.11 shows a short-circuit current I

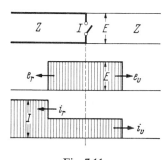

Fig. 7.11

in the left-hand portion of the line only, which is interrupted. At the point of interruption a voltage E is produced which is caused by equal traveling voltage pulses in the two lines,

$$e_v = e_r = E. \qquad (7.19)$$

The corresponding current pulses are equal and opposite in polarity and their sum, as shown in Fig. 7.11, is equal to I, that is,

$$i_v - i_r = \frac{e_v}{Z} + \frac{e_r}{Z} = I. \qquad (7.20)$$

Thus the current in the line after interruption is

$$i_v = -i_r = \tfrac{1}{2}I, \qquad (7.21)$$

and from Eq. (7.19) this gives the voltage upon interruption of the short circuit,

$$E = \frac{Z}{2} I = \frac{1}{2} \sqrt{\frac{l}{c}} I. \qquad (7.22)$$

Thus the right-hand portion of the line has reduced the voltage on the line to one-half of the value given by Eq. (7.16) and at the switch itself to one-quarter. If, as is the case in bus bars in large switching stations, several lines were to originate at the point of interruption, the voltage would be still further reduced because the rearward-traveling current pulse i_r would be only a fraction of the forward-traveling current pulses i_v in all the other lines.

If a short-circuit current $I = 2000\ \mathrm{A}$ is interrupted in an underground cable network of surge impedance $Z = 50\ \Omega$, a voltage

$$E = \frac{50}{2} 2000 = 50{,}000\ \mathrm{V}$$

is generated, which would still be large even if several lines were joined together. In practice, an arc will be formed at the point of interruption, preventing sudden changes in current and thus making the voltage rise a more gradual one, as shown in Fig. 7.10. In short lines, then, the voltage pulse may again not fully develop to its maximum value.

(f) Reignition. Step pulses of voltage increase in danger with the abruptness of their rise, that is, in inverse proportion to their front-rise length k, which by Eq. (7.18) is proportional to the switching time τ. As we have seen, the interruption of currents usually is gradual enough for the length to be several kilometers. In contrast, the switching on of lines is very rapid. Particularly for high-voltage systems, the process involves the formation of a spark across a gap and is completed in less than 1 μsec, so that it leads to a pulse with a front-rise length of the order of a few meters or even centimeters. In Ref. 1,

Chapters 39–44, it was shown that during the interruption of alternating currents there is almost always a sequence of reignitions or restrikes. These then always generate abrupt traveling pulses that are propagated into the lines.

Take the case of an alternating-current line of some appreciable capacitance that is slowly

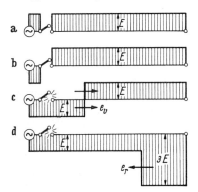

Fig. 7.12

disconnected from the current source. Then the arc carrying the line-charging current will be interrupted at an instant of zero current, which corresponds also to an instant of peak voltage on the line. On the disconnected line this peak value is stored as a charge voltage, as shown in Fig. 7.12*a*. No traveling pulses are generated during this process. The alternating voltage to the left of the switch now decreases sinusoidally and reaches its opposite polarity maximum a half period later, as shown in Fig. 7.12*b*. If the insulation resistance of the isolated line is high enough to maintain its voltage unaltered, there will be a voltage equal to twice the peak system voltage across the switch contacts at that instant. If these contacts have not moved very far since the instant of current interruption a half period ago, this large voltage will be enough to disrupt the oil or air gap between the contacts. Thus the line is now charged to the same voltage but with opposite polarity and, as shown in Fig. 7.12*c*, a charge pulse that has a step rise of amplitude $-2E$ advances into it.

This step pulse is reflected at the far open end of the line and its amplitude is doubled to $-4E$. As shown in Fig. 7.12*d*, the returning pulse charges the line to a voltage $-3E$, that

is, three times the peak voltage of the source. This step pulse now travels back to the switch and source and is reflected there; it is then reflected repeatedly at both ends, each time with some loss, until finally the entire line will only be at the voltage of the source, that is, close to $-E$. At the next passage of the current through zero the process will be repeated, and this will continue until the switch contacts have moved far enough apart that reignition can no longer be caused by a voltage $2E$. The step-pulse voltage reaches its full value $2E$ only toward the end of the interruption process. At its beginning the smaller contact separation leads to reignition for smaller values. Thus the amplitude of the step pulse increases gradually during the period of interruption.

Ignition or reignition almost always occurs when an open alternating-current line is disconnected. It gives rise to step pulses of voltage amplitude up to twice that connected with switching-on pulses. While connection

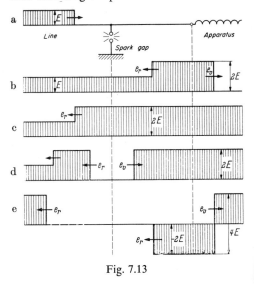

Fig. 7.13

of an open line may produce reflections up to twice the operating peak voltage, its disconnection may give rise to three times the operating peak voltage. To avoid this danger it is necessary to disconnect rapidly or introduce shunt resistances between the lines.

The processes here are similar to the ones described in Ref. 1, Chapter 42, for less rapid oscillations. If the arc is extinguished prematurely, say at the instant at which the returning pulse of amplitude $3E$ arrives at the switch, then here also the voltage on the line may rise to very high values by repeated reignition, as in the case treated in Ref. 1, Chapter 44.

(*g*) *Simple spark-gap arrester.* To protect the windings of machines or transformers against arc-over to ground it was at one time considered useful to introduce a simple spark gap across the line feeding them at some distance from the beginning of the winding, as shown in Fig. 7.13*a*. W. G. Hawley and H. M. Lacey drew attention to the danger of such an arrangement, which can be explained by the use of Fig. 7.13.

If a traveling pulse arrives along the line that is just insufficient to fire the spark gap, the pulse is reflected at the winding owing to its high surge impedance, giving almost twice the original voltage, or nearly $2E$ (Fig. 7.13*b*). The front of this pulse then travels back toward the spark gap with this increased voltage. Owing to the time required for ignition of the spark gap, this pulse now travels slightly past the gap (Fig. 7.13*c*). Shortly after the passage of the front of this pulse, which may have any shape, the spark gap fires rapidly, so that a very sharp step pulse of amplitude $-2E$ is propagated from it in both directions. When the pulse advancing toward the winding reaches it, after a small time interval, it too is reflected and increased to almost twice its value, that is, to $-4E$, so that a pulse of $-2E$ returns along the line (Fig. 7.13*e*). However, note that a step pulse of amplitude $4E$ is now propagated into the winding, instead of one of amplitude $2E$ originally caused without the spark gap (Fig. 7.13*b*). Thus the winding is stressed by a larger voltage between turns or coils.

Thus, owing to the ignition delay of the spark gap for some incident pulse amplitudes, the voltage incident upon the winding as a whole is hardly reduced while the voltage stresses within the windings, which are of considerable danger, are increased by a factor of 2. The introduction of a suitably large resistance in series with the spark gap avoids this danger and this arrangement, shown in Fig. 6.13, has in fact come into general use.

CHAPTER 8

NATURAL FREQUENCIES OF THREE-PHASE CIRCUITS

Three-phase circuits are generally used in power networks. Thus it is important to examine the oscillatory behavior of such circuits. Of particular importance are high-frequency oscillations of motors, generators, transformers, and lines due to the interruption of short-circuit currents, since they determine the interrupting capacity of the circuit breakers.

In networks many oscillatory modes can exist. Thus we must examine star- and delta-connected circuits with open and grounded terminations. We shall also note whether the grounded point is connected with the first or the last interrupted arc of the three-phase circuit breaker. The analysis will apply to overhead lines, underground cables, and windings of motors, generators, transformers, and inductors.

(a) *Method of analysis.* In Chapter 7 it was shown that switching processes in lines with open or shorted terminals give rise to waves or pulses that travel to and fro along the lines. This produces natural oscillations of the circuit. Whereas the velocity of propagation in open lines is of the order of 300 m/μsec and in cables 150 m/μsec, measurements show that in transformer windings it is generally below 100 m/μsec and it may drop to 30 m/μsec in motor or generator windings. In each case the velocity of propagation is

$$v = \frac{1}{\sqrt{lc}}, \qquad (8.1)$$

where l is the inductance per unit length and c is the capacitance per unit length. We shall assume that the velocity of propagation is known in a given case. For lines it was derived in Chapter 1 and for windings it is derived in Chapter 17.

The natural oscillations are generated by repeated reflection of the traveling pulses at the terminals of the lines or windings. Thus the natural frequency can be obtained from the length of the line or winding traversed between reflections. If the initial pulse has a very steep step front, we can also infer the frequency of higher harmonics. As the result of attenuation of these higher harmonics, the sharp features of the pulses are gradually flattened out.

In three-phase windings the single phases are coupled together mainly by their conductive connection, at either the star point or the delta corners. This coupling has a fundamental influence on the frequency of the oscillations. In addition to this conductive coupling, there is slight capacitive and inductive coupling between the phase windings, for example within the overhang of an alternator, or between the coils on different limbs of a transformer. We shall neglect this coupling as of secondary influence. Similarly we shall neglect here the influence of any additional capacitance at the terminals of the windings, which may be caused by terminal bushings, connecting cables, or similar elements, since this can easily be treated separately.

The natural oscillations and frequencies can be analyzed in a fairly simple manner by considering the traveling voltage pulses caused by the instantaneous interruption of a short-circuit current in the winding. This is also one of the most important operating conditions of interest and the solutions obtained agree closely with experimental results.

Figure 8.1 shows the method of analysis applied to the well-known case of interruption of the short circuit across an insulated single-phase winding. Because the short circuit has reduced the voltage across the winding to zero, it is sufficient to examine only voltage changes produced by the interruption of the short circuit. If the winding is symmetric, half the interruption voltage appears at each end but with opposite polarities. Each rectangular voltage step pulse travels over the whole winding, undisturbed by the other pulse. The solid lines in Fig. 8.1*a–e* represent the resultant voltage distribution over the winding after $\frac{1}{4}$, $\frac{3}{4}$, $\frac{5}{4}$, . . . of the time of travel

$$\tau = \frac{a}{v} \qquad (8.2)$$

that is required to cover the total length *a* of the winding. After a time *t* equal to $\frac{8}{4}\tau$,

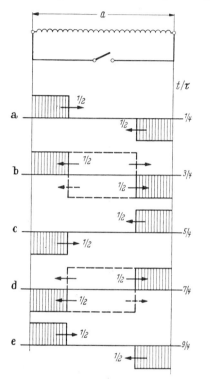

Fig. 8.1

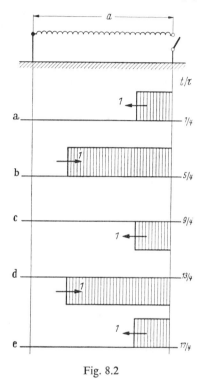

Fig. 8.2

pulse appears at the interrupted open end and travels with a steep front over the winding. The other end, which is grounded, now represents a short circuit and thus reflects the pulse with a reversal of polarity. The resultant voltage distribution over the winding is shown in Fig. 8.2*a–e* for *t*/τ equal to $\frac{1}{4}$, $\frac{5}{4}$, $\frac{9}{4}$, . . . and we see that only after *t*/τ = 4 does the process repeat itself. Thus the steep pulse

the entire process repeats itself, so that the period is 2τ and hence the frequency is

$$f_0 = \frac{v}{2a}. \qquad (8.3)$$

We see that this result is obtained simply by following any steep pulse front, which is reflected without change of polarity at the open ends of the winding. The pulse returns to its original position after traveling a distance $2a$ and this gives the natural frequency directly as in Eq. (8.3).

Figure 8.2 shows the single-phase winding grounded at one end. Now the entire voltage

front travels a distance of $4a$ and the natural frequency for this case is

$$f_0' = \frac{v}{4a}. \qquad (8.4)$$

Because voltage pulses cannot propagate through a short-circuit point, we may consider any other winding connected to the left-hand side of Fig. 8.2, for example two additional phases of a star-connected three-phase winding. Thus we see that Eq. (8.4) gives also the natural frequency of a three-phase system with grounded neutral star point.

We found in Chapter 1, Sec. (*d*), that in the case of two open-circuit terminations, as shown in Fig. 8.1, there are a number of higher-frequency harmonics h in addition to the fundamental frequency f_0 given in Eq. (8.3). These are simple multiples of f_0, according to

$$\frac{h}{f_0} = 1, 2, 3, 4, \ldots \qquad (8.5)$$

and the pulse shapes in Fig. 8.1 are composed of a complete set of harmonics.

For the case of the grounded winding shown

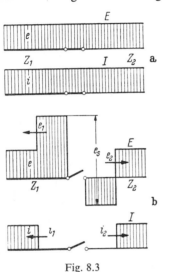

Fig. 8.3

in Fig. 8.2 only odd harmonics h of the fundamental frequency f_0' given by Eq. (8.4) can occur and so

$$\frac{h}{f_0'} = 1, 3, 5, 7, \ldots \qquad (8.6)$$

Again the complete set composes the rectangular step pulses shown in Fig. 8.2. In all of the following cases similar higher-frequency harmonics must be taken into consideration.

The initial distribution of voltage after interruption of a short circuit in three-phase systems is different from that for single-phase windings shown in Figs. 8.1 and 8.2, because several phase windings act in parallel and disturb the symmetry of the terminations. This leads to a composite heterogeneous circuit interrupted as shown in Fig. 8.3. Here E and I are the voltage and current immediately before interruption. In the neighborhood of the circuit breaker they are constant in space

(Fig. 8.3a). Immediately after the interruption the current I vanishes and therefore receding step pulses (Fig. 8.3b) travel away from both sides of the switch. The value of these current step pulses is given by

$$i_1 = i_2 = -I, \qquad (8.7)$$

and these are superimposed on the initial current I.

Now the voltage associated with the current in every traveling pulse is given by the product of the current and the surge impedance Z of the line, with equal polarity for pulses traveling forward and with opposite polarity for pulses traveling backward (Fig. 8.3b). We obtain therefore for the voltages on the lines immediately after interruption

$$e_1 = -Z_1 i_1 = Z_1 I,$$
$$e_2 = Z_2 i_2 = -Z_2 I, \qquad (8.8)$$

which are superposed on the initial voltage E. Figure 8.3b shows the resultant voltage distribution and we see that the voltage across the circuit-breaker contacts is

$$e_s = e_1 - e_2 = (Z_1 + Z_2)I. \qquad (8.9)$$

It depends only on the current and the sum of the surge impedances. This is the total interrupting voltage, which produces traveling pulses and oscillations within any winding. It divides between the two sides of the circuit breaker, so that by Eqs. (8.8)

$$\frac{e_1}{e_s} = \frac{Z_1}{Z_1 + Z_2}; \quad \frac{e_2}{e_s} = -\frac{Z_2}{Z_1 + Z_2}. \qquad (8.10)$$

The division thus depends only on the surge impedances of the lines or windings in the neighborhood of the switch. We can easily verify the initial distribution of voltages in Figs. 8.1 and 8.2 if we consider that in Fig. 8.1 $Z_1 = Z_2$ and in Fig. 8.2 $Z_1 = 0$ for the grounded end.

(b) *Three-phase short-circuited interruption.* If we interrupt the first phase of a short-circuited star-connected winding, as in Fig. 8.4, we can rearrange the parts of the winding adjacent to the circuit breaker as shown in Fig. 8.5 in order to get a diagram similar to Fig. 8.3. We realize that one side of the breaker is connected to $Z_1 = Z$, the surge impedance of each phase winding, while the other side is connected to $Z_2 = Z/2$, the surge impedance of two phase windings in parallel. We therefore obtain the surge voltages

$$\frac{e_1}{e_s} = \frac{Z}{Z + Z/2} = \frac{2}{3}; \quad \frac{e_2}{e_s} = -\frac{Z/2}{Z + Z/2} = -\frac{1}{3}.$$
$$(8.11)$$

The initial pulses, which travel away from both sides of the breaker, are therefore in the

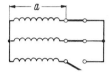

Fig. 8.4

ratio of -1 to 2, as shown graphically in Fig. 8.5.

These pulses travel through the various phase windings, all of the same length a, reach the star point at exactly the same time, and reunite there. The phenomena in the neighborhood of the star point are shown in Fig. 8.6. The incident pulses are represented in Fig. 8.6a; Fig. 8.6b shows the transmission

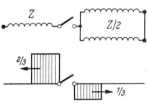

Fig. 8.5

of the left-hand pulse, which, as we saw in Chapter 6, Sec. (a), splits on passing the star point into two branches, each having $\frac{2}{3}$ of the incident voltage. Simultaneously, a pulse of $\frac{1}{3}$ of the incident voltage is reflected. The right-hand pulses, on the other hand, increase their voltage on passing the star point to $\frac{4}{3}$ of the incident voltage. Figure 8.6c shows the transmitted pulse and also the reflected pulse of $\frac{1}{3}$ of the incident voltage.

In Fig. 8.6d the two pulses are superposed. The voltage at the star point remains at zero, because the two transmitted pulses of Figs. 8.6b, c have equal but opposite values. As a

result there remain only discharge-voltage pulses on each side of the star point and these pulses have the same value as the original incident pulses. With this distribution of

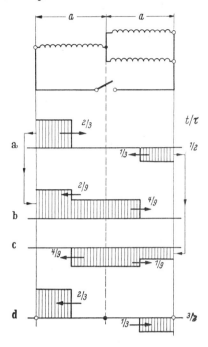

Fig. 8.6

voltages, the star point acts exactly like the center point in Fig. 8.1 or the left-hand short-circuit point in Fig. 8.2. The effect of two phase windings in parallel thus results in this case merely in an unequal subdivision of voltage. But as this is without influence on

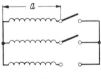

Fig. 8.7

the distance traveled and the time taken to cover it by the pulses, the natural frequency of each of the three-phase windings is given by

$$f_{\lambda 1} = \frac{v}{4a}. \quad (8.12)$$

The interruption of the last phase of the star-connected winding is shown in Fig. 8.7.

Here the division of voltage at the breaker is in the ratio $\frac{1}{2}$ to $-\frac{1}{2}$, the third phase terminal being inactive. Figure 8.8 shows the spatial behavior of the pulses in this case, which is identical to that of Fig. 8.1 but with a different distance of travel. The star point remains unaffected, its voltage always at zero, and thus the third phase winding does not receive any voltage. It is wholly inactive in this case. As the incident pulses are reflected at the star point with a change in polarity, this point

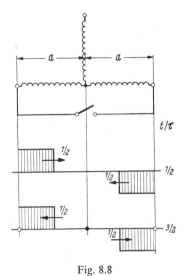

Fig. 8.8

acts as a short circuit for the pulses, although actually it is not grounded. Thus the natural frequency is

$$f_{\lambda 2} = \frac{v}{4a}. \qquad (8.13)$$

We see, therefore, that for star-connected windings the interruption of either the first phase or the last phase excites the same natural frequency, which is determined by a distance $4a$ traveled by the pulses and given in Eqs. (8.12) and (8.13).

The interruption of the first connection of a short-circuited delta-connected winding is shown in Fig. 8.9. We can rearrange the windings giving the linear diagram of Fig. 8.10 if we consider that one pole of the interrupter is connected to two ends of the windings, the other pole to four ends of the windings, all being free from ground and able to float.

While the phases y and z are in parallel and connected at both ends through the switch, the phase x is connected at both its ends to the same pole of the interrupting switch. Therefore we have drawn x in parallel halves in Fig. 8.10, so that the pulses enter both ends with the same sign. The center of this third

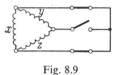

Fig. 8.9

phase, forming the left-hand end of Fig. 8.10, is naturally quite free and can float also.

We see that in this case we have two lines of winding, both of length $1.5a$, connected at three points with each other, which must oscillate completely in parallel and in synchronism, as if they were only one line of length $1.5a$ with open ends. This is indicated at the bottom of Fig. 8.10. As the distance of travel of any pulse in this structure up to its repetition is $3a$, the natural frequency for the delta winding is

$$f_{\Delta 1} = \frac{v}{3a}. \qquad (8.14)$$

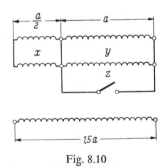

Fig. 8.10

To clarify the performance of the switching pulses further, Fig. 8.11 shows their distribution over the windings at a number of successive times until the distribution repeats. As one pole of the circuit breaker is connected to two ends and the other pole to four ends of the winding, the division of the breaker voltage is in the ratio $\frac{2}{3}$ to $\frac{1}{3}$, exactly as in

Fig. 8.5 and Eq. (8.11). The larger pulse is incident at the end of the winding, and the smaller pulse at $\frac{2}{3}$ of the resultant total length. After the opening of the breaker and the creation of the incident pulses, the entire winding is completely free from any external

as given by Eq. (8.14). The fact that the exciting points for the traveling waves in this case are not at both ends of the resultant line does not alter in the least the natural frequency, though it may influence the higher harmonics insofar as some of these may not be excited.

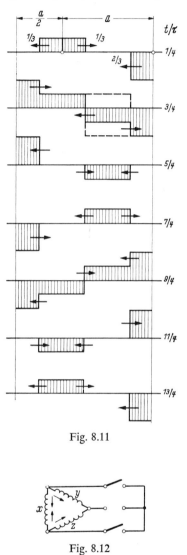

Fig. 8.11

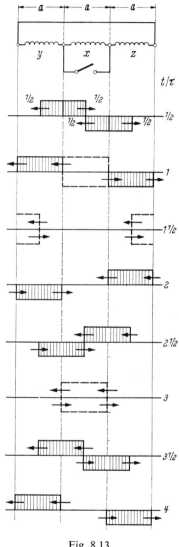

Fig. 8.13

Fig. 8.12

influence because the connecting cables to the breaker have only a negligible capacitance according to the assumptions made in the introduction. We see from Fig. 8.11 that the initial state is repeated after three periods τ of every pulse traveling over one phase length,

If we now interrupt the last two phases of the delta winding, as in Fig. 8.12, we have symmetry of both terminals of the circuit breaker, so that half the interrupted voltage appears on each side. The incident pulses pass into the two adjacent phase windings, dividing according to the arrows in Fig. 8.12, and so we have four pulses, which start to travel over the triangle. As the delta winding is entirely free from any discontinuities and forms a completely closed loop, each of the four pulses travels undisturbed through the whole winding, of total length $3a$, until it comes back to the starting point. It repeats this journey again and again without ever changing polarity. As the period is 3τ, the natural frequency is

$$f_{\Delta 2} = \frac{v}{3a}. \qquad (8.15)$$

This is exactly the same value as for the first interrupted pole according to Eq. (8.14). In Fig. 8.13 are shown the successive distributions of voltage over the delta winding, which is drawn out in a straight line. The various pulse

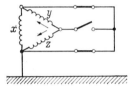

Fig. 8.14

shapes are all formed by the superposition of the four voltage pulses incident at the junction of the two phases of the line. The last junction remains free of incident voltage.

If we compare the natural frequencies of insulated star- and delta-connected windings, according to Eqs. (8.12) and (8.13) and Eqs. (8.14) and (8.15) respectively, we see that they are in the ratio $4:3$ and not perhaps $3:1$, as one might think at first sight.

From the foregoing considerations we see further that the natural frequencies of windings in ungrounded single-phase, delta, and star connections are in the ratios $\frac{1}{2}:\frac{1}{3}:\frac{1}{4}$ respectively, in accordance with the travel-time ratios $2:3:4$.

For zigzag connections of windings, which occur especially in transformers, the travel time of the pulses between the terminal and the star point is equal to that of the star connection, even though the two windings of a single phase are on different magnetic cores. Thus all the relations derived for star-

connected windings apply equally to zigzag-connected windings.

(*c*) *Interruption of a ground fault.* Because many short circuits in three-phase systems are formed by the action of already existing ground faults of a single-phase line, we shall consider the interruption of short circuits while the system is grounded at one pole. Since we measure all voltages between winding

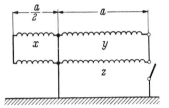

Fig. 8.15

and ground, every ground connection acts on the traveling pulses as a short-circuit point at which no voltage can exist.

The performance of star-connected windings under grounded conditions is rather complicated; since that of delta windings is much simpler, we shall consider them first.

A diagram of the interruption at the first pole of a grounded delta winding is shown in Fig. 8.14, and we see that this pole becomes completely insulated. The other two poles

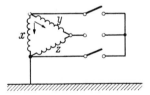

Fig. 8.16

remain connected to the ground, so we can replace Fig. 8.14 by the straight-line diagram of Fig. 8.15, which has exactly the same connections. In consequence the third phase on the left-hand side, whose action was important in Fig. 8.10, is excluded from any action in Fig. 8.15. It remains dead because no voltage can penetrate through a short-circuit point. Thus we have here exactly the same conditions as in Fig. 8.2, the fact that two windings are in parallel causing no change.

The natural frequency of the grounded delta winding for interruption at the first pole is therefore

$$f'_{\Delta1} = \frac{v}{4a}. \qquad (8.16)$$

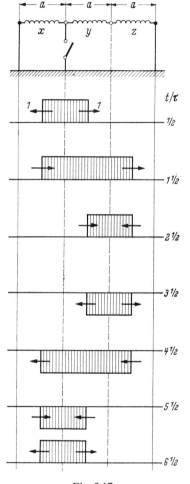

Fig. 8.17

This is different from the frequency in the ungrounded condition.

The interruption at the last poles of the same breaker is represented in Fig. 8.16. As the connecting point to ground acts as a short-circuit point, the entire interrupting voltage appears at the other end of the breaker. From its point of connection with the winding,

voltage pulses travel into both phase windings x and y. They are reflected at the grounded point with changed polarity, the one after traveling a distance a, the other a distance $2a$. When these pulses return, they travel over the connecting point with the breaker without any distortion until they reach the other end of the winding, where again they are reflected with changed polarity. Both waves therefore travel over a winding length $3a$ to and fro, and thus repeat their initial performance after traveling a distance $6a$. The natural frequency for this case is, therefore,

$$f'_{\Delta2} = \frac{v}{6a}, \qquad (8.17)$$

which is also the frequency of a line of length $3a$ with both ends short-circuited to ground.

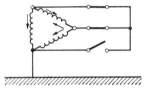

Fig. 8.18

The full course of the voltage pulses over the winding is shown in Fig. 8.17 for successive instants of time. The winding is fed at one third of its length by the initial pulses, which spread to both sides. By the reflection with changed polarity at the ends the whole voltage is concentrated after a time 3τ at the other phase junction; then it spreads again over the winding, and after a time 6τ the pulses converge once more at the initial exciting point. In Figs. 8.14 and 8.16 we have treated the case where the first interruption takes place at one of the nonfaulty or insulated poles, and the last interruption at the pole with the ground fault. However, the opposite sequence is equally likely and Fig. 8.18 shows the diagram for this condition with interruption at the grounded pole first, while the two others remain connected to each other. Here the full voltage appears at both of these insulated poles and four partial pulses travel over the windings. By comparison with Fig. 8.16 we see that each of the two insulated poles acts in this case exactly like the last insulated pole treated in Fig. 8.16. We have here, therefore, simply a doubling of all traveling pulses by the feeding of both free poles or phase junctions instead of only one as in Fig. 8.17.

Without drawing the whole combined diagram we see that the traveling distances and therefore the repetition periods are not

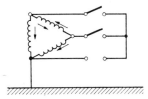

Fig. 8.19

altered. The natural frequency, if the ground fault is interrupted first, is therefore

$$f''_{\Delta 1} = \frac{v}{6a}. \tag{8.18}$$

Figure 8.19 shows the interruption at the last poles in this case. Since both are insulated and symmetric after the interruption, the corresponding phase junctions receive equal but opposite interruption voltages, $\pm\frac{1}{2}$ the total value. Both voltages spread over the winding in four pulses similar to those in Fig. 8.17, only with different magnitudes and signs. However, these pulses now being opposite each other, the halves of the complete delta winding act one against the other. We see this immediately if we impress $\pm\frac{1}{2}E$ on the two phase junctions in Fig. 8.17, which gives exactly the same course of pulses as in Fig. 8.13. The center of this diagram is now a point of symmetry between opposite halves and acts therefore like a short-circuit point. Thus we obtain for each half winding, and therefore also for the complete delta winding, a natural frequency

$$f''_{\Delta 2} = \frac{v}{3a}. \tag{8.19}$$

We saw in Sec. (*b*) that the natural frequencies of insulated or ungrounded three-phase windings remain the same for the first and last interrupting pole of the circuit breaker in both the star and delta connections. From Eqs. (8.16)–(8.19) we realize that in a grounded system it is merely a matter of chance whether

the first or the last excited frequency is the higher, since they are always different. Because we cannot predict whether the first quenched arc in the circuit breaker belongs to the line with the fault or to a sound line, the first frequency excited may vary as $3:2$, the last one as $1:2$, so that three of these possible frequencies are always different from the frequency for the insulated state of the entire winding.

There remains now the investigation of the interruption of star-connected windings with ground fault, and we begin again with the opening of one of the sound phases, as in Fig. 8.20. The straight-line diagram for this case is shown in Fig. 8.21 and we see that this is similar to Fig. 8.6, except that the terminals of the paralleled phase windings are grounded and thereby act as short-circuit points. Since the full interruption voltage appears at the open end of Fig. 8.21 and penetrates into the single winding, and since the laws of transmission and reflection at the star point are well known, we can draw the progressive pulses as in Fig. 8.22. Up to 8.5τ no repetition of the original pulse shape occurs, and even if we extend the picture further into time we do not find an exact repetition. The nearest to the original pulse form occurs at about 6.5τ, but the agreement is only fair.

The absence of exact repetition means that the overtones of this distribution of voltage pulses are not harmonic, and that the ratios

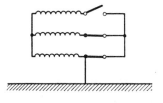

Fig. 8.20

of their frequencies to the fundamental frequency are irrational. Although we see that the travel time of the fundamental pulse may be in the neighborhood of 6.5τ, apparently there is no possibility of deriving an exact value for its frequency, and the more so for the frequencies of overtones, by the simple method of determining the distance traveled until repetition of the initial pulse form occurs.

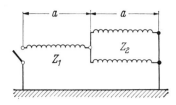

Fig. 8.21

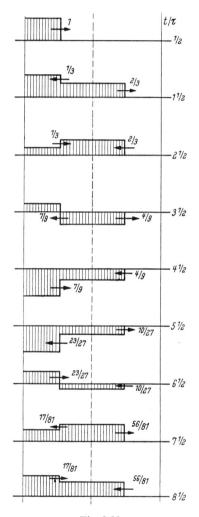

Fig. 8.22

Instead of using d'Alembert's method for considering traveling pulses, which we have followed until now, we shall apply Bernoulli's method, which consists in analyzing the phenomena by sine waves of various wavelengths, as treated in Chapter 1, Sec. (d).

If we apply this method to the general case of two lines with different characteristics connected at one point, as in Fig. 8.23, and open at one terminal while the other terminal is short-circuited, we obtain from the balance of voltage and current at the junction point the solution for the natural frequencies $\nu = 2\pi f$ from Eqs. (1.48)–(1.50),

$$Z_2 \tan \frac{\nu a_2}{v_2} = Z_1 \cot \frac{\nu a_1}{v_1}, \qquad (8.20)$$

where Z_1 is the surge impedance of the line with the open end and Z_2 that of the line with the short-circuited end. In this case we have within the phase windings equal values of

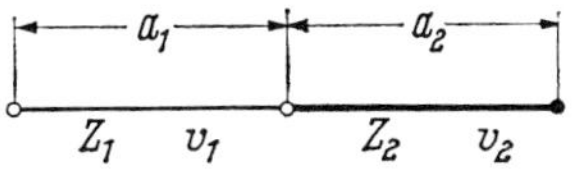

Fig. 8.23

the velocity v and of the length a. Therefore Eq. (8.20) becomes

$$\tan^2 \frac{\nu a}{v} = \frac{Z_1}{Z_2}, \qquad (8.21)$$

which transforms easily into

$$\cos \frac{\nu a}{v} = \pm \frac{1}{\sqrt{1 + Z_1/Z_2}}. \qquad (8.22)$$

For our diagram, Fig. 8.21, with $Z_2 = \frac{1}{2}Z_1$, we get

$$\cos \frac{\nu a}{v} = \pm \frac{1}{\sqrt{3}}, \qquad (8.23)$$

which is represented graphically in Fig. 8.24. From this relation we obtain the roots

$$\frac{\nu a}{v} = \text{arc cos} \left(\pm \frac{1}{\sqrt{3}} \right)$$

$$= 0.956,\ 2.186,\ 4.098,\ 5.328 \ldots \quad (8.24)$$

and from these we derive the natural frequencies $f = \nu/2\pi$ of the system,

$$f'_{\lambda 1} = \frac{v}{6.56a},\ \frac{v}{2.86a},\ \frac{v}{1.53a},\ \frac{v}{1.18a} \cdots \quad (8.25)$$

By synthesis of oscillations with these frequencies the whole picture of Fig. 8.22

can be described. We see that the fundamental frequency is in the neighborhood of our estimated value and that the overtones are far from being harmonic. The fundamental frequency is determined by an irrational number, in contrast to the integral values that have been obtained in all the previous cases.

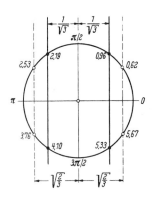

Fig. 8.24

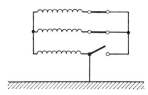

Fig. 8.25

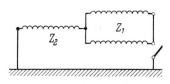

Fig. 8.26

If the pole of the circuit breaker that is connected to the ground fault opens first, as in Fig. 8.25, the straight-line diagram is given by Fig. 8.26. Now the single winding has the short-circuited end and the paralleled windings have the open end, which is common to both. Referring to Fig. 8.23, we see that now $Z_1 = \frac{1}{2}Z_2$ and Eq. (8.22) becomes

$$\cos\frac{va}{v} = \pm\sqrt{\frac{2}{3}}, \qquad (8.26)$$

which is represented by the dashed lines in Fig. 8.24. The roots of this equation are

$$v\frac{a}{v} = \arccos\left(\pm\sqrt{\frac{2}{3}}\right)$$

$$= 0.617, 2.527, 3.756, 5.668\ldots \quad (8.27)$$

and therefore the natural frequencies of this system are

$$f''_{\lambda 1} = \frac{v}{10.2a}, \frac{v}{2.49a}, \frac{v}{1.67a}, \frac{v}{1.11a}\ldots \quad (8.28)$$

Here also the overtones are not harmonic and the travel distance of the fundamental is

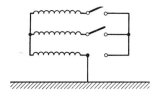

Fig. 8.27

an irrational multiple of the length of the phase windings.

In both cases of grounded windings that we have just treated, the fundamental frequency is considerably lower than for insulated windings as given by Eq. (8.12).

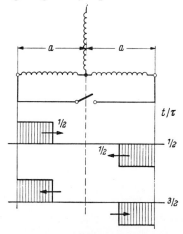

Fig. 8.28

If finally, as in Fig. 8.27, we interrupt those poles that are connected with the originally sound lines, the voltage divides into halves, of value $\pm\frac{1}{2}E$. The equivalent straight-lin diagram is shown in Fig. 8.28 and we see that the grounded phase winding is only an attachment to the center point. Since this point always has zero voltage, according to the course of the traveling pulses in Fig. 8.28, which is the same as that of Fig. 8.8 and Fig. 8.1, the third phase remains dead and has

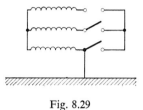

Fig. 8.29

no influence whatever on the natural frequency. The value of the frequency is therefore

$$f''_{\lambda 2} = \frac{v}{4a};$$ (8.29)

this is very different from the fundamental value of Eq. (8.28), which is valid for the pole opening first in this case.

Finally there remains the case of Fig. 8.29, in which the grounded pole of the circuit breaker opens last. The straight-line diagram of Fig. 8.30 shows that this example is similar

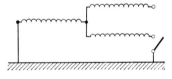

Fig. 8.30

to that of Fig. 8.26, except that the two open ends on the right-hand side are not connected with each other. In consequence, only one of the open phase windings is excited; the other one remains without external voltage impulse. Instead of pursuing the fairly complicated distribution of the traveling pulses in this system with three interlinked branches, we can assume it to be composed of two simpler distributions, as shown in Fig. 8.31a, b. Figure 8.31a represents the excitation of the two open branches with the voltages $\pm\frac{1}{2}E$. The oscillations of this system have already

been indicated by Fig. 8.28 and Eq. (8.29). Figure 8.31b, however, represents the excitation of the two ends with equal voltages $+\frac{1}{2}E$. The oscillations for this case were indicated by Fig. 8.26 and Eq. (8.28).

By addition of these two initial distributions of voltage we obtain Fig. 8.31c, which coincides exactly with Fig. 8.30. We see from this superposition that this system responds with two different series of oscillations, both of the same magnitude. Their fundamental natural frequencies are

$$f'_{\lambda 2} = \frac{v}{4a}, \frac{v}{10.2a},$$ (8.30)

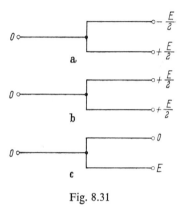

Fig. 8.31

and each is accompanied by the higher-frequency components, as in Eqs. (8.6) and (8.28) respectively.

The total voltage of interruption therefore divides into two oscillations of equal amplitudes but of very different frequencies. In this case we cannot consider the faster oscillations as overtones of the slower ones, since both have the same importance for the interrupting process.

In all previous examples we have drawn the diagrams so that they contain a well-determined ground connection immediately adjacent to the terminals of the star or delta winding. These schemes are widely used for tests of circuit breakers in high-power laboratories. In practice, however, the connections may be slightly different. Frequently the circuit breaker is adjacent to the winding of the machine or transformer, and the ground fault is situated on the line, so that the diagrams of

Figs. 8.32 and 8.33 are valid for the poles interrupting first and last respectively. We realize, however, that these are by no means different examples from those already discussed; rather they are merely different representations of Figs. 8.20 and 8.7. The same

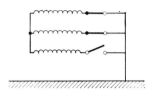

Fig. 8.32

remark naturally holds true for delta windings also. Our investigations therefore have covered the entire field of possible connections in three-phase systems.

(*d*) *Comparison with experiments.* The various examples that we have discussed are collected in Table 8.1, with the exception of the trivial

case of a neutral-grounded star system shown in Fig. 8.2. The series of multiplication factors for calculating the higher harmonics are also shown.

We see that for insulated three-phase windings, which means the interruption of a

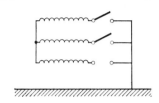

Fig. 8.33

mere short circuit, the natural frequencies are the same for the poles interrupted first and last in star as well as delta connections. For grounded windings, however, which means the interruption of a combined short circuit and ground fault, we have a wide variation of natural frequencies. The only general rule we can derive here is that in the

TABLE 8.1. Natural frequencies of three-phase windings under various internal and external conditions.

Winding	Interruption	λ	Δ
		Fig. 8.4	Fig. 8.9
Insulated	First pole	$f = \dfrac{v}{4a} \times 1, 3, 5, 7 \ldots$	$f = \dfrac{v}{3a} \times 1, 2, 3, 4 \ldots$
		Fig. 8.7	Fig. 8.12
	Last pole	$f = \dfrac{v}{4a} \times 1, 3, 5, 7 \ldots$	$f = \dfrac{v}{3a} \times 1, 2, 3, 4 \ldots$
		Fig. 8.20	Fig. 8.14
Grounded	First pole insulated	$f = \dfrac{v}{6.56a} \times 2.3, 4.3, 5.6 \ldots$	$f = \dfrac{v}{4a} \times 1, 3, 5, 7 \ldots$
		Fig. 8.29	Fig. 8.16
	Last pole grounded	$f_1 = \dfrac{v}{4a}$	$f = \dfrac{v}{6a} \times 1, 2, 3, 4 \ldots$
		$f_2 = \dfrac{v}{10.2a}$	
		Fig. 8.25	Fig. 8.18
	First pole grounded	$f = \dfrac{v}{10.2a} \times 4.1, 6.1, 9.2 \ldots$	$f = \dfrac{v}{6a} \times 1, 2, 3, 4 \ldots$
		Fig. 8.27	Fig. 8.19
	Last pole insulated	$f = \dfrac{v}{4a} \times 1, 3, 5, 7 \ldots$	$f = \dfrac{v}{3a} \times 1, 2, 3, 4 \ldots$

grounded condition the natural frequencies are always less than in the insulated state, the ratio varying between 1:2 and 1:2.6.

The natural frequencies for various connections were measured with a three-phase 50,000 kVA turbogenerator of 22 kV and 3000 rev/min by J. S. Cliff. The natural frequencies were determined not by the actual interruption of short circuits but by a resonance method using an external source of high frequency, thus exciting every possible frequency of the system under consideration without regard to amplitude.

The average test results on this generator were

$$\tau = 6.67 \ \mu\text{sec}$$

or

$$\frac{v}{a} = 150 \ \text{kc/sec},$$

which we consider as a measured constant

of the winding since we compare here only the influence of the different schemes of connection. The propagation velocity itself with a phase-winding length $a = 250$ m is calculated to be

$$v = 150 \cdot 10^{-3} \cdot 250 = 37.5 \ \text{m}/\mu\text{sec}.$$

In Table 8.2 every measured natural frequency is entered so that we can compare it easily with the calculated figures according to our formulas, including the harmonics, which are also entered. The agreement is exceedingly good, particularly for the fundamental frequencies, which are the most important ones.

The excellent agreement between measurement and calculation with so many different connections of the windings proves that the performance of machine windings corresponds under all circumstances with the formulas and the pulse pictures that we have derived.

TABLE 8.2. Comparison of measured and calculated natural frequencies (kc/sec) of a three-phase alternator.

Winding	Interruption	λ Measured Calculated	Δ Measured Calculated
Insulated	First pole	37.5, 97, 146 37.5, 113, 188	52, 98, 130 50, 100, 150
	Last pole	37.5, 100, 152 37.5, 113, 188	52, 98, 130 50, 100, 150
Grounded	First pole insulated	24, 55, 72, 97, 135 23, 52, 98, 127	37.5, 92, 152 37.5, 113, 188
	Last pole grounded	15, 37, 72, 84, 97, 142 14.7, 37.5, 60, 90, 113, 135	25, 37.5, 52, 95, 130, 150, 178 25, 50, 75, 100, 125, 150, 175
	First pole grounded	14.3, 52, 84, 98, 125 14.7, 60, 90, 135	24.5, 86, 175 25, 50, 75
	Last pole insulated	36.5, 99, 144 37.5, 113, 188	50, 98, 125 50, 100, 150

CHAPTER 9

PROTECTIVE RESISTANCES FOR LINE SWITCHING

Lumped ohmic resistances were seen earlier to produce considerable reduction in amplitudes of traveling pulses. This can be utilized most effectively if the resistances are designed and connected for use in anticipated switching operations on lines. In this way they can frequently damp from the very beginning traveling pulses produced by the connection, interruption, or short-circuiting of lines as considered in the previous chapters. Naturally, this is possible only for planned switching operations and automatic lightning arresters,

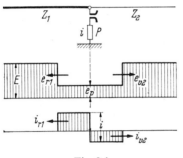

Fig. 9.1

whereas the effects due to random localized failures in insulators, coils, cables, or windings cannot be controlled in this way.

(a) *Spark arresters.* Spark arresters are used in transmission lines to prevent charging of the lines to very high voltages, considerably in excess of the regular system operating voltage. Calibrated spark gaps set to a given overvoltage, say twice the peak operating value, are introduced in several places. As soon as this set voltage is exceeded, the spark initiates an arc that can discharge the line to ground to prevent excessive overvoltage. The arc will be fed current by the system and must be rapidly extinguished, by means of horns or other quenching devices. With the ignition

of the spark the line suddenly assumes ground voltage at this point. Thus, according to Fig. 7.1, a voltage-discharge pulse of amplitude equal to the full breakdown voltage travels in each direction away from the arrester.

To reduce these dangerous step pulses and at the same time limit the current amplitude in the spark or arc, a resistance P is introduced between the spark arrester and ground (Fig. 9.1). The spark-arrester current i then produces a voltage drop

$$e_p = Pi \qquad (9.1)$$

across the resistance. This results in a discharge of the line from the breakdown voltage E to e_p instead of zero. Thus the two voltage-discharge pulses are reduced to

$$e = |E - e_p| = e_{v2} = e_{r1} \qquad (9.2)$$

and travel into both lines. The current in the damping resistance is the difference in currents between the two lines,

$$i = i_{v2} - i_{r1}. \qquad (9.3)$$

This current enters the ground, whose resistance should be included in P for the purposes of these calculations.

Figure 9.1 shows the voltages and currents immediately after spark ignition. To avoid constant use of negative sign for the voltage-discharge step pulses as given by Eq. (9.2), we shall consider them positive and let Fig. 9.1 represent them in their true polarity. Then

$$E = e + e_p, \qquad (9.4)$$

and by substitution from Eqs. (9.1) and (9.3) and use of the surge impedances Z_1 and Z_2, which we take to be different on the two sides, we obtain

$$E = e + P \left(\frac{e}{Z_2} + \frac{e}{Z_1} \right), \qquad (9.5)$$

74

Thus the amplitude of the voltage-discharge pulse is

$$e = \frac{E}{1 + P\left(\dfrac{1}{Z_1} + \dfrac{1}{Z_2}\right)}, \qquad (9.6)$$

and from Eq. (9.4) the voltage drop across the damping resistance P is

$$e_p = \frac{E}{1 + \dfrac{1}{P\left(\dfrac{1}{Z_1} + \dfrac{1}{Z_2}\right)}}. \qquad (9.7)$$

The current in P is then, from Eq. (9.1),

$$i = \frac{E}{P + \dfrac{Z_1 Z_2}{Z_1 + Z_2}}. \qquad (9.8)$$

Thus Eq. (9.6) shows that the amplitude of the voltage-discharge step pulses is governed by the ratio of P to the surge impedances Z_1 and Z_2 in parallel and can thus be limited by a choice of P that still allows discharge of the line to a safe voltage level.

As an example we shall take a spark arrester at the junction of an overhead line of $Z_1 = 500 \, \Omega$ and a transformer winding of surge impedance $Z_2 = 2000 \, \Omega$. The total parallel surge impedance is then

$$\frac{Z_1 Z_2}{Z_1 + Z_2} = \frac{500 \cdot 2000}{500 + 2000} = 400 \, \Omega.$$

Thus a damping resistance P of $400 \, \Omega$ will reduce the voltage-discharge step pulses, according to Eq. (9.6), to

$$\frac{e}{E} = \frac{1}{1 + \dfrac{400}{400}} = 0.5,$$

which is exactly half of the value without P. A damping resistance of $1600 \, \Omega$ would reduce the pulses to one-fifth of their amplitude without resistance. Thus the resistance is very effective in permitting the use of spark arresters that upon operation do not give rise to voltage step pulses of dangerous amplitudes.

At the same time the current entering the ground is seen from Eq. (9.8) to be reduced by P and this reduces danger in the vicinity of the grounding point. If the spark arrester

in the foregoing example is set for $E = 40{,}000 \, \text{V}$, the initial ground current without P would be $100 \, \text{A}$. For $P = 400 \, \Omega$ Eq. (9.8) gives for the current

$$i = \frac{40{,}000}{400 + 400} = 50 \, \text{A}$$

and for $P = 1600 \, \Omega$ it would be further reduced to $20 \, \text{A}$.

If a large damping resistance is used, the large reduction in voltage-pulse and current amplitude leads also to a less rapid and complete discharge of the line. Thus it is usual to insert a number of arresters in a system, which together can perform the necessary rapid discharge and at the same time do not individually give rise to dangerous voltage pulses.

If spark arresters are used in homogeneous lines, as is customary in overhead transmission lines, $Z_1 = Z_2 = Z$ and we obtain the simpler relations for the voltage-discharge step pulse,

$$e = \frac{E}{1 + \dfrac{2P}{Z}}, \qquad (9.9)$$

and for the current,

$$i = \frac{E}{P + \dfrac{Z}{2}}. \qquad (9.10)$$

Comparison with Eqs. (7.7) and (7.8) shows that the continuous line here acts exactly like a line of one-half the surge impedance discharged through a resistance at the end. Of course the two line segments joined at any point have exactly this half value of surge impedance when considered in parallel. Thus the line can be discharged most rapidly and without oscillations if we make

$$P = \tfrac{1}{2}Z. \qquad (9.11)$$

If m arresters are used along a line, their resistances are effectively in parallel and it is then adequate to make the individual resistances of higher value, namely,

$$P = m\frac{Z}{2}. \qquad (9.12)$$

Then simultaneous operation of all arresters will produce a rapid discharge just like that due to the resistance given by Eq. (9.11). However, instead of one large voltage step pulse we now obtain several smaller ones whose amplitude, by substitution of Eq. (9.12) in Eq. (9.9), is

$$e = \frac{E}{m+1}. \qquad (9.13)$$

These pulses arise at widely spaced points and thus reach the stations at very different times, which largely prevents a reinforcing summation of their effects.

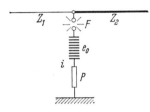

Fig. 9.2

At the same time the current through each arrester is reduced, according to Eqs. (9.10) and (9.12), to

$$i = \frac{2}{m+1}\frac{E}{Z}. \qquad (9.14)$$

Thus a large number of arresters m gives a lower arc current, which can be more easily interrupted, and reduces danger in the vicinity of the grounding point.

For overhead lines of surge impedance $Z = 500\,\Omega$ six arresters should each have, according to Eq. (9.12), a resistance

$$P = 6\frac{500}{2} = 1500\,\Omega.$$

For a flash-over voltage $E = 40,000$ V, Eq. (9.14) gives a current

$$i = \frac{2}{6+1}\frac{40,000}{500} = 23\ \text{A}$$

and voltage step pulses, by Eq. (9.13), of

$$\frac{e}{E} = \frac{1}{6+1} = \frac{1}{7} = 14\ \text{percent}$$

of the original breakdown voltage, which usually are not dangerous.

Figure 9.2 shows a spark arrester with a nonlinear characteristic like that of Fig. 6.12; after ignition the voltage drops to

$$e_p = e_0 + Pi, \qquad (9.15)$$

and this relation must now be used instead of Eq. (9.1). Equation (9.4) then leads to

$$E = e_0 + e + P\left(\frac{e}{Z_1} + \frac{e}{Z_2}\right) \qquad (9.16)$$

and thus the amplitude of the discharge pulse becomes

$$e = \frac{E - e_0}{1 + P\left(\dfrac{1}{Z_1} + \dfrac{1}{Z_2}\right)}. \qquad (9.17)$$

From Eq. (9.4) the voltage on the line is now given by

$$e_p = \frac{e_0 + P\left(\dfrac{1}{Z_1} + \dfrac{1}{Z_2}\right)E}{1 + P\left(\dfrac{1}{Z_1} + \dfrac{1}{Z_2}\right)} = \frac{e_0 + n\dfrac{P}{Z}E}{1 + n\dfrac{P}{Z}}, \qquad (9.18)$$

The last expression in Eq. (9.18) is for an arrester at a junction of n lines of equal surge impedance Z.

For $n = 2$, that is, for a simple switching station, this value caused by the discharge of a static charge is equal to that caused by an incident traveling pulse as developed in Eq. (6.47). For $n = 1$, that is, an end station, it is lower, and for $n \geqslant 3$ at a junction, it is higher than the former case. For a single nonlinear arrester at a bus-bar junction with $n = 5$, an ignition voltage $E = 60$ kV, a limiting voltage $e_0 = 20$ kV, and a resistance $P = 0.2Z$, we obtain a voltage residue on the line of

$$e_p = \frac{20 + \dfrac{5}{5}\cdot 60}{1 + \dfrac{5}{5}} = 40,000\ \text{V}.$$

This is a fairly high value because of the large number of lines and the considerable energy of their charge.

(b) *Circuit breakers with protective resistances.* Any switching on of lines leads to voltage step pulses that travel along the lines and may have amplitudes up to the full system operating voltage. In Fig. 7.12 we saw that even the switching off of an unloaded line produced step pulses of twice the system operating voltage and more, owing to reignition. We shall investigate the dependence of these traveling step pulses on the surge impedances of the lines to see if the insertion of a resistance at the circuit breaker can reduce the amplitude of these step pulses.

Figure 9.3 shows a line of surge impedance Z_1, charged to the voltage E, connected to another line of surge impedance Z_2 through a protective resistance R. As usual, we assume that R is a lumped circuit element of negligible dimensions. Prior to closing of the switch the second line had neither voltage nor current.

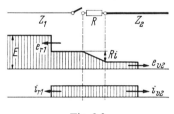

Fig. 9.3

Thus in the second line the closing of the switch can only give rise to a charging voltage pulse traveling into it. In line 1 we shall not consider all the component pulses but merely examine the total change superposed on the static charges by the switching process. Thus we need consider in line 1 only backward-traveling pulses originating at the switch. Figure 9.3 shows the voltage and current pulses immediately after closing of the switch. In line 1, which was originally charged, there travels a discharging voltage pulse and in line 2, which was originally uncharged, a charging voltage pulse. As the discharging pulse travels backward and the charging pulse forward, both give rise to positive current pulses.

The current amplitudes, after switching, must be equal in the two lines, and thus

$$i_{v2} = i_{r1} = i. \qquad (9.19)$$

If the current is expressed as the voltage divided by the surge impedance and the polarities shown in Fig. 9.3 are observed, we obtain

$$\frac{e_{v2}}{Z_2} = \frac{e_{r1}}{Z_1} = i, \qquad (9.20)$$

which can be rearranged as

$$e_{v2}\frac{Z_1}{Z_2} - e_{r1} = 0. \qquad (9.21)$$

The sum of the charging-voltage pulse in line 1, the discharging-voltage pulse in line 2, and the voltage drop across the protective resistance R is equal to the original voltage E at the switch, as seen in Fig. 9.3, and thus

$$e_{v2} + Ri + e_{r1} = E. \qquad (9.22)$$

Addition of Eqs. (9.21) and (9.22) gives

$$e_{v2}\left(1 + \frac{Z_1}{Z_2}\right) + Ri = E. \qquad (9.23)$$

If we now use Eq. (9.20) to express the current i in terms of the charging voltage pulse in line 2, we obtain

$$e_{v2} = \frac{Z_2}{Z_1 + Z_2 + R}E. \qquad (9.24)$$

The discharging voltage pulse in line 1 is now given by Eq. (9.21) as

$$e_{r1} = \frac{Z_1}{Z_1 + Z_2 + R}E. \qquad (9.25)$$

Finally, the charging current pulse that travels from the switch into both lines, given by Eq. (9.20), is

$$i = \frac{E}{Z_1 + Z_2 + R}. \qquad (9.26)$$

It is to be noted that the amplitude of the charging-voltage step pulse propagated into line 2 by closing of the switch is, according to Eq. (9.24), exactly half that produced by a pulse incident on the closed switch along line 1, as derived in Eq. (6.24). However, for the backward-traveling discharge pulse, Eq. (9.25) here has a quite different structure from Eq. (6.25). The decomposition of a static charge into equal pulses traveling in opposite directions, as shown in Fig. 2.1, accounts for this.

If the protective resistance is omitted, so

that $R = 0$, we obtain for the charging voltage pulse, from Eq. (9.24),

$$e_{v2} = \frac{Z_2}{Z_1 + Z_2} E, \qquad (9.27)$$

for the discharging pulse, from Eq. (9.25),

$$e_{r1} = \frac{Z_1}{Z_1 + Z_2} E, \qquad (9.28)$$

and for the switching current, from Eq. (9.26),

$$i = \frac{E}{Z_1 + Z_2}. \qquad (9.29)$$

These equations, for the simplest switching process, show that the current in the switch

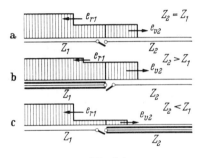

Fig. 9.4

immediately after closing is given simply by the voltage prior to switching divided by the sum of the surge impedances of the two lines. No other circuit parameters influence it. The voltage is divided into two pulses by the switching process. The voltage pulse is higher in the line of higher surge impedance. For $Z_1 = Z_2 = Z$, the two voltage pulses are equal and this is in agreement with Eqs. (2.19) and (2.20) for this symmetric case, given earlier.

For unequal surge impedances either the forward- or backward-traveling voltage pulse will be higher, and dangerous conditions are most likely to arise if very dissimilar lines are connected together. Figure 9.4 shows a few typical cases. In Fig. 9.4a equal lines are connected and so the voltage pulses are each equal to one-half. In Fig. 9.4b a cable feeds an open line. Since $Z_2 > Z_1$, Eq. (9.27) indicates that the charging-voltage pulse on the line is almost equal to the open-circuit voltage

and only a small discharging-voltage pulse is propagated into the cable. For $Z_1 = 50\ \Omega$ and $Z_2 = 500\ \Omega$,

$$\frac{e_{v2}}{E} = \frac{500}{50 + 500} = 91 \text{ percent}$$

and

$$\frac{e_{r1}}{E} = \frac{50}{50 + 500} = 9.1 \text{ percent.}$$

Thus the voltage at the end of the cable is only slightly reduced by the switching on of the line.

Figure 9.4c shows a cable switched on to an overhead line. Owing to the low surge impedance of the cable, it draws only a small charging-voltage pulse, while a large discharging pulse is propagated backward along the line. This reduces the voltage on the line almost to zero and travels back to the current source. Again for typical values of $Z_1 = 500\ \Omega$ and $Z_2 = 50\ \Omega$, we obtain

$$\frac{e_{v2}}{E} = \frac{50}{500 + 50} = 9.1 \text{ percent}$$

and

$$\frac{e_{r1}}{E} = \frac{500}{500 + 50} = 91 \text{ percent.}$$

This shows that the situation is the inverse of the preceding case.

The switching on of the cable acts almost like a short circuit on the overhead line. All these effects are produced by the different inductances and capacitances of the lines, which are sufficiently and most simply characterized by the surge impedance, which for this reason is also sometimes called the characteristic impedance.

If a protective resistance R is used at the switch, the ratio of the forward- and backward-traveling voltage pulses is unaltered; from Eq. (9.21), it is

$$\frac{e_{v2}}{e_{r1}} = \frac{Z_2}{Z_1}. \qquad (9.30)$$

Thus the larger pulse always occurs in the line of higher surge impedance. However, the amplitude of both voltage step pulses and the current pulses is reduced significantly by the resistance, as seen in Eqs. (9.24)–(9.26).

Because R is added to Z_1 and Z_2 in the denominators of all three equations, it must be of at least comparable magnitude and preferably several times larger.

For the last example above, of a cable switched on to an overhead line, we saw that the largest voltage step pulse was 91 percent of the original voltage. A protective resistance of 2000 Ω will reduce this very considerably, to

$$\frac{500}{500 + 50 + 2000} = 20 \text{ percent}$$

of the original voltage.

The pulses produced by the closing of the circuit breaker travel along both lines and after some time announce the closing at each end. At the ends and other discontinuities the pulses are transmitted and reflected and are gradually damped out owing to resistance losses in the lines. Thus a short time after closing of the switch these pulses have completely decayed and the steady-state operating distributions of voltages and currents exist on the lines.

If the line switched on is open at the far end, and draws no appreciable steady-state charging current from the source, then after decay of the switching pulses its voltage will be equal to that of the feeding line. Thus the protective

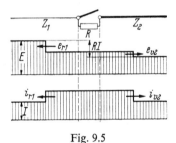

Fig. 9.5

resistance can then be short-circuited without giving rise to any further disturbances. However, if the line feeds a load at the far end that draws a steady-state current I, this causes a considerable voltage drop IR across the resistance, as shown in Fig. 9.5. Now a short-circuiting of the resistance gives rise again to switching pulses. These can be calculated from Eqs. (9.27)–(9.29) if E is taken as the voltage drop IR across the resistance. The total effect is shown in Fig. 9.5.

In practice this result limits the size of the protective resistance, for if it were chosen very large, so as to keep the first switching pulses very small, this second switching operation would give rise to large pulses. A good compromise is to produce switching pulses of equal amplitudes in both operations. This is possible for both the forward pulses, as in Eqs. (9.24) and (9.27), and the backward pulses,

Fig. 9.6

as in Eqs. (9.25) and (9.28). Both conditions have the common solution

$$\frac{E}{Z_1 + Z_2 + R} = \frac{RI,}{Z_1 + Z_2}, \qquad (9.31)$$

and thus Eqs. (9.26) and (9.29) also give equal current pulses for both operations.

If machines or transformers are connected to an unloaded line, the current I for moderate protective resistances is only slightly reduced if they draw predominantly reactive currents, as shown in Fig. 9.6. Thus it is possible to use the known steady-state current I_n as an approximation. Then, using Eq. (9.31) to determine the resistance, we obtain

$$R^2 + R(Z_1 + Z_2) = \frac{E}{I_n}(Z_1 + Z_2), \qquad (9.32)$$

and this gives for the best protective resistance

$$R = -\frac{Z_1 + Z_2}{2}$$

$$+ \sqrt{\left(\frac{Z_1 + Z_2}{2}\right)^2 + \frac{E}{I_n}(Z_1 + Z_2)}, \qquad (9.33)$$

omitting the negative root.

Here E/I_n is the reactive impedance of the line and its load. For high-voltage systems this can be considerably larger than the sum of the surge impedances under no-load or light-load conditions. Thus an approximate value for R will be

$$R = \sqrt{\frac{E}{I_n}(Z_1 + Z_2)}. \qquad (9.34)$$

This is the geometric mean of the reactive impedance and the sum of the surge impedances.

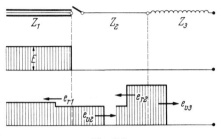

Fig. 9.7

Thus the larger the reactive load impedance, the larger may be the protective resistance.

A cable of $Z_2 = 50\ \Omega$ fed by an overhead line of $Z_1 = 500\ \Omega$, which draws a reactive charging current $I_n = 5$ A under the normal operating voltage $E = 20,000$ V, is best protected by switching through a resistance

$$R = \sqrt{\frac{20,000}{5}(500 + 50)} = 1480\ \Omega.$$

A transformer of $Z_2 = 5000\ \Omega$ fed by an overhead line of $Z_1 = 500\ \Omega$, which has a magnetizing current $I_n = 0.5$ A under the normal operating voltage $E = 50,000$ V, is best protected by switching through a resistance

$$R = \sqrt{\frac{50,000}{0.5}(5000 + 500)} = 23,400\ \Omega.$$

In both cases the approximation of Eq. (9.34) is quite valid.

In initial switching pulses decay quite rapidly owing to the fact that the resistance is in the circuit. Thus the second operation, shorting out the protective resistance, can follow after a very brief interval. It is usual to arrange the single continuous motion of the breaker contacts so that they touch first a preliminary contact that connects to the protective resistance and shortly thereafter the main contact, which shorts out the protective resist-

ance. In multipole breakers it is important to have all contacts close at the same time because the last contact to close will have to switch the highest voltage.

When electric motors or transformers are to be connected to an underground cable, the circuit breaker is sometimes arranged at the end of the cable with a length of open line between it and the machinery, as shown in Fig. 9.7. The line is constructed to have low capacitance between conductors, which gives it high surge impedance, and is of moderate length. Upon switching, only a moderate discharge voltage pulse travels into the cable. However, a high charging-voltage pulse travels along the open line and this is further almost doubled owing to the even higher surge impedance of the machinery. Thus a total charging-voltage step pulse of almost twice the voltage switched enters the winding of the machinery.

Similar effects occur, if, as shown in Fig. 9.8, a cable is fed by a generator or transformer over a short open line. Upon closing of the switch at the cable end only a moderate charging pulse enters the cable. However, a discharge pulse, of almost full-voltage amplitude, is sent along the open line to the feeding apparatus and again enters its windings doubled, owing to their high impedance.

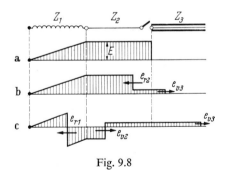

Fig. 9.8

This doubled voltage pulse can be observed at the first coils of transformer and machine windings connected as shown in Figs. 9.7 and 9.8. Motor or transformer-primary windings, as shown in Fig. 9.7, are exposed to these double voltage pulses and also have their voltage to ground raised to these high values. Generator or transformer-secondary windings are exposed to these double voltage pulses and at the same time have their voltage to ground reversed and thus raised to this double value at the star point.

The doubling of switching pulses, shown in Figs. 9.7 and 9.8, occurs to its full extent only if the open line between the ends of the cable and the windings is longer than the rise-time length of the pulses. In practice it can be observed for lengths as short as 10 m, so that pulse rise-time lengths of this order of magnitude do occur. Shorter connecting lines usually give only lower voltages.

Such increased voltage pulses can be completely avoided if the open line is omitted, or if the circuit breaker is connected between the open line and the winding, as shown in Fig. 9.9. In this case the step pulses in the

windings can at most attain the full system voltage E. Furthermore, reflection at the cable further reduces the pulse amplitude.

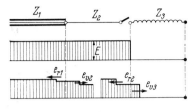

Fig. 9.9

If such an arrangement is impossible owing to physical layout problems, a protective resistance can be used to limit the switching pulses.

CHAPTER 10

DISTRIBUTED IMPEDANCE CHANGES

Not all conductor arrangements in use have uniform inductance and capacitance along their length, so that pulses can be propagated without disturbance. Long coils and windings, for example, have a lower surge impedance at the ends than in the middle sections. Power-line towers, when conducting lightning currents to ground, have a higher surge impedance at their top than at their base. In both these examples there is no sudden discontinuity between these values such as we have examined between different lines or windings in previous cases. Thus it is important to examine a gradual change in space of the surge impedance.

We saw in Chapter 6 that a pulse traveling from one line to another of different surge impedance undergoes a change of voltage and current. Because in each line of surge impedance Z the power of the pulse is given by

$$W = \frac{e^2}{Z} = i^2 Z, \qquad (10.1)$$

we can derive the laws of transmission and reflection at a junction easily from the conservation of power. The sum of the transmitted power and the reflected power must be equal to the incident power and thus

$$W_1 = \frac{e_{v1}^2}{Z_1} = \frac{e_{v2}^2}{Z_2} + \frac{e_{r1}^2}{Z_1} = W_2 + W_r. \quad (10.2)$$

On the other hand, the voltage balance at the junction gives

$$e_{v1} + e_{r1} = e_{v2}. \qquad (10.3)$$

By eliminating e_{r1} between these two equations we obtain for the transmitted voltage pulse

$$e_{v2} = \frac{2Z_2}{Z_1 + Z_2} e_{v1}, \qquad (10.4)$$

which of course agrees with Eq. (6.6). We shall here borrow some of the relations developed in Chapter 6.

For the transmitted power we obtain

$$W_2 = \frac{e_{v1}^2}{Z_2} \left(\frac{2Z_2}{Z_1 + Z_2} \right)^2, \qquad (10.5)$$

or as a fraction of the incident power

$$\frac{W_2}{W_1} = \frac{Z_1}{Z_2} \left(\frac{2Z_2}{Z_1 + Z_2} \right)^2 = \left(\frac{2}{\sqrt{Z_1/Z_2} + \sqrt{Z_2/Z_1}} \right)^2. \qquad (10.6)$$

This fraction is always less than unity because the transmitted power must be less than the

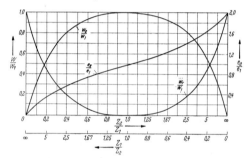

Fig. 10.1

incident power. Furthermore, it is invariant with interchange of Z_1 and Z_2, so that the direction of traversal of the junction has no effect on the power distribution. The reflected power is

$$W_r = \frac{e_{v1}^2}{Z_1} - \frac{e_{v2}^2}{Z_2} = \frac{e_{v1}^2}{Z_1} \left(\frac{Z_2 - Z_1}{Z_1 + Z_2} \right)^2, \qquad (10.7)$$

or, as a fraction of the incident power,

$$\frac{W_r}{W_1} = \left(\frac{Z_1 - Z_2}{Z_1 + Z_2} \right)^2. \qquad (10.8)$$

This fraction is always positive and less than unity. Both power ratios, from Eqs. (10.6) and (10.8), are plotted in Fig. 10.1 as functions of the ratio Z_1/Z_2. We see that when the difference between Z_1 and Z_2 is small, W_r becomes a negligible fraction of W_1. Even for some deviation of Z_1/Z_2 from unity, the losses due to reflection are quite low. Hence, to effect a transition between two lines of different surge impedances with the lowest possible losses due to reflection, it is best not to connect them directly, but to join them by a line of intermediate surge impedance.

Thus, for example, a direct transition from $Z_1 = 200\ \Omega$ to $Z_2 = 500\ \Omega$ gives a reflection loss of 19 percent, whereas an intermediate line of $Z = 300\ \Omega$ transmits $0.96 \cdot 0.94 = 90$ percent of the incident energy and thus reduces the reflection loss to 10 percent.

Still lower losses can be achieved if a series of suitably graded lines are introduced, and so one finally approaches the limiting case of a gradually changing transition.

(a) Reflectionless lines. The voltage changes along such an inhomogeneous line can be obtained from Eq. (10.4) if we apply it to a line element of small length in which only a

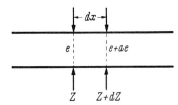

Fig. 10.2

small change dZ in surge impedance takes place. According to Fig. 10.2, then,

$$e + de = \frac{2(Z + dZ)}{Z + (Z + dZ)}e \qquad (10.9)$$

and thus

$$de = \frac{2Z + 2dZ - (2Z + dZ)}{2Z + dZ}e = \frac{dZ}{2Z}e. \qquad (10.10)$$

This gives the differential equation for the voltage,

$$\frac{de}{dZ} = \frac{e}{2Z}. \qquad (10.11)$$

The solution of this equation is

$$e = k\sqrt{Z}. \qquad (10.12)$$

The constant of integration k can be found from the initial values of surge impedance Z_0 and voltage e_0; this gives

$$e = e_0\sqrt{\frac{Z}{Z_0}}. \qquad (10.13)$$

Thus on a line with gradually changing surge impedance the general shape of the voltage distribution, given by e_0, remains unchanged. However, the amplitude changes as the square root of the ratio of the surge impedance. Increasing surge impedance produces higher, and decreasing surge impedance lower, voltage amplitude.

The power content of the traveling pulse is, according to Eqs. (10.1) and (10.13),

$$W = \frac{e^2}{Z} = \frac{e_0^2}{Z} \cdot \frac{Z}{Z_0} = \frac{e_0^2}{Z_0}. \qquad (10.14)$$

Thus it is constant at the initial value and no internal reflections occur. The total pulse energy travels forward, without loss, and the voltage changes according to Eq. (10.13) and the current, because of Eq. (10.1), according to

$$i = i_0\sqrt{\frac{Z_0}{Z}} \qquad (10.15)$$

in both time and space.

A line with a surge impedance that gradually changes from a value Z_1 to a value Z_2 thus acts on traveling pulses exactly like a transformer with a turns ratio

$$r = \sqrt{\frac{Z_2}{Z_1}} \qquad (10.16)$$

for alternating currents. Thus one can look upon the square root of the end impedances

Fig. 10.3

of a nonhomogeneous line as a transformation ratio.

We should note that these pulse phenomena are analogous to the surge of ocean waves onto gradually sloping beaches. In these waves also the energy content remains unchanged.

As the water becomes shallower the surface
displacement of the waves increases until,
owing to lack of stability, the wave breaks, as
shown in Fig. 10.3.

One can easily make a corresponding arrangement to produce the "breaking" of electrical
pulses or waves. As shown in Fig. 10.4, a
line in which pulses travel is connected to a
cable whose outer sheath is gradually enlarged,
so that the reduced capacitance gives an

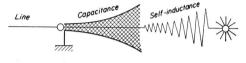

Fig. 10.4

increased surge impedance. In series with the
cable we further connect a line of increasing
inductance, such as a coil wound on a cone
of increasing diameter or a coil of increasing
density of turns. This continuous increase in
surge impedance will give a high transformation ratio, in accordance with Eq. (10.16).
Thus any incident voltage pulse will upon
traversal of the structure be transformed to
several times its original amplitude. Such
arrangements, which can also be constructed
simply out of lines with changing cross-
sectional dimensions, act to amplify pulse
voltages and thus are useful in the measurement or detection of traveling pulses. They
can also be used to dissipate such pulses, if
they are terminated in a suitably matched
large ohmic resistance, which will then absorb
their energy without reflection.

In communications technology, proper impedance matching between lines and equipment to avoid reflection losses is important.
It can be achieved with transformers, which
then must have the turns ratio given in Eq.
(10.16). However, as discussed here, it is also
possible to use transition lines with changing
inductance or capacitance, which will give a
gradual transition between impedances. Both
methods, which are shown schematically in
Fig. 10.5, were suggested by the earliest
pioneers in long-distance telephone engineering, but the latter is fully effective only under
certain circumstances.

We obtained Eq. (10.13), for the behavior
of voltage pulses on lines free of reflection, by

using a series or staircase of homogeneous
line elements. These all satisfied the usual
pulse equations and their solutions as in
Eq. (10.6) for the transmitted and in Eq. (10.8)
for the reflected power. If the line constants
change only slowly along the length of the
line in comparison with the extent of the
traveling pulse, this is valid. For example, a
short single pulse will not change in shape
during travel along the line. Only its amplitude
as a whole will change with the changing
impedance of the line.

However, if the changes in line impedance
occur over distances which are comparable
to or even smaller than the extent of the pulses,
it is clear that the pulses must be distorted
in shape. Thus, for example, the voltage at
the front of the pulse will have risen more than
that in its main body if it travels into a line of
abruptly increasing surge impedance. For
such cases, then, the foregoing equations are
no longer valid. They approach more closely
the cases previously treated of junctions between unequal lines, in which considerable
reflection losses were encountered.

Thus on lines with gradually changing surge
impedance the only pulses that propagate
without reflection are those that are short

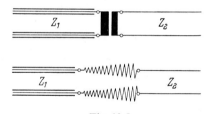

Fig. 10.5

compared with the distance over which
appreciable surge-impedance changes occur.
For this reason the overvoltage indicator at
the far right in Fig. 10.4 is activated only by
short traveling pulses, particularly step pulses,
and not by the more gradually changing
operating voltages of the system.

(b) *Conditions for zero reflection.* To obtain
a general solution for the voltage behavior on
nonhomogeneous lines we must reestablish
the differential equations on the line with the
added feature of changing inductance and
capacitance along its length. Just as in the case

of the homogeneous line, the laws of induction and dielectric displacement give us the two conditions

$$\frac{\partial e}{\partial x} = -l\frac{\partial i}{\partial t},$$

$$\frac{\partial i}{\partial x} = -c\frac{\partial e}{\partial t}. \qquad (10.17)$$

However, l and c now vary with x, though they are of course constant in time.

To eliminate one variable, for example the current i, we differentiate the second of Eqs. (10.17) with respect to t and the first with respect to x, and obtain

$$\frac{\partial^2 i}{\partial x\partial t} = -c\frac{\partial^2 e}{\partial t^2}, \qquad (10.18)$$

$$\frac{\partial^2 i}{\partial t\partial x} = -\frac{\partial}{\partial x}\left(\frac{1}{l}\frac{\partial e}{\partial x}\right) = -\frac{1}{l}\frac{\partial^2 e}{\partial x^2} + \frac{1}{l^2}\frac{\partial l}{\partial x}\frac{\partial e}{\partial x}.$$

Thus the differential equation for voltage on a line with changing l and c is

$$cl\frac{\partial^2 e}{\partial t^2} = \frac{\partial^2 e}{\partial x^2} - \frac{1}{l}\frac{\partial l}{\partial x}\cdot\frac{\partial e}{\partial x}, \qquad (10.19)$$

and similarly the differential equation for the current is

$$cl\frac{\partial^2 i}{\partial t^2} = \frac{\partial^2 i}{\partial x^2} - \frac{1}{c}\frac{\partial c}{\partial x}\cdot\frac{\partial i}{\partial x}. \qquad (10.20)$$

Comparison with Eq. (1.47) for homogeneous lines shows that we have now an additional term with $\partial/\partial x$. The strength of this term depends on the spatial change in l and c and it produces a spatial distortion of the pulses. A solution of the differential equations is now possible only if the changes in l and c as a function of x are given.

Let us find a solution for the case in which c is constant and l is a linear function of x, as shown in Fig. 10.6:

$$l = l_0\left(1 + \frac{x}{a}\right) = l_0 + (l_1 - l_0)\frac{x}{A}; \qquad (10.21)$$

here A is the actual length of the transition region and a the general length for a doubling of l. Then

$$\frac{1}{l}\frac{\partial l}{\partial x} = \frac{1}{a + x}. \qquad (10.22)$$

Further, for harmonic oscillations of circular frequency ω,

$$\frac{\partial^2 e}{\partial t^2} = -\omega^2 e, \qquad (10.23)$$

and thus the differential equation (10.19) becomes

$$\frac{\partial^2 e}{\partial x^2} - \frac{1}{a + x}\frac{\partial e}{\partial x} + \omega^2 cl_0\frac{a + x}{a}e = 0. \qquad (10.24)$$

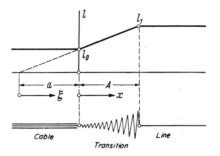

Fig. 10.6

If we now introduce the new variable

$$a + x = \xi$$

and put

$$\omega\sqrt{\frac{cl_0}{a}} = \beta, \qquad (10.25)$$

we obtain

$$\frac{\partial^2 e}{\partial \xi^2} - \frac{1}{\xi}\frac{\partial e}{\partial \xi} + \beta^2\xi e = 0, \qquad (10.26)$$

which we can recognize as a Bessel differential equation, which has the solution

$$e = \xi[k_1 J_{\frac{2}{3}}(\tfrac{2}{3}\beta\cdot\xi^{\frac{3}{2}}) + k_2 N_{\frac{2}{3}}(\tfrac{2}{3}\beta\cdot\xi^{\frac{3}{2}})] \qquad (10.27)$$

Here J and N are Bessel functions of the first and second kind respectively and of order $\frac{2}{3}$ as shown; k_1 and k_2 are constants of integration which can be evaluated by consideration of the boundary conditions at the two ends of the transition region, but we shall not carry out the numerical calculation.

However, without a detailed evaluation we can see by inspection that the shape of voltage pulses will be distorted along x or ξ owing to the nature of Bessel functions.

Because the peaks of the Bessel functions decrease only slowly with increasing ξ, the first factor ξ in Eq. (10.27) tends to give an increasing voltage along x, which is in accord with the corresponding increase in l along x. The complicated form of the Bessel functions, however, indicates internal reflections of voltage in the line. An expression for the power of the pulses would show this to be the case.

The process would have zero reflection if the solution had contained circular instead of Bessel functions, because these give undistorted pulses composed of harmonic oscillations. All Bessel functions approach circular functions for large values of their argument and thus a condition for zero reflection can be obtained in the form

$$\tfrac{2}{3}\beta\xi^{\frac{3}{2}} \gg 1; \qquad (10.28)$$

for the values in Eq. (10.25) this gives as an example for $x = 0$

$$\tfrac{2}{3}a\omega\sqrt{cl} \gg 1, \qquad (10.29)$$

and if we introduce

$$\lambda = \frac{v}{\omega} = \frac{1}{\omega\sqrt{cl}} \qquad (10.30)$$

as the wavelength on a homogeneous line with a velocity of propagation v we obtain

$$a \gg \tfrac{3}{2}\lambda. \qquad (10.31)$$

Thus the linear transition to twice the value of inductance (as shown in Fig. 10.6) must be considerably longer than $1\frac{1}{2}$ wavelengths to obtain zero reflection. This confirms in a quantitative manner the qualitative statements made earlier about the validity of Eqs. (10.13) and (10.14). If we actually let the Bessel functions in Eq. (10.25) approach asymptotically the circular functions we obtain, using Eq. (10.21),

$$e \approx k_3 \sqrt[4]{l} \cdot \sin\left[\tfrac{2}{3}\beta(a + x)^{\frac{3}{2}} + k_4\right]. \quad (10.32)$$

The increase of amplitude with the fourth root of l, that is, the square root of Z, agrees with Eq. (10.13). That the sine term is based on $x^{\frac{3}{2}}$, instead of x, indicates the progressive shortening of the wavelength owing to the increasing inductance.

If we have step pulses instead of harmonic waves we may assume equivalence between the front-rise length and a quarter wavelength. Thus the criterion for zero reflection in the

transition region is that the distance over which the inductance doubles must be more than six times the front-rise length of the step pulse. If this distance is longer, the voltage is transformed up or down according to $\sqrt{Z}$; if it is shorter, only a part of the voltage is propagated transformed, while the balance is reflected.

If the transition region shrinks to a point, or even a distance short compared with the front-rise length of the step pulse, maximum reflection will occur and the transmitted voltage pulse will be given by the simple relation in Eq. (10.4). Thus we must recognize that in an abrupt transition the voltage can at most be doubled, whereas in a gradual transition it can rise arbitrarily high (depending only on $\sqrt{Z}$) owing to the lack of reflection.

(c) *Exponential transitions.* Let us investigate numerically the particular arrangement of a

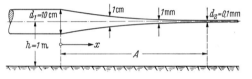

Fig. 10.7

line transition shown in Fig. 10.7. This has an exponential change in the diameter of the line conductor as given by

$$d = d_1\varepsilon^{-\delta x}. \qquad (10.33)$$

The diameter changes from $d_1 = 10$ cm to $d_2 = 0.1$ mm over a distance A on a center line at height $h = 1$ m above the ground. Thus δ is given by

$$\delta A = \ln\left(\frac{d_1}{d_2}\right). \qquad (10.34)$$

The surge impedance of such a conductor in air is given by Eq. (1.41),

$$Z = 60 \ln\left(\frac{4h}{d}\right) \Omega, \qquad (10.35)$$

and a corresponding dependence on cross-sectional dimensions is valid also for more complicated arrangements, for example multiple conductors. Using Eq. (10.33) for the diameter we obtain

$$Z = 60 \ln \left(\frac{4h}{d_1} \varepsilon^{\delta x} \right) = 60 \left[\ln \left(\frac{4h}{d_1} \right) + \delta x \right].$$

$$(10.36)$$

Thus the surge impedance increases as a linear function of x and the change over the length A becomes, by Eq. (10.34),

$$\Delta Z = 60 \, \delta A = 60 \ln \left(\frac{d_1}{d_2} \right), \quad (10.37)$$

which is now only a function of the two end diameters.

For our example this gives

$$\Delta Z = 60 \ln \left(\frac{10}{10^{-2}} \right) = 60 \cdot 6.9 = 413 \ \Omega,$$

while the initial and final values are

$$Z_1 = 60 \ln \left(\frac{4 \cdot 10^2}{10} \right) = 60 \cdot 3.69 = 222 \ \Omega,$$

$$Z_2 = 60 \ln \left(\frac{4 \cdot 10^2}{10^{-2}} \right) = 60 \cdot 10.6 = 635 \ \Omega.$$

Thus the voltage transformation ratio is

$$r = \sqrt{\frac{635}{222}} = 1.69.$$

And the general voltage transformation ratio is

$$r = \sqrt{\frac{Z_2}{Z_1}} = \sqrt{\frac{\ln (4h/d_1) + \ln (d_1/d_2)}{\ln (4h/d_1)}}$$

$$= \sqrt{1 + \frac{\ln (d_1/d_2)}{\ln (4h/d_1)}}. \quad (10.38)$$

Thus a considerable change in cross-sectional dimensions and close proximity to the ground are required to achieve appreciable transformation ratios.

The effect can be considerably increased if the line conductor is embedded at the beginning in a tapered cone made of a plastic insulator with a high dielectric constant and at the end in a tapered cone made of a highly magnetic insulator. Because the surge impedance Z, in its general formulation, contains $\sqrt{\mu/\varepsilon}$ in front of the logarithm, its effect is more pronounced than that of dimensional changes. For example, with $\varepsilon = 50$ at the beginning and $\mu = 500$ at the end of a transition zone, the transformation ratio

$$r = \sqrt{50 \cdot 500} = 158$$

for a conductor of constant diameter. This is almost 100 times the value for the arrangement shown in Fig. 10.8. However, because the space around the conductor (except in concentric cables) cannot be completely filled by such materials, this value represents merely an upper limit.

By winding the line into a coil it is also possible to increase considerably its inductance and thus the surge impedance. By means of a graded increasing density of windings it is then possible to obtain a high voltage-transformation ratio.

It is interesting to note that all these results depend on the initial and final values of the

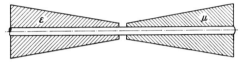

Fig. 10.8

surge impedance only and not on the form of the transition. Thus we need not, in the foregoing example, follow the exponential law for the diameter very closely, as long as the change in Z is smoothly continuous. The exponential law gives the optimum smoothness in continuity, but it should really be used for the surge-impedance changes instead of the dimensional ones. If an inhomogeneous transition line is open at its far end, a pulse is completely reflected. During reflection the voltage at the open end is twice the pulse voltage. During its return travel the voltage is reduced by the same ratio by which it was increased during its forward travel. Thus upon return to the beginning of the transition region it has again its original amplitude. Losses by internal reflection do not occur in this process if the conditions of Eq. (10.31) are fulfilled.

The velocity of propagation of the waves at any point in the transition region is merely determined by $\sqrt{\varepsilon \mu}$ and thus for conductors in air is constant at the velocity of light in air. However, if the conductor is embedded in dielectric or magnetic materials, the velocity decreases by a factor $\sqrt{\varepsilon \mu}$ and the time required to traverse the transition region increases correspondingly.

CHAPTER 11

INTERACTION BETWEEN ADJACENT LINES

Traveling pulses produce appreciable electro-magnetic-field effects in the vicinity of the lines along which they travel. These can produce strong interaction between adjacent lines. Because the energy content of the magnetic field of the pulses is equal to that of the electric field, we must here also consider the pulse-current effects and thus not only the electrostatic induction on adjacent lines as treated in Ref. 1, Chapter 29. We shall investigate the voltages and currents produced in a second line that, as shown in Fig. 11.1,

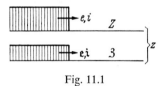

Fig. 11.1

runs in parallel with a power line on which a step pulse of voltage amplitude e and current i is traveling.

(*a*) *Mutual characteristics.* The power line has the surge impedance Z and the adjacent line the surge impedance 3. The electro-magnetic field of the traveling pulse then produces currents and voltages in the adjacent lines which in turn produce a back effect upon the pulse in the original line. We wish to examine this interaction between the pulses in the two lines, and as shown in Fig. 11.1 we shall designate quantities pertaining to the main line with Latin letters and those pertaining to the adjacent line with German letters.

The change of voltage with distance in each line was seen in Eq. (1.2) for a single line to depend only on the self-inductance of the line and time changes of current. Here, however, we must introduce an added term for the voltage induced by the magnetic field associ-ated with the current in the other line. Thus we obtain

$$-\frac{\partial e}{\partial x} = l\frac{\partial i}{\partial t} + m\frac{\partial i}{\partial t},$$
$$-\frac{\partial e}{\partial x} = \lambda\frac{\partial i}{\partial t} + m\frac{\partial i}{\partial t}. \tag{11.1}$$

Here l and λ are the self-inductances of the two lines and m is the mutual inductance of the lines per unit length.

The change of voltage with time in each line was seen in Eq. (1.4) for a single line to depend only on the self-capacitance of the line and space changes of the charging current. Here, however, we must introduce an added term for the voltage produced by charging current from the other line which flows into the mutual capacitance between the lines. Thus we obtain

$$-\frac{\partial e}{\partial t} = q\frac{\partial i}{\partial x} + g\frac{\partial i}{\partial x},$$
$$-\frac{\partial e}{\partial t} = \kappa\frac{\partial i}{\partial x} + g\frac{\partial i}{\partial x}. \tag{11.2}$$

Here q and κ are the reciprocals of the self-capacitances of the two lines and g is the reciprocal of the mutual capacitance between the two lines per unit length. For the simplest case of two single-conductor lines with a ground return, Ref. 1, p. 384, Eq. (4) gives q. The closer the two lines approach each other, the greater becomes the influence of the mutual coupling factors m and g.

The solution of these four differential equa-tions can be obtained, as in Chapter 1 for single lines, by assuming a linear relation for currents and voltages:

$$e = Zi + zi,$$
$$e = 3i + zi. \tag{11.3}$$

In addition to the surge impedances Z and 3 of the two lines we have here introduced a mutual surge impedance z between them which relates the current in one line to the voltage in the other line. Because the differential equations (11.1) and (11.2) and our trial solution (11.3) are linear in current and voltage, these effects superpose without interference. To determine the surge impedances as functions of the inductance and the capacitance we can simply let either one of the currents be zero.

If we first let the current $i = 0$ in the adjacent line the first members of Eqs. (11.1)–(11.3) give

$$-\frac{\partial e}{\partial x} = l\frac{\partial i}{\partial t},$$

$$-\frac{\partial e}{\partial t} = q\frac{\partial i}{\partial x}, \qquad (11.4)$$

$$e = Zi,$$

and the second members give

$$-\frac{\partial \mathrm{e}}{\partial x} = m\frac{\partial i}{\partial t},$$

$$-\frac{\partial \mathrm{e}}{\partial t} = g\frac{\partial i}{\partial x}, \qquad (11.5)$$

$$\mathrm{e} = zi.$$

If we let the main current i vanish, we obtain from the first of Eqs. (11.1)–(11.3),

$$-\frac{\partial e}{\partial x} = m\frac{\partial \mathrm{i}}{\partial t},$$

$$-\frac{\partial e}{\partial t} = g\frac{\partial \mathrm{i}}{\partial x}, \qquad (11.6)$$

$$e = z\mathrm{i}$$

and from the second

$$-\frac{\partial \mathrm{e}}{\partial x} = \lambda\frac{\partial \mathrm{i}}{\partial t},$$

$$-\frac{\partial \mathrm{e}}{\partial t} = \kappa\frac{\partial \mathrm{i}}{\partial x}, \qquad (11.7)$$

$$\mathrm{e} = 3\mathrm{i}.$$

The four sets of equations (11.4)–(11.7) all have the same structure. This is identical to that developed in Chapter 1 for single lines as actually given by Eq. (11.4). If then, as in

Chapter 1, we take the third equation of each set, substitute it in the first two equations, and multiply these, we obtain the equations for the surge impedances:

$$Z = \pm\sqrt{\frac{l}{1/q}},$$

$$3 = \pm\sqrt{\frac{\lambda}{1/\kappa}}, \qquad (11.8)$$

$$z = \pm\sqrt{\frac{m}{1/g}}.$$

The third of these equations follows from both Eqs. (11.5) and (11.6). In Eq. (11.8) the reciprocals are used so as to give directly capacitances per unit length, as in earlier expressions. We see that all surge impedances are given by inductance and capacitance values which must be constants for a given arrangement of the two lines. The mutual surge impedance z will be small compared with the surge self-impedances Z and 3 if the lines are far apart because the mutual inductance and capacitance coefficients will be small.

Just as in Eq. (1.12) we can here determine the velocity of propagation of the undistorted pulse by the ratio of the spatial to the temporal changes in each line. This gives

$$v = \frac{\pm 1}{\sqrt{l \cdot 1/q}} = \frac{\pm 1}{\sqrt{\lambda \cdot 1/\kappa}} = \frac{\pm 1}{\sqrt{m \cdot 1/g}}. \qquad (11.9)$$

The equality of the three expressions is confirmed by the known relation between inductance and capacitance coefficients.

The positive signs in Eqs. (11.8) and (11.9) correspond to forward-traveling partial pulses and the negative signs to backward-traveling partial pulses of voltage and current. The complete electromagnetic conditions on the two lines are obtained by superposition of all these partial pulses according to Eq. (11.3).

(*b*) *Mutual coupling coefficients.* For the purpose of calculations let us relate the surge mutual impedance between lines to the surge impedance proper of the lines. From Eq. (11.8) we obtain the ratio

$$\frac{z}{Z} = \sqrt{\frac{mg}{lq}}. \qquad (11.10)$$

Now by using the first and third members of Eq. (11.9) we obtain

$$\sqrt{\frac{m}{l}} = \sqrt{\frac{g}{q}}.$$ (11.11)

Thus the ratio of the surge mutual impedance to that of the main line is

$$\frac{z}{Z} = \frac{g}{q}.$$ (11.12)

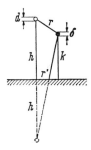

Fig. 11.2

Similarly Eqs. (11.8) and (11.9) give us the ratio to that of the adjacent line,

$$\frac{z}{3} = \frac{g}{\kappa}.$$ (11.13)

The reciprocal self- and mutual capacitances g, q, and κ can always be determined from the cross-sectional areas and distances apart of the lines.

Let us obtain some values for pulses that travel between a line and ground. In this case the mutual effects are large owing to the large loop formed by the two main current paths. Because of the high frequency of the partial-pulse components, we may regard the ground surface as a perfect mirror.

Figure 11.2 shows two single-conductor lines with ground return. By Ref. 1, p. 383, Eq. (2), the reciprocal of the capacitance of the two lines is

$$g = 2v^2 \ln\left(\frac{r'}{r}\right)$$ (11.14)

and according to Eq. (4) the reciprocal of the capacitance of the main line of height h and diameter d is

$$q = 2v^2 \ln\left(4\frac{h}{d}\right).$$ (11.15)

Similarly, the reciprocal of the capacitance of the adjacent line of height k and diameter δ is

$$\kappa = 2v^2 \ln\left(4\frac{k}{\delta}\right).$$ (11.16)

This gives the ratio of the surge mutual impedance to that of the main line,

$$\frac{z}{Z} = \frac{\ln (r'/r)}{\ln (4h/d)} \approx \frac{\ln [(h+k)/r]}{\ln (4h/d)}$$ (11.17)

and to that of the adjacent line,

$$\frac{z}{3} = \frac{\ln (r'/r)}{\ln (4k/\delta)} \approx \frac{\ln [(h+k)/r]}{\ln (4k/\delta)}.$$ (11.18)

The approximate expressions are valid for small distances r between the lines.

For lines on the same tower with $h = 10$ m, $k = 5$ m and hence $r = 5$ m, $d = 8$ mm and $\delta = 4$ mm, the ratio with respect to the main line is

$$\frac{z}{Z} = \frac{\ln 15/5}{\ln 4 \cdot 10/0.008} = 13 \text{ percent},$$

and with respect to the adjacent line is the same,

$$\frac{z}{3} = \frac{\ln 15/5}{\ln 4 \cdot 5/0.004} = 13 \text{ percent}.$$

If either line is made up of several conductors, the reciprocal of mutual capacitance

Fig. 11.3

remains the same if the distances r and r' are taken to the equivalent center of each line system. However, the reciprocals of self-capacitance q and κ become smaller, according to Eqs. (11.15) and (11.16), respectively, for multiple conductors. For two conductors, as shown in Fig. 11.3, Ref. 1, p. 371, Eq. (13) gave for the equivalent diameter

$$\Delta_2 = \sqrt{2\,ds}, \qquad (11.19)$$

where s is the distance between the centers of the two conductors.

Similarly, Ref. 1, p. 390, Eq. (24) shows the equivalent diameter for three conductors arranged as in Fig. 11.4a to be

$$\Delta_3 = \sqrt[3]{4\,ds^2}, \qquad (11.20)$$

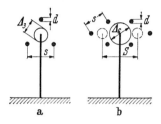

Fig. 11.4

Thus for six conductors arranged as in Fig. 11.4b the sequential use of the last two expressions, with the distance S between equivalent centers as shown, gives

$$\Delta_6 = \sqrt{2S\sqrt[3]{4\,ds^2}} = d\sqrt{\frac{2S}{d}}\sqrt[3]{\frac{2s}{d}}. \quad (11.21)$$

For linear arrangements of three-phase lines, as shown in Fig. 11.5, the effective

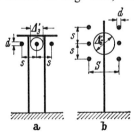

Fig. 11.5

diameter is somewhat larger because the arrangement is spread out more in space. Considerations similar to those used in Ref. 1, p. 390, Eq. (24) give for the effective diameter for three conductors, as shown in Fig. 11.5a,

$$\Delta_3' = \sqrt[3]{4\,ds^2\sqrt[3]{4}}, \qquad (11.22)$$

and thus for six conductors arranged as in Fig. 11.5b it is

$$\Delta_6' = \sqrt{2S\sqrt[3]{4\,ds^2\sqrt[3]{4}}} = d\sqrt[9]{2}\,\sqrt{\frac{2S}{d}}\sqrt[3]{\frac{2s}{d}}.$$

$$(11.23)$$

These equations are valid for the main lines of diameter d and adjacent lines of diameter δ. Thus the effective diameter of groups of conductors is much larger than that of single conductors in the group and is determined by both the diameters of the single conductors and the distances between them.

Figure 11.6 shows a 100,000-V main line of six conductors, which has, according to Eq. (11.23), an equivalent diameter of

$$\Delta_6' = 1.2 \cdot 1.08 \sqrt{\frac{2 \cdot 600}{1.2}} \sqrt[3]{\frac{2 \cdot 300}{1.2}} = 325 \text{ cm},$$

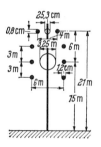

Fig. 11.6

topped by an adjacent two-conductor line, which has, according to Eq. (11.19), an equivalent diameter of

$$\Delta_2 = \sqrt{2 \cdot 0.8 \cdot 400} = 25.3 \text{ cm}.$$

Thus according to Eq. (11.17) the surge mutual impedance is

$$\frac{z}{Z} = \frac{\ln\left[(15+21)/6\right]}{\ln\left(4 \cdot 15/3.25\right)} = 62 \text{ percent}$$

of the main-line surge impedance and

$$\frac{z}{\mathcal{Z}} = \frac{\ln\left[(15+21)/6\right]}{\ln\left(4 \cdot 21/0.253\right)} = 31 \text{ percent}$$

of the adjacent-line surge impedance. However, for one of the six main conductors, say the center one, the surge mutual impedance between it and the adjacent line is only

$$\frac{z}{Z} = \frac{\ln\ [(15 + 21)/6]}{\ln\ (4 \cdot 15/0.012)} = 21 \text{ percent}$$

of its own surge impedance. Thus it is coupled only about one-third as strongly as all six conductors combined.

We should note that the ratios of surge impedances in Eqs. (11.17) and (11.18) are very similar to the shielding effect derived in Ref. 1, p. 370, Eq. (10) for the electrostatic shielding by ground conductors. From Eqs. (11.19)–(11.23) we could then similarly derive these shielding factors for three-phase lines.

(c) *Perpendicular approach of an adjacent line.* Figure 11.7 shows an adjacent line that

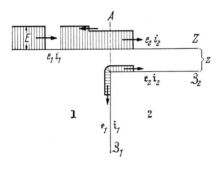

Fig. 11.7

approaches the main line at right angles. We wish to examine the effects of pulses on the main line in such an arrangement. To the left of the perpendicular A, that is, in sector 1, the pulse $e_1 i_1$ travels unchanged on the main line and the adjacent line experiences no effects. To the right of A, that is, in sector 2, the two lines run in parallel and are thus coupled. On the lines in sector 1 the last of Eqs. (11.4) applies to the main line and the last of Eqs. (11.7) applies to the adjacent line, since they are not coupled. However, in sector 2 the complete set of Eqs. (11.3) must be used. Let us first differentiate between the adjacent line surge impedances $\mathfrak{z}_1$ and $\mathfrak{z}_2$ to the left and right respectively of the point A. Because each line is continuous at A, we can draw up continuity equations for the main line,

$$\begin{aligned} e_1 &= e_2, \\ i_1 &= i_2, \end{aligned} \qquad (11.24)$$

and for the adjacent line,

$$\begin{aligned} \mathfrak{e}_1 &= \mathfrak{e}_2, \\ \mathfrak{i}_1 &= \mathfrak{i}_2. \end{aligned} \qquad (11.25)$$

In sector 2 we can have only forward-traveling pulses, since we consider it free of any earlier pulses. In sector 1, however, in addition to the incident pulse along the main line, we can also have in both lines reflected pulses that originate at A. Thus, by resolution into forward and backward partial pulses of the first of Eqs. (11.24), the voltage on the main line is composed of

$$e_{v1} + e_{r1} = e_{v2}. \qquad (11.26)$$

If we now express these as currents, using for the left-hand member the last of Eqs. (11.4) and for the right-hand member the first of Eqs. (11.3), and if we take into account the negative sign of Eq. (11.8) for rearward-traveling pulses, we obtain

$$Zi_{v1} - Zi_{r1} = Zi_{v2} + zi_{v2}. \qquad (11.27)$$

From the second of Eqs. (11.24) the partial current pulses in the main line are

$$i_{v1} + i_{r1} = i_{v2}. \qquad (11.28)$$

In the adjacent line, which has no incident pulse, the partial voltage pulses are, from the first of Eqs. (11.25),

$$\mathfrak{e}_{r1} = \mathfrak{e}_{v2}, \qquad (11.29)$$

and after insertion of the currents in sector 1 from the last of Eqs. (11.7) and in sector 2 from the second of Eqs. (11.3) we obtain

$$-\mathfrak{z}_1 \mathfrak{i}_{r1} = \mathfrak{z}_2 \mathfrak{i}_{v2} + z i_{v2}. \qquad (11.30)$$

From the second of Eqs. (11.25) we obtain the further condition for the currents in the adjacent line,

$$\mathfrak{i}_{r1} = \mathfrak{i}_{v2}. \qquad (11.31)$$

The four equations, (11.27), (11.28), (11.30), and (11.31), are sufficient to determine the four current pulses traveling on the two lines from the starting point A. By addition of

Eqs. (11.27) and (11.28) and insertion of the incident pulse on the main line,

$$i_{v1} = I,$$
$$e_{v1} = E, \qquad (11.32)$$

we obtain

$$2I = 2i_{v2} + \frac{z}{Z} i_{v2}, \qquad (11.33)$$

and by addition of Eqs. (11.30) and (11.31),

$$0 = \left(1 + \frac{\mathfrak{Z}_2}{\mathfrak{Z}_1}\right) i_{v2} + \frac{z}{\mathfrak{Z}_1} i_{v2}. \qquad (11.34)$$

Solution of these last two equations gives for the transmitted current pulse on the main line

$$i_{v2} = \frac{I}{1 - \dfrac{z^2}{2Z(\mathfrak{Z}_1 + \mathfrak{Z}_2)}} \approx I, \qquad (11.35)$$

and for the currents on the adjacent line, with due regard for Eq. (11.31)

$$i_{r1} = i_{v2} = \frac{-I}{\dfrac{\mathfrak{Z}_1 + \mathfrak{Z}_2}{z} - \dfrac{z}{2Z}} \approx -\frac{z}{2\mathfrak{Z}} I. \qquad (11.36)$$

Thus, according to Eq. (11.35), the current in the main line is somewhat increased by the presence of the adjacent line. At the same time, according to Eq. (11.36), a current of opposite sign appears in the adjacent line, which outwardly compensates for the increase and travels in both directions from A as a secondary current pulse. For lines far apart, z is only small compared with Z and $\mathfrak{Z}$, so that the second term in the denominators of Eqs. (11.35) and (11.36) can be neglected. In this case we obtain the approximations given, which apply also to homogeneous adjacent lines with $\mathfrak{Z}_1 = \mathfrak{Z}_2 = \mathfrak{Z}$.

The voltage pulse in the adjacent line is obtained by multiplying the current i_{r1} from Eq. (11.36) by its surge impedance and then using Eq. (11.29), which gives

$$e_{v2} = e_{r1} = -\mathfrak{Z}_1 i_{r1} =$$
$$\frac{\mathfrak{Z}_1}{Z} \frac{E}{\dfrac{\mathfrak{Z}_1 + \mathfrak{Z}_2}{z} - \dfrac{z}{2Z}} \approx \frac{z}{2Z} E. \qquad (11.37)$$

Here the incident current is expressed as a voltage-dependent quantity, and an approximation is also given. Thus upon arrival of the main-line pulse at A, voltage pulses of the same polarity are propagated in the adjacent line from A. The voltage pulses in the adjacent line have approximately the same ratio to the incident main-line pulse as one-half the surge mutual impedance has to the main-line surge impedance.

The reflected current pulse in the main line is, by Eqs. (11.28), (11.32), and (11.35),

$$i_{r1} = \frac{-I}{\dfrac{2Z(\mathfrak{Z}_1 + \mathfrak{Z}_2)}{z^2} - 1} \approx \frac{z^2}{4Z\mathfrak{Z}} I, \qquad (11.38)$$

and thus after multiplication by $-Z$ the reflected voltage pulse in the main line is

$$e_{r1} = \frac{-E}{\dfrac{2Z(\mathfrak{Z}_1 + \mathfrak{Z}_2)}{z^2} - 1} \approx -\frac{z^2}{4Z\mathfrak{Z}} E. \qquad (11.39)$$

Both pulses are small if the lines are far apart.

Finally, the transmitted voltage pulse on the main line is, according to Eqs. (11.26), (11.32), and (11.39),

$$e_{v2} = \frac{1 - \dfrac{z^2}{Z(\mathfrak{Z}_1 + \mathfrak{Z}_2)}}{1 - \dfrac{z^2}{2Z(\mathfrak{Z}_1 + \mathfrak{Z}_2)}} E \approx \left(1 - \frac{z^2}{4Z\mathfrak{Z}}\right) E, \qquad (11.40)$$

and thus only slightly smaller than the incident pulse.

From these relations we conclude that traveling pulses on one line induce voltage and current pulses of the same form on an adjacent line. They start at the point of approach and travel in both directions on the adjacent line. The power of the pulses in the adjacent line is the product of current and voltage, as given by Eqs. (11.36) and (11.37), respectively, and thus

$$\mathfrak{w}_{v2} = \mathfrak{w}_{r1} = \frac{\mathfrak{Z}_1}{Z} \frac{W}{\left(\dfrac{\mathfrak{Z}_1 + \mathfrak{Z}_2}{z} - \dfrac{z}{2Z}\right)^2} \approx \frac{z^2}{4Z\mathfrak{Z}} W. \qquad (11.41)$$

Thus for a homogeneous adjacent line the power is determined by the product of the ratios of surge mutual to self-impedances.

The amplitude e of the induced voltage can become sizable for close lines. If, as shown in Fig. 11.8, the lines are extremely close, then in the limit the values of surge mutual and self-impedance may be considered equal. Then, according to Eq. (11.8), we can put

$$z = Z = \mathfrak{Z}_1 = \mathfrak{Z}_2, \qquad (11.42)$$

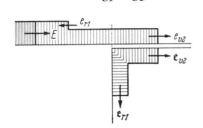

Fig. 11.8

so that, by the exact equations (11.37) and (11.40), the induced voltage approaches the limiting value

$$\mathfrak{e}_{v2} = \mathfrak{e}_{r1} = e_{v2} = \tfrac{2}{3}E. \qquad (11.43)$$

This is the same voltage that was shown in Fig. 6.3 to result if a pulse is incident on a bifurcating junction.

Thus traveling pulses are propagated, even without conductive connections, only in accordance with the surge impedances. They can enter adjacent lines with considerable amplitudes. This is due to their high-frequency components, which can produce coupling exceeding that due to direct connection paths.

If the adjacent line is somewhat more distant, the induced voltages rapidly decrease. For homogeneous adjacent lines with $\mathfrak{Z}_1 = \mathfrak{Z}_2$ Eqs. (11.37) and (11.12) give for the voltage

$$e = \frac{z}{2Z} E = \frac{1}{2} \frac{g}{q} E. \qquad (11.44)$$

Comparison with Ref. 1, p. 384, Eq. (5) shows that this induced pulse voltage is exactly one-half of that induced electrostatically on a line. This is because only a part of the induced pulse travels along with the main pulse. An equal pulse travels in the opposite direction also in the adjacent line.

We have based our calculations and figures on simple step pulses. But, since we used only linear relations for the induced effects, we can expect other pulse shapes to be induced in their true form. Thus our conclusions apply in general for pulses or pulse trains of any shape.

Strong step pulses are produced in adjacent lines when an insulator in a high-voltage line fails to ground. For an open 20,000-V line, for example, a threefold breakdown voltage results in a voltage step pulse of 60,000-V amplitude. If this pulse travels toward a point of approach of a low-voltage line, then, according to Eq. (11.44) and the values of our first example,

$$\frac{e}{E} = \frac{1}{2} \, 13 \text{ percent} = 6.5 \text{ percent}.$$

According to Eq. (11.36) the current pulse for the same case is

$$\frac{i}{I} = \frac{z}{2\mathfrak{Z}} = \frac{1}{2} \, 13 \text{ percent} = 6.5 \text{ percent},$$

and finally the transferred power is

$$\frac{w}{W} = 0.065^2 = 0.42 \text{ percent}$$

of the incident pulse power. If this pulse has a voltage $E = 60 \text{ kV}$, a main-line surge impedance $Z = 500 \, \Omega$, and thus a current $I = 120 \text{ A}$ and a power $W = 7200 \text{ kW}$, the pulse in the adjacent line has a voltage $e = 3900 \text{ V}$, current $i = 7.8 \text{ A}$, and power $w = 30 \text{ kW}$. These are values that, even for the short time duration involved, can cause serious disturbances.

In telephone lines the disturbances induced by traveling pulses are particularly unpleasant. Not only are the voltages and currents of excessive amplitude, but their high-frequency content also produces interfering noises that are far more noticeable than those due to the low operating frequency of the power lines.

Effects analogous to those for an approaching line are produced if an adjacent line recedes from the main line. Here too the contact point A serves to generate pulses traveling in both directions. Which of the two lines actually makes a right-angle turn is immaterial. Only the change in surge mutual impedance produces the effects. If the approach is not at right angles, as shown in Fig. 11.7, but is more gradual, we can regard the induced effects as a sequence of elementary pulses that superpose, each determined by the changing surge mutual impedance z. Thus the induced pulses will have a rise length corresponding to the transition length of the approaching or diverging lines. Step pulses on the main line thus will induce only pulses with spreadout rise lengths.

(d) *Linkages at transformers and ground wires.* Strong disturbances can be created by traveling pulses that are transferred between the primary and secondary windings of transformers. We shall deal with this more explicitly in Chapter 22. Figure 11.9 shows a step pulse incident on the high-voltage winding. If the low-voltage winding is very close to it, the ratio of surge mutual to self-impedance z/Z is, according to Eq. (11.37), relatively high. Thus traveling pulses are induced in the low-voltage winding, which partly penetrate into the winding and partly propagate into the line connected with it.

The limiting value, never actually reached, for very closely coupled or linked windings would be

$$z = Z = \mathfrak{Z}_2. \qquad (11.45)$$

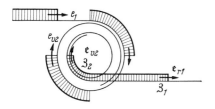

Fig. 11.9

The surge impedances inside the transformer are mostly considerably higher than those like $\mathfrak{Z}_1$ for the connecting lines. Thus the induced voltage becomes, according to Eq. (11.37),

$$e_{v2} = e_{r1} = \frac{2\mathfrak{Z}_1}{2\mathfrak{Z}_1 + \mathfrak{Z}_2} E. \qquad (11.46)$$

For $\mathfrak{Z}_1 = 500\ \Omega$ and $\mathfrak{Z}_2 = 5000\ \Omega$, we obtain

$$e = \frac{2 \cdot 500}{2 \cdot 500 + 5000} E = \frac{1}{6} E = 0.167E.$$

In general, the voltage is lower because the windings are separated by their insulating distances. For 30- to 60-percent coupling, 5 to 10 percent of the incident pulse-voltage amplitude can be transmitted through the transformer.

The voltage transfer appears here independent of the turns ratio and, from Eq. (11.46), is determined only by the surge impedances in the vicinity of the ends of the winding. Thus a voltage pulse that is quite harmless

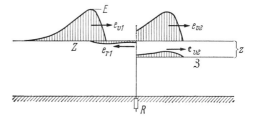

Fig. 11.10

to the high-voltage winding can enter the low-voltage side with an amplitude that can cause considerable damage in it and along the lines connected to it. To avoid this effect completely, a metallic shield must be placed between the two windings.

Even between adjacent telephone and telegraph lines, cross-coupling causes disturbances. Especially in telephone cables with closely packed conductors the energy transferred, according to Eq. (11.14), is appreciable. These effects are avoided by twisting and transposition of conductors in such systems.

Figure 11.10 shows a traveling pulse that arrives at a given tower along a line without a ground wire and then continues with a ground wire. At the tower the pulse voltage is reduced. The ground wire, which is grounded through the tower and footing resistance R, acts at the tower exactly as if it were receding from the line with a surge impedance

$$\mathfrak{Z}_1 = R. \qquad (11.47)$$

Thus for relatively small z a good approximation for the voltage pulse in the ground wire, from Eq. (11.37), is

$$e_{v2} = \frac{R}{R + \mathfrak{Z}} \frac{z}{Z} E. \qquad (11.48)$$

The subscript 1 is omitted here from the ground-wire surge impedance. For $R = 0$ the voltage becomes zero and for $R = \infty$ it assumes the static value.

For the reflected pulse in the main line Eq. (11.39) gives

$$e_{r1} = -\frac{z^2}{2Z(\mathcal{3} + R)}E, \qquad (11.49)$$

and for the transmitted pulse in the main line, which is reduced by the effect of the ground wire, Eq. (11.40) gives

$$e_{v2} = \left[1 - \frac{z^2}{2Z(\mathcal{3} + R)}\right]E. \qquad (11.50)$$

We see from Eq. (11.48) that the ground wire can be raised to considerable voltages, depending on the ground resistance R. For the earlier example of the 100,000-V six-conductor line and two top ground wires of $\mathcal{3} = 300\ \Omega$, a tower resistance $R = 100\ \Omega$, and an atmospheric pulse of double the operating voltage, that is, $E = 200,000$ V on all six conductors in common, Eq. (11.48) gives

$$e_{v2} = \frac{100}{100 + 300} \cdot \frac{62}{100} \cdot 200 = 31 \text{ kV},$$

which is a remarkably high value for a grounding wire. The reflected pulse on the main line is given by Eq. (11.49) and is also, by Eq. (11.50), equal to the reduction in the transmitted pulse. With the same values it is

$$\frac{e_{r1}}{E} = -\frac{0.62 \cdot 0.31}{2\left(1 + \dfrac{100}{300}\right)} = -7.2 \text{ percent}$$

of the incident voltage pulse. Thus the overvoltage has been reduced from 200,000 V to 186,000 V.

This relatively small reducing effect of the ground wire is due to the fact that, according to Eq. (11.49), it is governed by two coupling factors. The first coupling transfers energy from the main line to the ground wire and the second coupling transfers it back to the main line. Even if in the last example we push the ground-wire current to its maximum by making $R = 0$, the reduction in the voltage pulse on the main line would be only 9.6 percent. If the ground wire is not grounded at

all, that is, if $R = \infty$, all effects on the main line disappear, as seen in Eqs. (11.50) and (11.49). The voltage pulse in the ground wire reaches, according to Eq. (11.48), its static value

$$e_{v2} = \frac{z}{Z}E = \frac{g}{q}E, \qquad (11.51)$$

which agrees with Ref. 1, p. 384, Eq. (5) and in the present example is

$$e_{v2} = \frac{62}{100}\,200 = 124 \text{ kV}.$$

If the incident pulse travels along only one of the six conductors, the coupling factor z/Z, as we have seen, is only 21 percent, or one-third of the value used above.

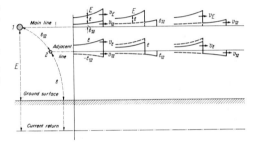

Fig. 11.11

Thus the voltage pulses produced in the ground wires and the changes in pulse voltage produced in the main line will be only one-third of the previously calculated values. Hence in all cases the reduction of incident traveling pulses is far less than if, as developed in Ref. 1, Chapter 28, the ground wire statically shields the line against the creation of a traveling pulse.

(e) *Different velocities of propagation.* A remarkable phenomenon is observed when the voltage on the adjacent line is measured at some distance from its point of origin, as shown in Fig. 11.11. At the origin, from voltage E on the main line a voltage e is transferred to the adjacent line, as previously calculated. The voltage e is between the adjacent line and ground.

Because the ground is a poor conductor, the high-frequency current components of the traveling pulses are carried at some "skin" depth and this, as shown in Ref. 1, p. 407, Eq. (48), increases the inductance above that for a good metallic conducting surface. Thus the velocity of propagation of the voltage pulses E and e is decreased slightly below that obtaining in air, owing to this ground penetration. However, the voltage e_{12}, which exists between the two metallic line conductors 1 and 2 and which is negative on conductor 2, is propagated with the velocity of light in air and thus travels faster than e. L. V. Bewley showed that as a result the voltage e_{12} arrives first at a distance from the origin and couples in part statically to ground. It has been frequently observed on the adjacent line as a brief negative forerunner and on the main line as a positive forerunner. Figure 11.11 shows these forerunners.

Similar phenomena can occur if the main pulse is slowed down by corona effects or if the ground-wire pulse is slowed down by the skin effect of its steel cable.

In spiral-clad cables (Fig. 11.12) the helical nature of the outer conductors leads to a lower propagation velocity. Thus similar forerunners are observed. Furthermore, they cause considerable disturbances, since the outer shield

Fig. 11.12

is ineffective for the grounding of traveling pulses.

In transformers, the differences in number of turns between primary and secondary windings lead to different velocities of propagation along the common axis of the windings. These complicated phenomena will be further examined in Chapter 22.

PART III

COILS AND CONDENSERS

CHAPTER 12

PULSE SHAPING

Choke coils and bypass condensers have long been used to give protection against over-voltages. High-frequency components of voltage and current will on the one hand not penetrate coils and thus will be choked off, and on the other hand will easily flow through condensers and thus can be made to bypass circuits in which they might otherwise cause disturbances. The coils and condensers also act as shock absorbers by temporarily storing the energy contained in the more sustained pulses. In addition, traveling pulses frequently encounter coils and condensers on power lines in the form of instrument transformers, relay coils, bus bars, and insulators. First we shall investigate the effect of these devices on step pulses caused by sudden switching processes. Later we shall consider the effect of single pulses and pulse trains.

(a) *Effect of self-inductance.* Figure 12.1 shows two lines, of surge impedances Z_1 and Z_2, connected by a choke coil L. A step pulse of voltage E and current I is incident from the left along line 1. The choke coil is assumed to have no appreciable physical length, no capacitance, and no resistance, and thus it is considered simply as a lumped self-inductance. This implies a uniform current throughout the coil and thus equal currents at the terminals connected to the two lines, that is,

$$i_1 = i_2. \qquad (12.1)$$

The voltages on the two lines differ by the voltage drop across the choke coil. This is given by the self-inductance L and the change with time of the current. Thus

$$e_1 = e_2 + L\frac{di_2}{dt}. \qquad (12.2)$$

As a result of the incident pulse in line 1 we can have only a forward-transmitted pulse in line 2 and also an additional reflected

rearward pulse in line 1. By separation of voltages and currents into partial pulses, we obtain from Eq. (12.2) the voltage equation

$$e_{v1} + e_{r1} = e_{v2} + \frac{L}{Z_2}\frac{de_{v2}}{dt}. \qquad (12.3)$$

Here the current is given as the voltage divided by the surge impedance. Similarly, Eq. (12.1) multiplied by Z_1 gives

$$e_{v1} - e_{r1} = \frac{Z_1}{Z_2}e_{v2}. \qquad (12.4)$$

Addition of Eqs. (12.3) and (12.4) eliminates the reflected voltage pulse and multiplication

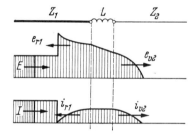

Fig. 12.1

by Z_2/L gives an equation for the two forward voltage pulses,

$$\frac{de_{v2}}{dt} + \frac{Z_1 + Z_2}{L}e_{v2} = 2\frac{Z_2}{L}e_{v1}. \qquad (12.5)$$

This is a linear differential equation of first order, with constant coefficients and known perturbation function, for the voltage at the beginning of line 2 as a function of time. For a given behavior as a function of time of the voltage e_{v1} at the beginning of the choke coil we can then calculate e_{v2} at its end connection to line 2.

If a rectangular step pulse of amplitude E is incident upon the choke coil, then, from the moment of arrival on,

$$e_{v1} = E, \qquad (12.6)$$

101

and thus the right-hand member of the differential equation is constant. In this case the solution is simple and was given in Ref. 1, p. 10, Eq. (14) for the switching on of a coil with direct current; it is

$$\frac{di}{dt} + \frac{R}{L} i = \frac{E}{L}, \qquad (12.7)$$

except that the coil resistance R is replaced by $Z_1 + Z_2$ and the perturbation function is quite different, as seen in Eq. (12.5). Corresponding to the time rise of the current given in Ref. 1, p. 11, Eq. (22),

$$i = \frac{E}{R} (1 - \varepsilon^{-(R/L)t}), \qquad (12.8)$$

we obtain here the rise of voltage on line 2 as a function of time,

$$e_{v2} = \frac{2Z_2}{Z_1 + Z_2} E (1 - \varepsilon^{-(Z_1 + Z_2)t/L}). \quad (12.9)$$

We see that step pulses of constant amplitude that arrive at a choke coil enter the second line, not abruptly, but with a voltage that rises gradually from zero to reach its final value only after some time. The speed of increase is governed by the exponent in Eq. (12.9), which thus has the usual nature of a time constant,

$$T_L = \frac{L}{Z_1 + Z_2}. \qquad (12.10)$$

It increases with L and decreases with the sum $Z_1 + Z_2$ of the surge impedances of the lines. Thus the lines act just like energy-dissipating resistances owing to the pulse energy transmitted into them.

For a choke coil of inductance $L = 3$ mH between an overhead line of $Z_1 = 500\ \Omega$ and a transformer winding of $Z_2 = 5000\ \Omega$, the time constant is

$$T_L = \frac{3 \cdot 10^{-3}}{500 + 5000} = 0.55 \cdot 10^{-6} \text{ sec.}$$

During one time constant the pulse travels a distance

$$vT_L = 3 \cdot 10^8 \cdot 0.55 \cdot 10^{-6} = 163 \text{ m.}$$

As shown in Chapter 1, the spatial shape of the voltage pulse along line 2 will correspond to its time change at the beginning of the

line. Thus the rectangular step pulse incident along line 1 on the choke coil is transmitted into line 2 as an exponentially flattened pulse, shown in Fig. 12.1.

The forward-traveling current pulse in line 2, derived from Eq. (12.9) by division by Z_2, is

$$i_{v2} = \frac{2Z_1}{Z_1 + Z_2} I(1 - \varepsilon^{-t/T}). \quad (12.11)$$

Here E has been replaced by the product of I and Z_1.

Thus the current in line 2 also rises not abruptly but gradually. The time dependence of Eqs. (12.9) and (12.11) shows that the factor in parenthesis is 0 at $t = 0$ but reaches a final value of 1. Thus voltage and current reach the same final values as in Eqs. (6.6) and (6.7) for the transmission of a pulse through a junction between two lines. Thus the choke coil acts only to flatten the front rise of the pulse. After several time constants T_L, the values of the voltages and currents are independent of the choke coil. The effect of the choke coil is that it stores magnetic energy drawn from the sharp front rise of the pulse. However, it has no effect on the flat main portion or body of the step pulse.

In line 1 the incidence of the pulse on the choke coil produces a reflected voltage pulse. Its amplitude, obtained from Eq. (12.4) by substituting e_{v1} as given in Eq. (11.6) and e_{v2} as given in Eq. (11.9), is

$$e_{r1} = E - \frac{2Z_1}{Z_1 + Z_2} E(1 - \varepsilon^{-t/T}). \quad (12.12)$$

Dividing by $-Z_1$ for the rearward-traveling current pulse we obtain

$$i_{r1} = -I + \frac{2Z_1}{Z_1 + Z_2} I(1 - \varepsilon^{-t/T}). \quad (12.13)$$

At $t = 0$ the second terms of both equations are also zero. Thus at the first instant of arrival of the pulse at the choke coil the voltage on line 1 is doubled and the current in it is reduced to zero. Hence the choke coil acts on the steep front of the pulse as an open line end. However, as time increases, the second terms of Eqs. (11.12) and (11.13) become larger and so the total voltage drops from $2E$ and the total current increases from zero. After several time constants the parentheses in the second terms have the value 1 and thus the reflected voltage at the line-1 terminal of the choke approaches

$$(e_{r1})_\infty = \frac{Z_2 - Z_1}{Z_1 + Z_2} E, \qquad (12.14)$$

and the current approaches

$$(i_{r1})_\infty = \frac{Z_1 - Z_2}{Z_1 + Z_2} I. \qquad (12.15)$$

Thus, after some time constants have elapsed, the pulses reach the values that were seen in Eqs. (6.8) and (6.9) to apply to a junction without a choke coil. Figure 12.1 shows the spatial distribution of the pulses derived from the above time behavior.

We see that lumped self-inductance can reshape the most abrupt features of voltage step pulses, caused by switching or flash-over processes, so that they can be kept away from sensitive parts of the system, such as machinery and transformers. The flattening of the transmitted pulse increases with increased self-inductance. However, the voltage of the reflected pulse has twice the amplitude of that of the incident pulse and drops more slowly to its final value as the self-inductance is increased.

If the lines on both sides of the choke coil have equal surge impedances, the equations become somewhat simpler. The essentially transient behavior retains its character; however, the final values are exactly those of the incident pulse owing to the homogeneity of the lines.

Figure 12.2 shows a single square pulse incident upon a choke coil in a line. We can regard this as composed of a positive step pulse followed after a short time by a negative step pulse of equal magnitude. The resulting transmitted and reflected pulses at different time intervals are shown for identical lines on either side of the choke coil. We note that the transmitted pulse is still a single one, flattened and stretched out. However, the reflected pulse now consists of two very sharp spike pulses of opposite polarity.

Choke coils were formerly in general use as protection against traveling pulses. They fell out of favor because with air cores they

could not be built small enough to really form a lumped self-inductance, and steel cores could not respond to the essential high-frequency components of the pulses owing to eddy currents.

However, with the advent of ferrites, which have negligible eddy currents up to very high frequencies and are also good insulators, choke

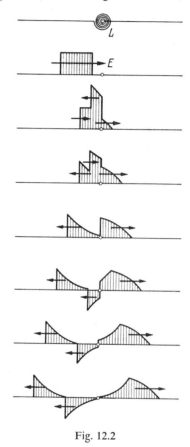

Fig. 12.2

coils with ferrite cores will become increasingly competitive with capacitors. In particular, they appear useful to activate overvoltage devices on the incident line due to the sharp voltage rises or spikes shown in Figs. 12.1 and 12.2.

(*b*) *Effect of condensers.* Figure 12.3 shows a condenser C connected between the junction of two dissimilar lines and ground. We shall consider it as a lumped capacitance without distributed self-inductance or resistance. It does not interfere with the equality of voltage at the two line ends joined; thus

$$e_1 = e_2. \qquad (12.16)$$

However, the condenser draws current from the lines that is the product of its capacitance and the change of voltage across it with time:

$$i_1 = i_2 + C\frac{de_2}{dt}. \qquad (12.17)$$

Here also line 2 carries only a forward transmitted pulse and line 1 an additional reflected

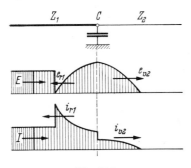

Fig. 12.3

pulse. Thus Eq. (12.16) when split into partial pulses leads to

$$e_{v1} + e_{r1} = e_{v2}, \qquad (12.18)$$

and Eq. (12.17), with the currents replaced by voltages and then multiplied by Z_1, leads to

$$e_{v1} - e_{r1} = \frac{Z_1}{Z_2}e_{v2} + Z_1 C\frac{de_{v2}}{dt}. \qquad (12.19)$$

Addition of the last two equations gives

$$\frac{de_{v2}}{dt} + \frac{\frac{1}{Z_1} + \frac{1}{Z_2}}{C}e_{v2} = \frac{2}{Z_1 C}e_{v1}. \qquad (12.20)$$

This is again a differential equation of first order for the voltage pulse in line 2.

Just as in Eqs. (12.7) and (12.8), the incidence of a step pulse of constant amplitude

$$e_{v1} = E \qquad (12.21)$$

gives the solution of the differential equation as

$$= \frac{2Z_2}{Z_1 + Z_2}E(1 - \varepsilon^{-[(1/Z_1)+(1/Z_2)]t/C}). \qquad (12.22)$$

We see by comparison with Eq. (12.9) that the time variation of the voltage in line 2 behind a condenser is very similar to that behind a choke coil. Here, too, the voltage rises only gradually and approaches a final value equal to that in the absence of a condenser.

The rapidity of the rise is also governed by a traveling-pulse time constant T_C, which is seen in the exponent of Eq. (12.22) to be

$$T_C = \frac{C}{(1/Z_1) + (1/Z_2)} = \frac{Z_1 Z_2 C}{Z_1 + Z_2}. \qquad (12.23)$$

Comparison with Ref. 1, p. 20, Eq. (11) shows that this time constant is of the same structure as that of a capacitive circuit with the resistance replaced by the parallel surge impedance of the two lines. The lines here carry away energy that in the lumped circuit is dissipated by the resistance. The larger the capacitance of the condenser, the larger the time constant and thus the longer the time taken for the voltage to rise in line 2.

A condenser of capacitance $C = 0.003\ \mu\text{F}$ connected between a line of $Z_1 = 500\ \Omega$ and a winding of $Z_2 = 5000\ \Omega$ has a time constant of

$$T_C = \frac{500 \cdot 5000}{500 + 5000} \cdot 0.003 \cdot 10^{-6}$$
$$= 1.36 \cdot 10^{-6}\ \text{sec}.$$

During 1 time constant the pulse travels a distance of

$$vT_C = 3 \cdot 10^8 \cdot 1.36 \cdot 10^{-6} = 410\ \text{m}.$$

Figure 12.3 shows the spatial increase of voltage along line 2, which is equal to the time increase at the junction. We see that a condenser in parallel with the junction produces the same transmitted pulse in the second line as a series-connected choke coil.

The current in line 2 is derived from Eq. (12.22) by division by Z_2 and introduction of the incident pulse current:

$$i_{v2} = \frac{2Z_1}{Z_1 + Z_2}I(1 - \varepsilon^{-t/T_C}). \qquad (12.24)$$

This agrees completely with Eq. (12.11) for the case of the choke coil except for the different value of the time constant T. The reflected voltage pulse can be derived most easily from Eq. (12.18); it is

$$e_{r1} = -E + \frac{2Z_2}{Z_1 + Z_2} E(1 - \varepsilon^{-t/T_c}),$$

$$(12.25)$$

and this gives for the reflected current pulse

$$i_{r1} = I - \frac{2Z_2}{Z_1 + Z_2} I(1 - \varepsilon^{-t/T_c}). \quad (12.26)$$

At time $t = 0$ the terms in the parentheses are zero and thus the voltage at the junction drops to zero while the total current increases to $2I$. The condenser thus acts on the front of the pulse as a short circuit. The voltage gradually rises and the current drops to the final values as calculated in Eqs. (12.14) and (12.15). Thus like the choke coil the condenser initially stores energy from the front of the pulse. Only later can the voltages and currents pass beyond it.

Figure 12.3 shows the spatial behavior of all pulses. We see by comparison between Figs. 12.1 and 12.3 as well as the pertinent equations that there is equivalence between the voltage pulses for choke coils and the current pulses for condensers and also between the current pulses for choke coils and the voltage pulses for condensers.

For equal lines on either side of the condenser the equations again become simpler, but without any significant change in the phenomena.

Figure 12.4, like Fig. 12.2, shows the behavior of a single square voltage pulse. Here, too, a flattened pulse is transmitted into line 2 while a double spiked pulse is reflected into line 1.

In the case of the choke coil the reflected voltage pulse in line 1 had the same polarity as the incident one and thus temporarily increased the voltage. However, in the case of the condenser it has the opposite polarity and thus reduces the voltage initially to zero. This absorption, so to speak, of the voltage makes the condenser effective for the protection of the incident line against overvoltages. However, as stated earlier, choke coils are useful in activating discharge devices, particularly in the case of pulses that are danger-

ous because of steep fronts and not because of their amplitude and that thus cannot directly activate such devices.

If the sole purpose is to flatten out steep pulse fronts in transmission, choke coils and condensers are equally effective if they are chosen to have the same time constant. Thus according to Eqs. (12.10) and (12.23) equal

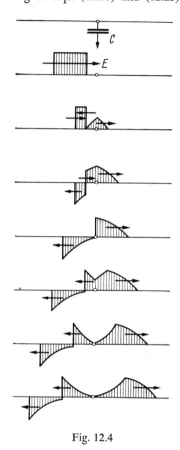

Fig. 12.4

protection is afforded by inductance or capacitance if the ratio

$$\frac{L}{C} = Z_1 Z_2 \quad (12.27)$$

is maintained. Thus the equivalent inductance and capacitance are related only by the product of the line surge impedances and not by any other parameters.

Thus the equivalence condition for protection of a winding of $Z_2 = 2000 \ \Omega$ connected to a line of $Z_1 = 500 \, \Omega$ by an inductance or a capacitance is

$$L = 10^6 C.$$

Hence for equal protection the choke coil must have the same size in henries as the condenser has in microfarads.

(c) *Single pulses and pulse trains.* If single pulses or pulse trains are incident upon a junction that is protected by an inductance or a capacitance, they are also in part reflected and in part transmitted. We can obtain their shape by repeated superposition of positive and negative step pulses according to the method shown in Fig. 3.12 and used already for single pulses in Figs. 12.2 and 12.4. However, we wish to treat a few special cases here in detail but restrict our attention to the transmitted pulses.

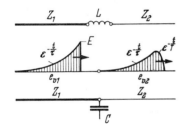

Fig. 12.5

Figure 12.5 shows a single pulse with a step front and an exponential back of time constant τ given by

$$e_{v1} = E\varepsilon^{-t/\tau}. \qquad (12.28)$$

The transmitted voltage pulse for inductance at the junction follows Eq. (12.5) and for capacitance Eq. (12.20). If we introduce in these the time constant T given by Eq. (12.10) or (12.23), respectively, we obtain for both cases the same differential equation

$$T\frac{de_{v2}}{dt} + e_{v2} = \frac{2Z_2}{Z_1 + Z_2} e_{v1}. \qquad (12.29)$$

The solution consists · of an impressed voltage component e'_{v2} which is based on the incident pulse form given by Eq. (12.28),

$$e'_{v2} = K_1\varepsilon^{-t/\tau}. \qquad (12.30)$$

By substitution in Eq. (12.29) we obtain

$$-\frac{T}{\tau} K_1 + K_1 = \frac{2Z_2}{Z_1 + Z_2} E, \qquad (12.31)$$

and for the amplitude,

$$K_1 = \frac{2Z_2}{Z_1 + Z_2} \frac{E}{1 - T/\tau}. \qquad (12.32)$$

In addition there is a free oscillation derived from Eq. (12.29) by putting the forcing function equal to zero and thus

$$e''_{v2} = K_2\varepsilon^{-t/T}. \qquad (12.33)$$

We saw earlier that at $t = 0$ the voltage is zero and cannot rise abruptly, and thus the sum of Eqs. (12.30) and (12.33) must be zero, which gives us

$$K_2 = -K_1. \qquad (12.34)$$

Now the complete transmitted voltage pulse is

$$e_{v2} = e'_{v2} + e''_{v2} =$$
$$\frac{2Z_2}{Z_1 + Z_2} \frac{E}{1 - T/\tau} (\varepsilon^{-t/\tau} - \varepsilon^{-t/T}). \qquad (12.35)$$

Figure 12.5 shows the shape of this pulse for $T < \tau$. The front is flattened by a rise according to the smaller time constant T of the protected junction, while its back drops exponentially according to the larger time constant τ of the incident pulse. If $T > \tau$, the total amplitude of the transmitted pulse will become very small. The initial voltage rise of the pulse is, by differentiation with respect to time,

$$\left(\frac{de_{v2}}{dt}\right)_0 = \frac{2Z_2}{Z_1 + Z_2} \frac{E}{T}, \qquad (12.36)$$

and thus a function only of the time constant of the junction. The peak value of the transmitted voltage pulse is always smaller than the peak value of the incident pulse, and increasingly so for larger time constants.

Figure 12.6 shows a rising exponential pulse which is the sum of a positive step pulse and a negative exponential pulse of time constant τ. Thus the transmitted pulse follows from Eq. (12.9) or (12.22) and Eq. (12.35):

$$e_{v2} = \frac{2Z_2}{Z_1 + Z_2} E \left[(1 - \varepsilon^{-t/T}) - \frac{\varepsilon^{-t/\tau} - \varepsilon^{-t/T}}{1 - T/\tau} \right]. \quad (12.37)$$

The shape now has several exponential functions shown composed in Fig. 12.6 for $T < \tau$. The transmitted pulse rises only very gradually and also approaches its final value

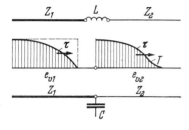

Fig. 12.6

gradually. The final value is again given purely by the surge impedances of the connected lines. As seen for $T < \tau$, the slow rise of the pulse is only slightly further flattened in transmission.

If the incident pulse has a sharp rise with the small time constant τ and a broad back with the large time constant θ, then it is given by

$$e_{v1} = E(\varepsilon^{-t/\theta} - \varepsilon^{-t/\tau}). \quad (12.38)$$

Then the transmitted pulse is made up of two parts, each structured in accordance with Eq. (12.35), and thus becomes

$$e_{v2} = \frac{2Z_2}{Z_1 + Z_2} E \left[\frac{\varepsilon^{-t/\theta} - \varepsilon^{-t/T}}{1 - T/\theta} - \frac{\varepsilon^{-t/\tau} - \varepsilon^{-t/T}}{1 - T/\tau} \right]. \quad (12.39)$$

Figure 12.7 shows this case. In both the last two cases the initial rise of the transmitted pulse is shallow and governed by the time constant T.

A periodic pulse train is shown in Fig. 12.8 incident upon a junction. If we ignore the decay in time and space of the pulse train, which is usually present and is shown in the figure, we can assume a constant peak value E which will represent the first and hence most damaging peak and let

$$e_{v1} = E \cos (\omega t + \psi). \quad (12.40)$$

The solution of the differential equation (12.29) was carried out in Ref. 1, p. 12, for Eq. (25). It also contains a forced harmonic and a free

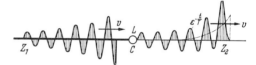

Fig. 12.8

oscillatory part, and from inspection of Ref. 1, p. 13, Eqs. (29) and (34) we obtain

$$e_{v2} = \frac{2Z_2}{Z_1 + Z_2} \frac{E}{\sqrt{1 + (\omega T)^2}}$$
$$\times [\cos (\omega t + \varphi) - \cos \varphi \varepsilon^{-t/T}].$$
$$(12.41)$$

Here φ is the phase with which the pulse train enters line 2. In Fig. 12.8 $\varphi = 180°$ and the transmitted voltage rises initially to about $2E$. This will always be the case if the incident pulse train has a steeply rising front (that is, for $\varphi = 0, 180°, 360°, \ldots$). Because the transmitted voltage pulse must always start at zero, this produces a transient overshoot that can cause dangerous overvoltages.

The amplitude of the transmitted steady-state harmonic pulse is of course always smaller owing to the presence of the inductance or capacitance as given by the square-root term in the denominator of Eq. (12.41). For example, a pulse train of 10^6 cy/sec is reduced by a choke coil of $L = 3$ mH and a time constant $T = 0.55 \cdot 10^{-6}$ sec to

$$\frac{1}{\sqrt{1 + (2\pi 10^6 \cdot 0.55 \cdot 10^{-6})^2}} = 28 \text{ percent}$$

of the value without the choke coil. For lower frequencies this reduction is far less effective, so that a frequency of 10^5 cy/sec, for example, produces only a reduction to 95 percent, which is almost negligible. The absolute value of the transmitted voltages must of course be calculated in detail using all factors in Eq. (12.41).

(d) *Series capacitance and parallel inductance.* The upper portion of Fig. 12.9 shows a condenser in series between lines 1 and 2, of surge impedances Z_1 and Z_2. Only the front

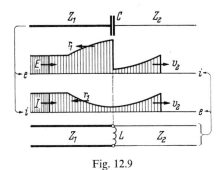

Fig. 12.9

of a step pulse can now be transmitted. For the currents again

$$i_1 = i_2, \qquad (12.42)$$

or, after resolution into partial pulses,

$$e_{v1} - e_{r1} = \frac{Z_1}{Z_2} e_{v2}. \qquad (12.43)$$

The difference in voltage between the lines is equal to the voltage drop across the condenser, so that

$$e_1 = e_2 + \frac{1}{C} \int i_2 \, dt. \qquad (12.44)$$

Again using partial pulses, we obtain

$$e_{v1} + e_{r1} = e_{v2} + \frac{1}{CZ_2} \int e_{v2} \, dt, \qquad (12.45)$$

and added to Eq. (12.43) this gives

$$\frac{1}{C} \int e_{v2} \, dt + (Z_1 + Z_2)e_{v2} = 2Z_2 e_{v1}. \qquad (12.46)$$

This is an integral equation which we can transform by a single differentiation into a

differential equation of the form treated earlier. If we let the time constant of the series condenser be

$$T_c = C(Z_1 + Z_2), \qquad (12.47)$$

then we can write the integral equation in the form

$$\frac{1}{T_c} \int e_{v2} \, dt + e_{v2} = \frac{2Z_2}{Z_1 + Z_2} e_{v1}, \qquad (12.48)$$

which is easy to handle. For an incident step pulse of constant amplitude E the solution is

$$e_{v2} = \frac{2Z_2}{Z_1 + Z_2} E\varepsilon^{-t/T_c}, \qquad (12.49)$$

as can be seen at once by substitution into Eq. (12.48) with integration from the limit $t = 0$.

Thus, as shown in Fig. 12.9, only a single voltage pulse, which has a step front and an exponentially decaying back, is transmitted. The amplitude of the front depends on the ratio of the surge impedances, and the time constant of the back depends on the product of the capacitance and the sum of the surge impedances. In this way it is easy to produce single pulses of this kind for experiments.

For a condenser of capacitance $C = 0.005 \, \mu\text{F}$ in a homogeneous open line of $Z_1 = Z_2 = 500 \, \Omega$ the time constant is

$$T_c = 0.005 \cdot 10^{-6}(500 + 500) = 5 \cdot 10^{-6} \text{sec}.$$

The total length of the single pulse produced by such an arrangement is about 4500 m (at about 3 time constants).

The reflected pulse in line 1, according to Eq. (12.43), is

$$e_{r1} = E\left(1 - \frac{2Z_1}{Z_1 + Z_2} \varepsilon^{-t/T_c}\right). \qquad (12.50)$$

The currents can be determined simply by dividing the voltages by the surge impedances. All pulse shapes are shown in Fig. 12.9 for $Z_1 = Z_2$.

The lower portion of Fig. 12.9 shows a coil connected across the lines at the junction. An incident step pulse will also transmit to line 2 a single sharp-front pulse with exponentially decaying back. However, the voltage and current shapes are now reciprocally inverted, as indicated in the figure. Thus a current increase in line 1 now takes the place of the former voltage increase because the coil stores energy in its magnetic field while the condenser stored energy in its electrostatic field. The time constant T_l of the coil, in parallel with the two line surge impedances, is

$$T_l = L \left(\frac{1}{Z_1} + \frac{1}{Z_2} \right) = L \frac{Z_1 + Z_2}{Z_1 Z_2}. \quad (12.51)$$

This case occurs in practice when transformers are connected to a continuing line. We see that they cannot change the front of traveling pulses, merely their back.

The self-inductance of a transformer is determined by its leakage flux. If, for a medium-sized transformer, it is $L = 100$ mH, then for a line of $Z_1 = Z_2 = 500 \ \Omega$ we obtain the time constant

$$T_l = \frac{100 \cdot 10^{-3} \cdot 2}{500} = 4 \cdot 10^{-4} \ \text{sec}.$$

Assuming again 3 time constants for the effectively complete decay of the transmitted pulse, we see that its length is 360 km. Thus the front of step or similar pulses is hardly changed by the transformer.

CHAPTER 13

EFFECTS OF CHANGE OF SHAPE

The changes of shape that traveling pulses undergo owing to inductances or capacitances are of considerable importance in the operation of high-voltage power systems. Such changes can be produced by switching devices or by arrangements introduced for this purpose, or they occur in the normal course of propagation along the lines or in the machinery of the system. Thus they can be manipulated to produce useful effects but may also be a source of danger.

(a) *Protective inductances and capacitances for line switching.* Just as we introduced a series resistance to protect a line against switching pulses of undesirable shape, it is possible, as shown in Fig. 13.1, to introduce

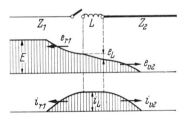

Fig. 13.1

a series inductance as a choke coil.

Because both lines are without current prior to switching, only pulses originating at the switch are produced and the current in the choke coil after switching is

$$i = i_{v2} = i_{r1}. \qquad (13.1)$$

The voltage, as shown in Fig. 13.1, is

$$E + e_{r1} = e_{v2} + L\frac{di}{dt}. \qquad (13.2)$$

Conversion of current to voltage gives

$$i = \frac{e_{v2}}{Z_2} = -\frac{e_{r1}}{Z_1}, \qquad (13.3)$$

and substitution into Eq. (13.2) gives the differential equation

$$\frac{de_{v2}}{dt} + \frac{Z_1 + Z_2}{L}e_{v2} = \frac{Z_2}{L}E. \qquad (13.4)$$

The complete solution is

$$e_{v2} = \frac{Z_2}{Z_1 + Z_2}E(1 - \varepsilon^{-t/T}), \qquad (13.5)$$

where the time constant T is that of the choke coil given by

$$T_L = \frac{L}{Z_1 + Z_2}. \qquad (13.6)$$

The solution of Eq. (13.4) in the form of Eq. (13.5) is very similar to that of Eq. (12.5) in the form of Eq. (12.9) for the behavior of the junction upon arrival of a step voltage pulse E. However, in that case the transmitted voltage pulse had twice the amplitude of the pulse here propagated into line 2. This is because in the former case line 1 carried an arriving current pulse i_{v1} as well as a departing one i_{r1}. As this is independent of the arrangement at the junction, we can deduce the general rule that the pulse propagated into a line by switching on has always one-half the amplitude of the pulse transmitted owing to an incident step pulse.

The voltage pulse traveling rearward into line 1 from the switch is, according to Eq. (13.3),

$$e_{r1} = -\frac{Z_1}{Z_1 + Z_2}E(1 - \varepsilon^{-t/T}). \qquad (13.7)$$

It has the same shape as the pulse in line 2 but a different amplitude. Again there is a similarity with Eq. (12.12) for the case of an arriving step pulse; however, it is not as simple as in the case of the transmitted pulse e_{v2}.

Figure 13.1 shows the pulses in both lines due to switching. We see that they have a gradually rising front and that the rise time of the front can be selected by the time constant of the choke coil T_L. Thus such choke coils are useful for suppressing the sharp pulse fronts that can occur in switching. At the moment of switching, the full voltage is impressed upon the choke coil, which must be able to withstand it. However, once the steady state is established, the choke coil no longer produces a voltage drop, as was the case with a protective resistance. Thus there is no need to short-circuit it after switching provided it can carry the steady-state current permanently. Similar switching phenomena occur in connection with choke coils introduced into networks to limit short-circuit currents, which have been treated in Ref. 1, Chapter 14 (*b*).

Figure 13.2 shows a capacitance connected as a bypass condenser across the switched

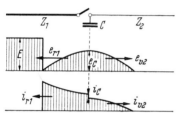

Fig. 13.2

line. By analogy with Chapter 12 we obtain here also Eq. (13.5); however, the time constant is now that of the condenser,

$$T_C = \frac{Z_1 Z_2}{Z_1 + Z_2} C. \tag{13.8}$$

The pulse traveling into line 1, however, has the opposite exponential term, so that the entire process is as shown in Fig. 13.2. We see that the charging pulses propagated into line 2, the fed line, are exactly the same as in the case of the choke coil and thus free of steep-front features. However, a step pulse with a very steep front is now propagated rearward in line 1, the feeding line, because the condenser produces a short circuit across line 1 at the instant of switching. Thus the arrangement is less favorable in respect to line 1.

If the condenser is connected across the feeding line instead of the fed line, as above, the situation is reversed as shown in Fig. 13.3,

line 1 now is free of steep pulse fronts but a full voltage step pulse is propagated into line 2. This arrangement is frequently used to produce such step pulses essentially by the discharge of the condenser.

Figure 13.4 shows the condenser connected in series with the switch and line 2. Now the

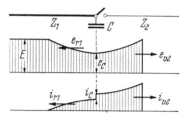

Fig. 13.3

first switching pulses arise as if the condenser were a short circuit. However, the current gradually charges the condenser, so that the pulse is flattened and has a time duration given by the time constant of the condenser in the series connection,

$$T_C' = (Z_1 + Z_2)C. \tag{13.9}$$

The pulse shape is shown in Fig. 13.4 and the charging voltage pulse in line 2 is given by

$$e_{v2} = \frac{Z_2}{Z_1 + Z_2} E\varepsilon^{-t/T'_C}. \tag{13.10}$$

A similar voltage discharge pulse of opposite sign is produced in line 1 as shown. Thus such

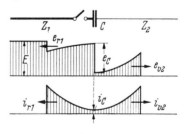

Fig. 13.4

an arrangement can be used to produce a single pulse or impulse of desired proportions.

(*b*) *Pulse flattening due to tower capacitance.* At the towers from which overhead lines are suspended there is an increased capacitance to ground owing to the proximity between the line and grounded tower conductors and also the capacitance of the insulators used to support the line conductors. This can be represented as shown in Fig. 13.5 by a sequence of periodically spaced condensers connected across the line. Let us examine the flattening of pulse fronts that travel along such a periodic chain of condensers on a line.

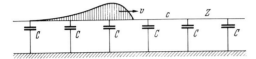

Fig. 13.5

For a single tower regarded as a junction the voltages are

$$e_{v1} + e_{r1} = e_{v2}, \quad (13.11)$$

and the current balance is

$$i_{v1} + i_{r1} = i_{v2} + C\frac{de_{v2}}{dt}, \quad (13.12)$$

which multiplied by the surge impedance is

$$e_{v1} - e_{r1} = e_{v2} + CZ\frac{de_{v2}}{dt}. \quad (13.13)$$

The sum of the two equations is

$$\frac{de_{v2}}{dt} = \frac{2}{CZ}(e_{v1} - e_{v2}) = v\frac{de_{v2}}{dx}. \quad (13.14)$$

Here we have introduced in the right-hand member the space differential instead of the

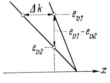

Fig. 13.6

time differential of the voltage rise of the transmitted pulse leaving the tower.

Figure 13.6 shows the pulse front for the incident and transmitted pulses. We see that the slope of the transmitted pulse is

$$\frac{de_{v2}}{dx} = \frac{e_{v1} - e_{v2}}{\Delta k} \quad (13.15)$$

if Δk represents the increase in pulse-front length for the front-rise portion of the pulse. If we compare these geometric relations as shown in Fig. 13.6 with Eq. (13.14), we obtain for the change in pulse-front length at each tower

$$\Delta k = \frac{CZv}{2} = \frac{C}{2c}, \quad (13.16)$$

where c is the capacitance of the line per unit length. We see that the ratio of tower plus insulator capacitance C to twice the distributed line capacitance c determines the lengthening of the front-rise portion of the pulse. The shape of the incident pulse is not important; pulses of any given shape are lengthened, stretched, or in effect flattened in the same proportion.

For a suspension insulator chain of capacitance $C = 10\ \mu\mu\text{F}$, an increase in distributed line capacitance due to the tower that is 50 percent of the insulator capacitance, and a line capacitance $c = 6.7\ \mu\mu\text{F/m}$, the additional front-rise length at each tower is

$$\Delta k = \frac{1.5 \cdot 10}{2 \cdot 6.7} = 1.1\ \text{m.}$$

If there are four towers per kilometer, the stretching amounts to 4.4 m/km. For a total of n towers on a line, the front rise is stretched by

$$\Delta k = \frac{nC}{2c}. \quad (13.17)$$

For a line 10 km long this is a stretching of a pulse front by 44 m, which is a significant reduction if the front was initially very steep. Comparison with Chapter 5 shows that this stretching is greater than that due to skin effects in the line conductor but smaller than that due to return currents in the ground or ground conductors.

Figure 13.7 shows the time rise of an initial ($n = 0$) voltage step pulse after passage of n towers ($n = 1$ to 50), calculated by B. G. Gates for this phenomenon. It shows the increased flattening of the rise caused by the stretching. In addition, the stretching produces a time delay in the onset of the voltage rise, because the added capacitance introduced by the tower and insulator must reduce the velocity of propagation of the pulse along the line according to the fundamental Eq. (1.13).

(c) *Inductance or capacitance at the end of a line.* From our general relations we can derive the laws of reflection at a line terminated in a pure inductance L or capacitance C as shown in Fig. 13.8.

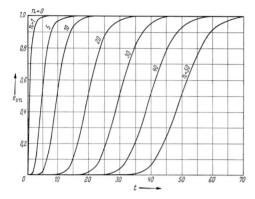

Fig. 13.7

If in Fig. 12.1 we reduce the surge impedance of line 2 to zero, we obtain in essence line 1 terminated in a choke coil. Thus let us put

$$Z_2 = 0,$$
$$Z_1 = Z, \qquad (13.18)$$

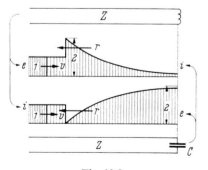

Fig. 13.8

and we obtain for an incident voltage step pulse E, according to Eq. (12.12), the reflected voltage pulse

$$e_r = -E + 2E\varepsilon^{-t/T}, \qquad (13.19)$$

and for the reflected current pulse, according to Eq. (12.13),

$$i_r = I - 2I\varepsilon^{-t/T}. \qquad (13.20)$$

Here the time constant T of the line termination for traveling pulses is

$$T_L = \frac{L}{Z}, \qquad (13.21)$$

which has the form usual for inductances with Z in place of R.

The total voltage at the end of the line, the sum of the incident and reflected pulses, is

$$e_L = E + e_r = 2E\varepsilon^{-t/T}, \qquad (13.22)$$

and the total current is

$$i_L = I + i_r = 2I(1 - \varepsilon^{-t/T}). \qquad (13.23)$$

The two equations also give the voltage across and the current in the choke coil.

Figure 13.8 shows the total pulses. We see that at the instant of arrival the voltage is doubled but the current is reduced to zero. Both then change exponentially, the voltage to zero and the current to twice the value in the incident pulse. Thus an inductive line termination acts at first as an open circuit but later as a short circuit. Only the steep front of the pulse is reflected without change. The body of the pulse undergoes pronounced changes in shape and amplitude.

If in Fig. 12.3 we let the surge impedance of line 2 become infinite, then again we have in essence terminated line 1 in a condenser, as shown in the lower portion of Fig. 13.8, and we have now

$$Z_2 = \infty,$$
$$Z_1 = Z. \qquad (13.24)$$

The arrival of a step voltage pulse E at the condenser C now gives rise, by Eq. (12.25), to a reflected voltage pulse

$$e_r = E - 2E\varepsilon^{-t/T} \qquad (13.25)$$

and, by Eq. (12.26), to a reflected current pulse

$$i_r = -I + 2I\varepsilon^{-t/T}. \qquad (13.26)$$

Here the time constant T of the line termination for traveling pulses is

$$T_C = ZC, \qquad (13.27)$$

which has the form usual for capacitance with Z in place of R.

Comparison of Eqs. (13.25) and (13.26) for a condenser-terminated line with Eqs. (13.19) and (13.20) for a choke-coil-terminated line shows the reciprocal nature of current and voltage pulses in the two cases, as indicated by the arrows for the two portions of Fig. 13.8.

The total voltage at the end of the condenser-loaded line is

$$e_C = E + e_r = 2E(1 - \varepsilon^{-t/T}), \quad (13.28)$$

and the total current is

$$i_C = I + i_r = 2I\varepsilon^{-t/T}. \quad (13.29)$$

The voltage initially drops to zero and then rises exponentially to twice the step value. The current jumps initially to twice the step value and then decays exponentially to zero. Thus a condenser at the line end acts initially as a short circuit and later as an open circuit. Here, also, only the steep pulse front is reflected without change, while the body is exponentially reduced.

It should be noted that voltage and current in coils or condensers at the ends of long lines behave as though they were switched to a voltage source of amplitude $2E$ in series with an ohmic resistance of value Z. This often allows a simple examination of the effects upon the arrival of a traveling pulse at more complicated apparatus.

If the left-hand end of the line in Fig. 13.8 is open or short-circuited or connected to a constant-voltage or constant-current source, the waves incident there are reflected without distortion in shape but in some of the cases

with a reversal of polarity. They then arrive at the coil- or condenser-loaded end for a second time. The step front is again fully reflected and the body of the pulse is further modified as before. This can be calculated according to the pattern set by Eqs. (12.28) and (12.35) and leads to reflected pulses which decay more rapidly so that the voltage and current after the step rise change increasingly quickly to their final values.

With repeated pulse traversal of the line between the two ends, only the steep step

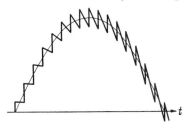

Fig. 13.9

front of the original step pulse is preserved. The body of the pulse is changed and thus a series of increasingly sharp spikes is formed, as shown in Fig. 13.9. These spikes have a period given by the time taken to make one round trip along the line. The shape of the total envelope is also determined by an oscillation between the reactive load at the end of the line and the line itself terminated in an open or short circuit at the opposite end. This oscillation has usually a longer natural period. Thus, as shown in Fig. 13.9, the voltage can rise to several times the amplitude of the original step pulse.

CHAPTER 14

PROTECTIVE VALUE OF COILS AND CONDENSERS

Series choke coils and bypass condensers produce changes in traveling pulses on lines in several ways, which have been treated in Chapters 12 and 13. On the one hand, they flatten the front rise of steep pulses or of other steep features of pulses. On the other hand, they stretch out single pulses and thus reduce their voltage amplitude. Finally, they reduce the intensity of pulse trains considerably. All these features help to protect the lines lying beyond these devices against overvoltages. Furthermore, the reduction of sharp voltage discontinuities leads to reduced voltages between turns of windings in machinery. Thus the distortion and breaking of the pulse trains produce lower peak voltages on lines and in windings.

(a) *Charging by a single pulse.* We shall examine the overvoltage caused, for example,

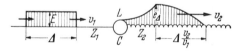

Fig. 14.1

by an atmospheric discharge in the form of a single pulse in the winding of a transformer protected by a choke coil or a bypass condenser. For simplicity, let the single pulse have the rectangular shape shown in Fig. 14.1. It is also possible, of course, to carry out the analysis for exponentially shaped pulses or to compose a more complex pulse shape out of several rectangular components. We shall assume that the amplitude E and duration or length Δ are typical of lightning discharges.

Before the pulse can penetrate from the line, of surge impedance Z_1, into the winding, of surge impedance Z_2, it encounters the protective coil of inductance L or condenser of capacitance C and stores energy in it.

Only after some energy storage in the protective device can the pulse propagate into the winding. Because we wish to examine the transmitted and not the reflected pulse the shape is the same for the two cases of coil or condenser, as seen in Figs. 12.2 and 12.4. The shape is shown in Fig. 14.1; it has a voltage peak e_Δ, which occurs when the energy stored in the protective device from the incident pulse has reached a maximum. This maximum voltage must be withstood by the transformer insulation. From Eqs. (12.9) and (12.22), the amplitude of the transmitted voltage pulse in line 2 is

$$e_{v2} = \frac{2Z_2}{Z_1 + Z_2} E(1 - \varepsilon^{-t/T}), \quad (14.1)$$

where the time constant T has the value

$$T_L = \frac{L}{Z_1 + Z_2} \text{ for a series coil}$$

and $\qquad (14.2)$

$$T_C = \frac{Z_1 Z_2 C}{Z_1 + Z_2} \text{ for a parallel condenser.}$$

The energy stored in the protective device has reached a maximum when the rectangular pulse of length Δ, traveling with velocity v_1, has entered it fully, that is, at a time

$$t = \frac{\Delta}{v_1} \quad (14.3)$$

after arrival of the front of the pulse. Substitution in Eq. (14.1) gives the peak voltage beyond the protective device,

$$e_\Delta = \frac{2Z_2}{Z_1 + Z_2} E(1 - \varepsilon^{-\Delta/v_1 T}), \quad (14.4)$$

which travels with velocity v_2 into the transformer winding. Because this velocity in the winding is in general quite different from that on the line, the length of the transmitted pulse is changed in proportion to the velocity ratio and is given by

$$\Delta_2 = \frac{v_2}{v_1}\,\Delta. \qquad (14.5)$$

The voltage increases according to an exponential law until the peak e_Δ is reached and then decreases again exponentially. If we wish to limit e_Δ in the winding, L and C must be chosen to give the desired value of T. From Eq. (14.4),

$$T = \frac{\Delta/v_1}{\ln\left(1 - \dfrac{e_\Delta}{E}\dfrac{Z_1 + Z_2}{2Z_2}\right)^{-1}}, \qquad (14.6)$$

and the required value of L or C is obtained from Eqs. (14.2).

For very strong protection, e_Δ/E should be very small and, since Z_2 is generally much larger than Z_1, the natural logarithm in Eq. (14.6) can be approximated by the first term of a series, to give

$$T = \frac{\Delta}{v_1}\frac{E}{e_\Delta}\frac{2Z_2}{Z_1 + Z_2}, \qquad (14.7)$$

and thus from Eq. (14.2) approximately for the inductance,

$$L = \frac{2\Delta Z_2}{v_1}\frac{E}{e_\Delta}, \qquad (14.8)$$

and for the capacitance,

$$C = \frac{2\Delta}{v_1 Z_1}\frac{E}{e_\Delta} = 2c_1\Delta\frac{E}{e_\Delta}. \qquad (14.9)$$

Here c_1 is the capacitance per unit length of the line as derived in Eq. (1.21). The product $c_1\Delta E$ gives the total charge in the incident pulse.

We note that the same protection can be achieved by either a coil or a condenser of the proper value. The capacitance of the condenser depends only on the constants of the line, while the inductance of the coil is in addition dependent on the constants of the protected winding.

Let us protect a transformer winding of $Z_2 = 2000\,\Omega$ against a pulse of length $\Delta = 1$ km and voltage $E = 100,000$ V that has a velocity of propagation $v_1 = 300,000$ km/sec on a line of $Z_1 = 500\,\Omega$ so as to reduce the peak voltage to one-half, that is, to make

$e_\Delta = 50,000$ V. To provide this protection we need, by Eq. (14.8), a coil of inductance

$$L = \frac{2\cdot 1\cdot 2000}{3\cdot 10^5}\cdot\frac{100,000}{50,000} = 27\text{ mH},$$

or by Eq. (14.9) a condenser of capacitance

$$C = \frac{2\cdot 1}{3\cdot 10^5\cdot 500}\cdot\frac{100,000}{50,000} = 0.027\ \mu\text{F}.$$

The more exact Eq. (14.6) would reduce these two values by about 15 percent. These values are quite high and are larger than those usually found in systems. For a three-phase station operating at 25,000 V per phase and 60 cy/sec, Table 14.1 gives the voltage drop in the cal-

TABLE 14.1. Effect of the calculated protective devices on normal operation

Station load rating (kVA)	Current (A)	Voltage drop in coil (percent)	Current in condenser (percent of normal)
5000	67	2.8	0.38
500	6.7	0.28	3.8

culated inductance or the reactive current in the calculated capacitance as percentages of the normal operating values for two load ratings.

In general, the coils are easier to implement for high voltages and low currents, while the condensers are easier to implement for high currents and low voltages. This is due to the equivalence equation (12.27).

(b) *Flattening of step-pulses.* The stress on the insulation between the turns or the layers

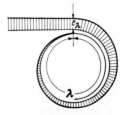

Fig. 14.2

of a coil or between neighboring coils depends on the voltage distribution that an incident traveling pulse has produced in a winding, a layer, or a coil.

Figure 14.2 shows an element of a winding of length λ. The voltage stress of the insulation across the element as shown is e_λ. If a flattened step pulse, as shown in Fig. 14.3, is traveling into this winding, the maximum voltage stress across the insulation at the beginning takes place when the front of the pulse has just traversed the distance λ, that is, the first turn, say. The voltage amplitude can be seen in Fig. 14.3 as a function of λ. This maximum voltage stress now propagates into the winding

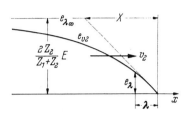

Fig. 14.3

with the same velocity as the pulse front. We can calculate it from Eq. (14.1) if we substitute for the time t the time required to traverse the distance λ, namely,

$$t_\lambda = \frac{\lambda}{v_2},\qquad(14.10)$$

and we then obtain

$$e_\lambda = \frac{2Z_2}{Z_1 + Z_2} E(1 - \varepsilon^{-\lambda/v_2 T}),\quad(14.11)$$

where the exponents are for inductance

$$\frac{\lambda}{v_2 T_L} = \frac{Z_1 + Z_2}{v_2}\frac{\lambda}{L},$$

and for capacitance $\qquad(14.12)$

$$\frac{\lambda}{v_2 T_C} = \frac{Z_1 + Z_2}{v_2 Z_1 Z_2}\frac{\lambda}{C}.$$

Thus, apart from the constants of the line and winding, the exponent decreases with increasing L or C. Hence the larger L or C the smaller will be e_λ, the voltage stress in the winding.

Figure 14.4 shows the value of the exponent in the parentheses of Eq. (14.11), which gives the effectiveness of protective coils or condensers. We see that a sufficiently large device can reduce the voltage stress to a small fraction of the value that would be produced by

a step pulse without such a device. The abscissa of Fig. 14.4 is scaled in terms of the reciprocal values of Eqs. (14.12), namely, $v_2 T/\lambda$, which is a dimensionless number directly proportional to L or C. For actual determinations the values of Z_1, v_1, Z_2, and v_2 can be obtained quite accurately, as will be shown for the last two in Chapter 17.

Figure 14.5 shows an actual measurement that is in good agreement with Eq. (14.11)

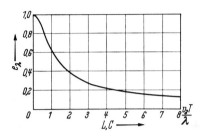

Fig. 14.4

and Fig. 14.4. It was conducted on a transformer of 30,000-V rating and 2000 turns. The voltage stress was measured across 200 turns so that a very large inductance was needed to produce significant reduction.

For ordinary windings the voltage stresses e_λ of the single turns, layers, or coils should

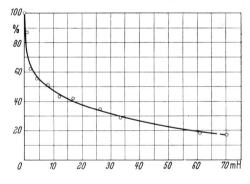

Fig. 14.5

be only a small fraction of the incident voltage step E. To obtain this reduction the front rise of the pulse must be flattened or stretched out so that e_λ is at the tip of the rise, hence in the approximately linear rising portion of the curve of Fig. 14.3. In this case we can deduce an approximation by introducing a space constant X for the winding corresponding to the time constant T simply as

$$X = v_2 T, \qquad (14.13)$$

in which case X gives the distance occupied by the pulse front if it were to rise linearly at its initial rate of rise to its final value as shown in Fig. 14.3. Then, as shown in Fig. 14.3,

$$\frac{e_\lambda}{e_{\lambda\infty}} = \frac{e_\lambda}{2Z_2 E/(Z_1 + Z_2)} = \frac{\lambda}{X} = \frac{\lambda}{v_2 T}. \qquad (14.14)$$

This then gives, upon insertion of values from Eq. (14.12), the voltage stress on a winding protected by a choke coil,

$$\frac{e_\lambda}{E} = \frac{2Z_2 \lambda}{v_2 L}, \qquad (14.15)$$

and by a bypass condenser,

$$\frac{e_\lambda}{E} = \frac{2\lambda}{v_2 Z_1 C}. \qquad (14.16)$$

Since the permissible voltage stress e_λ between turns, layers, or coils is given by the strength of the insulating material, we can determine the required inductance from Eq. (14.15),

$$L = \frac{2Z_2 \lambda}{v_2} \frac{E}{e_\lambda}. \qquad (14.17)$$

and the required capacitance from Eq. (14.16),

$$C = \frac{2\lambda}{v_2 Z_1} \frac{E}{e_\lambda}. \qquad (14.18)$$

Thus the protective device needed will increase with a decrease in the permissible voltage per unit length e_λ/λ of the winding.

A transformer of surge impedance $Z_2 = 5000 \ \Omega$, a velocity of propagation $v_2 = 3 \cdot 10^8$ m/sec, and a winding length $\lambda = 10$ m, in which a voltage $e_\lambda = 10,000$ V per turn is to be allowed for a step pulse of amplitude $E = 100,000$ V incident along a line of surge impedance $Z_1 = 500 \ \Omega$, requires a series coil of inductance

$$L = \frac{2 \cdot 5000 \cdot 10}{3 \cdot 10^8} \frac{100,000}{10,000} = 3.3 \text{ mH}$$

or a bypass condenser of capacitance

$$C = \frac{2 \cdot 10}{3 \cdot 10^8 \cdot 500} \frac{100,000}{10,000} = 0.0013 \ \mu\text{F}.$$

Protective coils and condensers have long been used to keep rapidly changing voltages away from machinery windings. To keep away voltage pulses of the most dangerous kind, namely step pulses, the size of the devices is determined by Eqs. (14.17) and (14.18). As we shall see later, if such devices are to operate effectively other conditions must be satisfied and they may not in fact be made arbitrarily small.

(c) *Protection against pulse trains.* Periodically structured pulse trains produce high-voltage stresses between parts of the winding that are separated by a half wavelength of the periodic train if they also are in close proximity within the structure of the winding.

Figure 14.6 shows a winding into which a pulse train is moving that has unilaterally

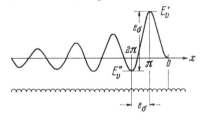

Fig. 14.6

displaced peak values as produced by choke coils or bypass condensers. The highest voltage with respect to ground occurs one-half period after the beginning of the pulse train, when $t = \pi/\omega$, if the initial phase angle φ is $0°$ or $180°$, according to Eq. (12.41), which then gives

$$E'_{v2} = \frac{2Z_2}{Z_1 + Z_2} \frac{-1 - \varepsilon^{-\pi/\omega T}}{\sqrt{1 + (\omega T)^2}} E. \qquad (14.19)$$

A still larger voltage can occur between two unfavorably placed locations in the winding of which the second is another half period π/ω behind the one just calculated, where again from Eq. (12.41) the voltage is

$$E''_{v2} = \frac{2Z_2}{Z_1 + Z_2} \frac{+1 - \varepsilon^{-2\pi/\omega T}}{\sqrt{1 + (\omega T)^2}} E. \qquad (14.20)$$

The highest possible winding voltage difference e_σ is thus

$$e_\sigma = E''_{v2} - E'_{v2}$$
$$= \frac{2Z_2}{Z_1 + Z_2} \frac{2 + \varepsilon^{-\pi/\omega T}(1 - \varepsilon^{-\pi/\omega T})}{\sqrt{1 + (\omega T)^2}} E. \qquad (14.21)$$

The second factor in this equation is purely a function of the product of the time constant T of the coil or condenser and the period ω of the pulse train, which is for the inductance

$$\omega T_L = \frac{\omega L}{Z_1 + Z_2}$$

and for the capacitance $\qquad (14.22)$

$$\omega T_C = \frac{Z_1 Z_2}{Z_1 + Z_2} \omega C.$$

Figure 14.7 shows the factor as a function of ωT, which is dimensionless and directly

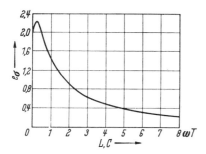

Fig. 14.7

proportional to L or C. We see that sufficiently large protective devices can reduce the voltage stress in the winding. However, below $\omega T = 0.5$ the devices produce an increase rather than a decrease in voltage. This is confirmed by an actual measurement made on a transformer protected by a series inductance against incident pulse trains. Figure 14.8 shows the results with changing inductance.

To obtain low winding stresses, ωT must be chosen large enough for its influence in the numerator of Eq. (14.21) to disappear while it predominates in the denominator. Then one approaches as an approximation the value

$$e_\sigma = \frac{2Z_2}{Z_1 + Z_2} \frac{2}{\omega T} E, \qquad (14.23)$$

which, on substitution of the values in Eqs. (14.22), gives for the voltage in a winding protected by an inductance

$$\frac{e_\sigma}{E} = \frac{4Z_2}{\omega L}, \qquad (14.24)$$

and by a capacitance

$$\frac{e_\sigma}{E} = \frac{4}{Z_1 \omega C}. \qquad (14.25)$$

Because the allowable voltage is known from the structure and insulation used, we then obtain the necessary protective inductance,

$$L = \frac{4Z_2}{\omega} \frac{E}{e_\sigma}, \qquad (14.26)$$

and the necessary protective capacitance,

$$C = \frac{4}{Z_1 \omega} \frac{E}{e_\sigma}. \qquad (14.27)$$

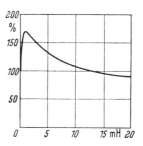

Fig. 14.8

Thus the protective device increases in size with decrease in the frequency of the wave trains against which it must protect. Furthermore, the magnitudes of the surge impedances have a marked influence—in the case of the series coil only the impedance of the winding protected and in the case of the bypass condenser only that of the line along which the pulse train arrives.

To protect a transformer winding of $Z_2 = 5000 \ \Omega$ against a pulse train of $E = 100,000$ V and $\omega = 2\pi \cdot 10^6$ so that a maximum voltage $e_\sigma = 50,000$ V is produced in the winding, we must connect in series with a line of $Z_1 = 500 \, \Omega$ an inductance of

$$L = \frac{4 \cdot 5000}{6.3 \cdot 10^6} \frac{100,000}{50,000} = 6.4 \text{ mH}$$

or in parallel a capacitance of

$$C = \frac{4}{500 \cdot 6.3 \cdot 10^6} \frac{100,000}{50,000} = 0.0025 \, \mu\text{F}.$$

For lower frequencies of the pulse train larger devices are needed, which in some cases may be more expensive than a strengthening of the insulation in the transformer winding.

CHAPTER 15

SURGE IMPEDANCE OF LUMPED CIRCUIT ELEMENTS

Up to now we have treated inductances and capacitances in the form of coils and condensers as lumped circuit elements in their effect on traveling pulses. We have assumed that the entire effect of these devices was concentrated at two terminals infinitely close together, so that currents and voltages arose instantaneously throughout the device, no consideration being given to propagation velocity.

In fact, however, any real coil will have some capacitance in addition to its inductance and any real condenser will have some inductance

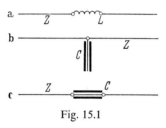

Fig. 15.1

in addition to its capacitance. Thus, because these devices are of necessity distributed structures of finite dimensions, pulses must produce effects that have finite propagation velocity within them. Essentially, then, they act in a more exact second approximation as lines or structures with distributed inductance and capacitance and have a surge impedance that is very large for coils because inductance is preponderant and very small for condensers because capacitance is preponderant.

We wish here to examine the effect of this finite surge impedance of such lumped circuit elements upon their behavior in connection with step pulses. It is customary to connect coils in series, as shown in Fig. 15.1a, and condensers in parallel, as shown in Fig. 15.1b. However, in the form of, say, a short length of cable a condenser can also be "series" connected as shown in Fig. 15.1c. The effects in the series connections of Fig. 15.1a, c can be treated together, but those of the parallel

connection of Fig. 15.1b require separate treatment.

(a) *Series connection.* If a voltage step pulse of amplitude E is incident upon a coil L or condenser C of surge impedance Z at the junction between two lines of surge impedances Z_1 and Z_2, a transmission and reflection of

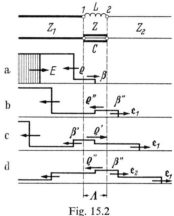

Fig. 15.2

the pulse occurs at both the entrance and the exit terminals of the coil or condenser. Figure 15.2 shows the arrangement for the case in which Z is smaller than either Z_1 or Z_2, so that considerable reflection takes place. At the entrance terminal, point 1, the pulse is split into a reflected partial pulse governed, according to Eq. (6.8), by the reflection factor

$$\rho = \frac{Z - Z_1}{Z + Z_1} = \frac{1 - Z_1/Z}{1 + Z_1/Z} \quad (15.1)$$

and a transmitted partial pulse governed, according to Eq. (6.6), by the refraction factor

$$\beta = \frac{2Z}{Z + Z_1} = \frac{2}{1 + Z_1/Z}. \quad (15.2)$$

The actual pulse amplitudes are simply obtained by multiplying the amplitude of the incident voltage step pulse by these factors.

121

At the exit terminal, point 2, the pulse is again split into two partial pulses, according to a reflection factor

$$\rho'' = \frac{Z_2 - Z}{Z_2 + Z} = \frac{1 - Z/Z_2}{1 + Z/Z_2} \quad (15.3)$$

and a refraction factor

$$\beta'' = \frac{2Z_2}{Z_2 + Z} = \frac{2}{1 + Z/Z_2}. \quad (15.4)$$

Thus the voltage step pulse transmitted into line 2 has an amplitude of

$$\frac{e_1}{E} = \beta\beta'' = \frac{4}{(1 + Z_1/Z)(1 + Z/Z_2)}, \quad (15.5)$$

while a reflected pulse of amplitude

$$\frac{r_1}{E} = \beta\rho'' \quad (15.6)$$

travels backward from point 2 into the protective device. This is split upon arrival at the entrance terminal, point 1, into two partial pulses according to a reflection factor

$$\rho' = \frac{Z_1 - Z}{Z_1 + Z} = \frac{1 - Z/Z_1}{1 + Z/Z_1} \quad (15.7)$$

and a refraction factor

$$\beta' = \frac{2Z_1}{Z_1 + Z} = \frac{2}{1 + Z/Z_1}. \quad (15.8)$$

At point 1, then, a transmitted pulse travels backward into line 1 and will not be followed further. A reflected pulse travels forward again into the protective device and has the amplitude

$$\frac{v_1}{E} = \frac{r_1}{E}\rho' = \beta\rho''\rho'. \quad (15.9)$$

This pulse again arrives at point 2 and is split for the third time according to Eqs. (15.3) and (15.4). Thus it is transmitted in amplitude as

$$\frac{e_2}{E} = \frac{v_1}{E}\beta'' = \beta\rho''\rho'\beta'' \quad (15.10)$$

and reflected in amplitude as

$$\frac{r_2}{E} = \frac{v_1}{E}\rho'' = \beta\rho''\rho'\rho''. \quad (15.11)$$

The transmitted pulse enters line 2 to produce a second charging effect. The reflected pulse again traverses the protective device and is split at point 1 into a transmitted pulse traveling backward again into line 1 and a reflected pulse traveling forward again to point 2, of amplitude

$$\frac{v_2}{E} = \frac{r_2}{E}\rho' = \beta(\rho''\rho')^2. \quad (15.12)$$

At point 2 a further charging pulse is transmitted into line 2, of amplitude

$$\frac{e_3}{E} = \frac{v_2}{E}\beta'' = \beta(\rho''\rho')^2\beta'', \quad (15.13)$$

while a reflected portion is again propagated toward point 1.

The travel of pulses in the protective device between points 1 and 2 continues indefinitely. The voltage in the protective device is gradually raised by these traveling pulses and upon arrival of the front at either end they give rise

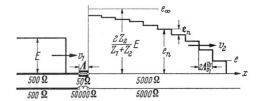

Fig. 15.3

also to a series of step pulses propagated in succession into lines 1 and 2. The amplitudes of the charging pulses in line 2 are given in sequence by Eqs. (15.5), (15.10), and (15.13). The following ones can be obtained by reiterated multiplication by the reflection factors ρ' and ρ''. Because these two factors are, as shown in Eqs. (15.3) and (15.7), always less than 1, the successive pulses have amplitudes that become smaller in a geometric series. They form a total transmitted voltage pulse shown in Fig. 15.3 for the arrangement of Fig. 15.2.

As seen in Fig. 15.3, the voltage pulse transmitted through the protective device into line 2 no longer has an exponential rise shape. Instead it consists of a series of step increases forming a staircase whose average shape approximates an exponential rise. For step n, a continuation of Eqs. (15.5), (15.10), and (15.13) gives the incremental voltage step pulse amplitude,

$$\frac{e_n}{E} = \beta(\rho''\rho')^{n-1}\beta''. \qquad (15.14)$$

Thus the total voltage on line 1, as the sum of all pulses up to n, becomes

$$\frac{e_n}{E} = \sum_1^n \frac{e_n}{E} =$$

$$\beta\beta''[1 + \rho'\rho'' + (\rho'\rho'')^2 + \cdots + (\rho'\rho'')^{n-1}]$$

$$= \beta\beta'' \frac{1 - (\rho'\rho'')^n}{1 - \rho'\rho''},$$

$$(15.15)$$

in which the geometric series is also expressed as a sum.

After a long time, as n becomes very large, the numerator of the last term becomes 1, because ρ' and ρ'' are both less than 1. Thus the final value that the voltage in line 2 gradually approaches is

$$\frac{e_\infty}{E} = \frac{\beta\beta''}{1 - \rho'\rho''}. \qquad (15.16)$$

If in this equation we substitute from Eqs. (15.2), (15.3), (15.4), and (15.7), we obtain simply

$$\frac{e_\infty}{E} = \frac{2Z_2}{Z_1 + Z_2}. \qquad (15.17)$$

Thus Eq. (15.15) can be rewritten as

$$e_n = \frac{2Z_2}{Z_1 + Z_2} E[1 - (\rho'\rho'')^n], \quad (15.18)$$

to give the voltage on line 2 in step n. Of course the pulse reflected back into line 1 has a correspondingly stepwise staircase shape.

Comparison of Eq. (15.18) with Eqs. (12.9) and (12.22) for the voltage as a function of time in a line protected by a coil or condenser shows that the final value given here by Eq. (15.17) is the same in both cases, only now the former gradual rise is replaced by a series of steps of geometrically decreasing amplitude. The example in Fig. 15.3 shows the transmitted pulse in a transformer winding of $Z_2 = 5000\ \Omega$ for a step pulse that arrives on the overhead line of $Z_1 = 500\ \Omega$ and is transmitted via a short length of cable of

$Z = 50\ \Omega$ that acts as a bypass condenser. Then the ratio of the first step pulse in the winding to the incident step-pulse voltage is

$$\frac{e_1}{E} = \frac{4}{\left(1 + \dfrac{500}{50}\right)\left(1 + \dfrac{50}{5000}\right)} = 0.36.$$

The identical voltage pulse is transmitted if the cable of $Z = 50\ \Omega$ is replaced by a coil of $Z = 50{,}000\ \Omega$ surge impedance connected in series as shown also in Fig. 15.3.

If the reflection factors ρ' and ρ'' are both positive or both negative, the product raised to any power is positive and the voltage in line 2 will increase monotonically with increasing n. However, if the two have opposite signs, then successive powers of the product have opposite signs, and successive pulses into line 2 alternate between positive and negative polarity as seen in Eq. (15.14). Inspection of Eq. (15.18) also shows that the first step e_1 for $n = 1$ is considerably greater in this case than the final value e_∞.

This unfavorable oscillatory behavior arises if the protective device has a surge impedance of value between that of the adjoining lines, because for

$$Z_1 < Z < Z_2 \qquad (15.19)$$

the signs are opposite according to Eqs. (15.3) and (15.7). Earlier, Fig. 6.8 showed that in

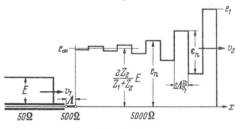

Fig. 15.4

such cases strong voltage rises are produced. Here we see that the oscillations consist of trains of rectangular-shaped pulses whose period or frequency is given by the length of the protective device, which acts almost as an open circuit at one and a short circuit at the other end.

Figure 15.4 shows the transmitted pulse in a transformer winding of $Z_2 = 5000\ \Omega$ connected to a cable of $Z_1 = 50\ \Omega$ over a short overhead line of $Z = 500\ \Omega$. As a fraction of the incident voltage pulse the first step is

$$\frac{e_1}{E} = \frac{4}{\left(1 + \dfrac{50}{500}\right)\left(1 + \dfrac{500}{5000}\right)} = 3.3,$$

a very high first step, which is furthermore followed by a series of large negative and positive steps that decay gradually. If the intermediate line has a length $\Lambda = 75$ m, the pulse train has a frequency given by Eq. (4.18),

$$\nu = \frac{\pi}{2}\frac{v}{\Lambda} = \frac{\pi \cdot 3 \cdot 10^8}{2 \cdot 75} = 6.3 \cdot 10^6 \text{ in } 2\pi \text{ sec}$$

or $f = 10^6$ cy/sec.

Traveling-pulse phenomena of the kind shown in Fig. 15.4 may occur if relay coils, current transformers, or choke coils of insufficient size are used in series with machinery windings or if these are connected to a cable by a short length of open line. In fact, whenever a circuit arrangement has a sequence of three monotonically different surge impedances, such rectangular natural oscillations arise and are propagated into the network at both junction points. These high-frequency oscillations, which owing to their step nature are rich in higher harmonics, can excite resonant modes elsewhere in the network. They can place particularly high stresses on the insulation of the machinery windings and should therefore be avoided wherever possible.

To obtain the smallest possible pulses in line 2, we must keep the first step e_1 of the voltage pulse, as given by Eq. (15.5), as small as possible. This can be achieved in two ways. One is to make Z as large as possible, so that it is several times as large as Z_2; this makes the first term in the denominator of Eq. (15.5) just a little more than 1 and the second term very large. Alternatively, we can make Z very small so that it is only a fraction of Z_1, in which case the role of the two terms in the denominator of Eq. (15.5) is reversed and we again obtain a small total fraction for the first voltage step pulse transmitted. So in both cases the protective device functions as desired.

Thus the larger we make the surge impedance of coils or the smaller we make the surge impedance of condensers, the smaller are the step pulses that can penetrate these devices.

In actual installations these pulses must be limited to the strength of the insulation of the turns, layers, or windings of the machinery protected. Thus the first and largest penetrating voltage step e_1 must be less than or equal to the safe winding or layer voltage e_λ. Equation (15.5) then gives for the large surge impedance of a choke coil

$$Z_L \geq 4Z_2\frac{E}{e_\lambda}, \qquad (15.20)$$

and for the small surge impedance of a bypass condenser cable,

$$Z_C \leq \frac{1}{4}Z_1\frac{e_\lambda}{E}. \qquad (15.21)$$

Thus the protective devices must have a definite surge impedance given by the surge impedance of the protected or feeding line and the permissible voltage stress in the winding protected.

Equal effects of coil and condenser are obtained according to Eq. (15.5) by the condition

$$Z_L Z_C = Z_1 Z_2, \qquad (15.22)$$

as can be seen by inspection.

For a voltage step pulse of amplitude $E = 100{,}000$ V arriving on an open line of $Z_1 = 500\ \Omega$ that is to be limited to a first step pulse of $e_\lambda = 10{,}000$ V in a transformer winding of $Z_2 = 5000\ \Omega$, Eq. (15.20) gives for the surge impedance of a choke coil needed at least

$$Z_L = 4 \cdot 5000\,\frac{100{,}000}{10{,}000} = 200{,}000\ \Omega,$$

or Eq. (15.21) gives for the surge impedance of a bypass condenser cable at most

$$Z_C = \frac{1}{4} \cdot 500\,\frac{10{,}000}{100{,}000} = 12.5\ \Omega.$$

It is difficult to construct a choke coil with such high surge impedance, while bypass condensers or cables of the required low surge impedance can be easily constructed. Furthermore, Eq. (15.21) shows that in the case of the bypass condenser the surge impedance of the feeding line enters the size determination. As this is more generally known than that of machinery windings to be protected, which must be considered in the case of choke coils by Eq. (15.20), it has become general practice to protect machinery windings by bypass-condenser cables rather than choke coils because the protective effects can be predicted with more assurance. Furthermore, changes in machinery can probably be made with greater freedom without a change in the protective device.

The capacitances of coils and the inductances of condensers impose quite serious limitations on the protection offered by these devices. They cause the front of step pulses to remain steep in the successive steps of the propagated staircase instead of producing an overall flattening of the propagated pulse, as shown in Figs. 12.1 and 12.3.

The selection of the right value of surge impedance of the protective device is a necessary but not sufficient condition for the protection of machinery windings against incoming step pulses. This is because the device reduces the amplitude of the individual pulses, whereas, as shown in Fig. 15.3, two or more step pulses can add up across a given portion of the winding. The distance between two successive steps is determined by the time taken for the internally reflected pulse to traverse the length of the protective device Λ twice. This must be at least equal to the time taken by the first pulse to traverse the length λ of the critical portion of the protected winding. Now because the velocities of propagation v in the protective device and v_2 in the protected device differ, we obtain for the time difference

$$t_\lambda = \frac{2\Lambda}{v} \geq \frac{\lambda}{v_2}. \qquad (15.23)$$

Thus the length of the protective device must be at least

$$\Lambda \geq \frac{1}{2} \frac{v}{v_2} \lambda, \qquad (15.24)$$

if only one transmitted pulse step at a time can be present in the length of winding λ to be protected. This condition for the physical length of the protective device must now be added to Eqs. (15.20) and (15.21) for the surge impedance as giving the minimum physical dimensions for these devices.

If both conditions are considered, it is possible to calculate the required total

inductance or capacitance of the protective device. We must again use from Eqs. (1.20) and (1.21) the self-inductance per unit length of a line l as given by the surge impedance divided by the velocity of propagation and the capacitance c per unit length as given by the reciprocal of the product of the surge impedance and the velocity of propagation. Then, to meet the conditions set by Eqs. (15.20) and (15.24), the choke coil must have a minimum inductance of

$$L = l\Lambda = \frac{Z_L}{v} \Lambda = \frac{2Z_2\lambda E}{v_2 e_\lambda}, \qquad (15.25)$$

and to meet the conditions of Eqs. (15.21) and (15.24) the condenser must have a minimum capacitance of

$$C = c\Lambda = \frac{\Lambda}{vZ_C} = \frac{2\lambda E}{v_2 Z_1 e_\lambda}. \qquad (15.26)$$

These two equations are exactly the same as Eqs. (14.17) and (14.18) for lumped coils and condensers. We now see that they alone are not sufficient for the correct design of protective devices and that we must add the conditions for surge impedance as given by Eqs. (15.20) and (15.21) or the conditions for the minimum transit time as given by Eqs. (15.23) and (15.24).

It is desirable to construct protective devices of large inductance or capacitance in a form that gives them a single well-defined surge impedance that is independent of the shape and frequency components of the traveling pulses or periodic waves that must be handled. This can be most easily accomplished in cables and similar uniformly structured distributed devices. In contrast, wound coils have interactions between different parts of their windings that produce a frequency dependence of the surge impedance. Often this leads to less protection against steep-front step pulses or other high-frequency pulse components than against more slowly changing pulse features. This will be treated in a later chapter.

(b) *Parallel connection.* Figure 15.5 shows a length of line with the surge impedance Z connected at the junction between two lines of surge impedances Z_1 and Z_2 as a protective device. A step pulse arriving along line 1 is now split at the junction of the three lines; a reflected pulse is produced along line 1 while two pulses of equal voltage amplitude are transmitted into line 2 and the parallel protective line. For the transmitted pulses the parallel surge impedance P_1 of the two lines is important and is given by

$$\frac{1}{P_1} = \frac{1}{Z} + \frac{1}{Z_2}. \qquad (15.27)$$

Thus according to Eq. (15.2) the refraction factor for traveling pulses arriving along line 1 is

$$\beta = \frac{e_1}{E} = \frac{2}{1 + Z_1/P_1} = \frac{2}{1 + Z_1/Z + Z_1/Z_2}. \qquad (15.28)$$

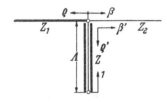

Fig. 15.5

This gives at the same time the fractional amplitude of the first voltage step transmitted into line 2.

The voltage step pulse of equal amplitude that enters the protective line is completely reflected without change of polarity at its open end and returns again to the junction after traversing the length Λ twice. At the junction it splits into two equal transmitted voltage pulses into lines 1 and 2 and a pulse reflected back into the protective line. The surge impedance P of lines 1 and 2 in parallel is given by

$$\frac{1}{P} = \frac{1}{Z_1} + \frac{1}{Z_2}, \qquad (15.29)$$

and thus according to Eq. (15.3), the reflection factor for the reflected pulse is

$$\rho' = \frac{1 - \dfrac{Z}{P}}{1 + \dfrac{Z}{P}} = \frac{1 - \dfrac{Z}{Z_1} - \dfrac{Z}{Z_2}}{1 + \dfrac{Z}{Z_1} + \dfrac{Z}{Z_2}} \qquad (15.30)$$

and the refraction factor for the transmitted pulses is according to Eq. (15.4)

$$\beta' = \frac{2}{1 + \dfrac{Z}{P}} = \frac{2}{1 + \dfrac{Z}{Z_1} + \dfrac{Z}{Z_2}}. \qquad (15.31)$$

The second step pulse into lines 2 and 1 is thus

$$\frac{e_2}{E} = \frac{e_1}{E}\beta' = \beta\beta', \qquad (15.32)$$

and the pulse reflected back into the protective line is

$$\frac{r_2}{E} = \frac{e_1}{E}\rho' = \beta\rho'. \qquad (15.33)$$

This is again fully reflected at the far end without change of polarity and after traversing Λ twice returns to the junction, where it gives rise to the third step pulse transmitted into lines 2 and 1,

$$\frac{e_3}{E} = \frac{r_2}{E}\beta' = \beta\rho'\beta', \qquad (15.34)$$

and a reflected pulse transmitted into the protective line,

$$\frac{r_3}{E} = \frac{r_2}{E}\rho' = \beta\rho'^2. \qquad (15.35)$$

This process is repeated and a series of decreasing step pulses are propagated into lines 2 and 1. For the nth step pulse the sequence of Eqs. (15.28), (15.32), and (15.34) leads to

$$\frac{e_n}{E} = \beta(\rho')^{n-2}\beta'. \qquad (15.36)$$

Thus here too the transmitted pulse into the protected line is a series of step pulses in staircase form. The total voltage in line 2 and at the junction is the sum of all the foregoing steps:

$$\frac{e_n}{E} = \sum_1^n \frac{e_n}{E}$$

$$= \beta + \beta\beta'[1 + \rho' + \rho'^2 + \cdots + (\rho')^{n-2}]$$

$$= \beta + \beta\beta'\frac{1 - (\rho')^{n-1}}{1 - \rho'}, \qquad (15.37)$$

where as before the geometric series is given as a sum.

After some time the numerator of this fraction, for n very large, approaches 1, because by Eq. (15.30) ρ' is always less than 1. Hence the final value of the voltage becomes

$$\frac{e_\infty}{E} = \beta \left(1 + \frac{\beta'}{1 - \rho'}\right) = \frac{2}{1 + Z_1/Z_2}. \quad (15.38)$$

This is exactly the same as the final value obtained if a lumped condenser is connected or if two dissimilar lines are joined. The total behavior of the staircase charging pulse into lines 2 and 1 is obtained by introducing Eq. (15.38) into Eq. (15.37), which gives

$$e_n = e_1 + (e_\infty - e_1)[1 - (\rho')^{n-1}], \quad (15.39)$$

where e_1 is the first step as obtained in Eq. (15.28). Figure 15.6 shows the total transmitted

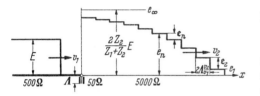

Fig. 15.6

pulse shape if a short cable of $Z = 50\ \Omega$ is connected at the junction of an open line of $Z_1 = 500\ \Omega$ and a transformer winding of $Z_2 = 5000\ \Omega$. The voltage increases in a series of steps that, except for the first, form a geometric series.

Periodic oscillations can also occur in this arrangement if the surge impedance of the protective line is greater than the parallel surge impedance of lines 1 and 2 as given by Eq. (15.29). In this case ρ' becomes negative by Eq. (15.30) and thus successive steps will be of opposite polarity in Eq. (15.36). This condition can occur at line junctions or bus bars. However, for protective lines, Z will always be chosen small to keep the first step e_1 given by Eq. (15.28) small. In this case no oscillations occur and for $Z < P$ Eq. (15.31) shows that $\beta' > 1$ always; hence by Eq. (15.32) the second voltage step e_2 is always greater than the first, as shown in Fig. 15.6. Thus in

the parallel connection the second step of the transmitted pulse is most dangerous for the winding insulation instead of the first, as in the earlier case of series connection. Using Eqs. (15.28) and (15.31) in Eq. (15.32), we get for the amplitude

$$\frac{e_2}{E} = \beta\beta'$$

$$= \frac{4}{\left(1 + \dfrac{Z_1}{Z} + \dfrac{Z_1}{Z_2}\right)\left(1 + \dfrac{Z}{Z_1} + \dfrac{Z}{Z_2}\right)}. \quad (15.40)$$

Inspection of this equation shows that only a very small value of Z can make this second voltage step small if at the same time the first step given by Eq. (15.28) is also to be kept small. Thus a good parallel-connected protection is possible only with condensers of low surge impedance.

If the greatest coil or layer voltage of the protective winding

$$e_\lambda = e_2 \quad (15.41)$$

must be a small fraction of the incident voltage step E, then by Eq. (15.40) the surge impedance of the condenser for small values of Z is approximately

$$Z_C \leq \frac{1}{4} Z_1 \frac{e_\lambda}{E}. \quad (15.42)$$

This is the same value as found in Eq. (15.21) for the series-connected condenser cable. Thus for step-pulse protection series or parallel connection of the same cable condenser is equally effective.

Because the distance between successive steps is here given by the time taken to traverse Λ twice, just as in Fig. 15.5, we must again make the length of the protective cable greater than the length of a coil or layer of the protected winding. Taking into account the velocities of propagation in the two devices, we find that the cable length is given here also by Eq. (15.24) and its total capacitance by Eq. (15.26). Also again this total value is in agreement with that calculated earlier for a lumped parallel condenser. Thus for effective protection by a parallel condenser or cable, in addition to the correct total capacitance we must also fulfill the two further conditions of sufficiently low surge impedance and sufficient physical length.

Less effective than a condenser connected at the junction is one that has a connecting branch line of appreciable length, as shown in Fig. 15.7. This is because the condenser

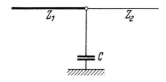

Fig. 15.7

cannot act upon the first step pulse transmitted owing to the time delay on the junction line. Thus, self-inductance due to either a series connection or internal structure is quite detrimental to the proper functioning of a bypass condenser.

(c) *Star-point oscillations of transformers.* We saw in Fig. 6.9 that the star point of a three-phase transformer winding can be charged to a high voltage by an atmospheric discharge pulse that travels simultaneously on all three lines. We wish here to examine this process more exactly by considering the changes in pulse shape that occur when the pulses enter the transformer windings, of surge impedance Z, from the lines, of surge impedance Z_1. If the star point is isolated we have, as shown in Fig. 15.8, a surge impedance

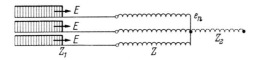

Fig. 15.8

$Z_2 = \infty$ for the second line. If a grounding coil is connected as Z_2, it too can be considered of almost infinite surge impedance when compared with Z of the loaded transformer, which is entirely due to its small leakage flux inductance.

Equation (15.18) gives the voltage that the star point reaches owing to the staircase nature of the repeated charging pulses. If then Z_2 is considered very large, ρ'' by Eq. (15.3)

approaches 1 and from Eq. (15.7) the star-point voltage is given by

$$e_n = 2E\left[1 - \left(\frac{Z_1 - Z}{Z_1 + Z}\right)^n\right]. \quad (15.43)$$

For high-voltage transformers Z is always greater than Z_1 for the feeder lines and thus the fraction in parentheses is negative, which produces a reversal of sign between successive steps. Thus the star-point voltage oscillates about the mean value $2E$ with rectangular pulses, as shown in Fig. 15.9 for the case of $Z_1 = 500\ \Omega$ and $Z = 5000\ \Omega$.

Inspection of Eq. (15.43) shows that, if the difference between these two surge impedances is large, the oscillations are also large and decay only slowly because the fraction in Eq. (15.43) approaches 1. Thus high-voltage windings, on account of their high

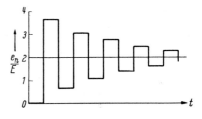

Fig. 15.9

surge impedance, undergo much higher star-point oscillations than low-voltage transformers or machinery. The damping between two successive oscillations of the same polarity is simply

$$\delta = \left(\frac{1 - Z_1/Z}{1 + Z_1/Z}\right)^2. \quad (15.44)$$

This is 0.67 for $Z = 10\ Z_1$ and 0.96 for $Z = 100\ Z_1$. Note also that the first step already comes close to four times the value of the incident voltage pulse E. If the star point is isolated inside the transformer case, this high voltage exists only between the end windings and the case, and this path, which lies in the insulating oil, has a high breakdown voltage. However, if the star point is brought out of the transformer case by a terminal bushing or other means, breakdowns at these terminals are often observed.

The damping resistances and the mutual effects between coils of transformer and machine windings frequently change the shape of the star-point oscillations from the sharp rectangular one shown to a more sinusoidal one. They also decay somewhat faster.

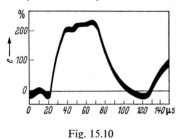

Fig. 15.10

Figure 15.10 shows an oscillogram of the star-point voltage of a three-phase machine upon arrival of a step pulse.

To avoid such oscillations it is possible to ground the star point through resistance R instead of the hitherto omitted Z_2. This reduces ρ'' in Eq. (15.3), which increases the damping. However, R must not be much larger than Z if it is to have appreciable effect.

If we wish then further to avoid the disadvantage of such a low star-point-to-ground impedance, which could give rise to high short-circuit currents, we can connect a series condenser as shown in Fig. 15.11. This will bypass the rapid step-pulse fronts without appreciably affecting steady-state currents. In very large distribution and transmission

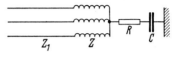

Fig. 15.11

networks the capacitance currents become so large that an inductive extinction of ground faults, as shown in Ref. 1, Chapter 26*b*, is no longer possible. In such cases it is usual to ground the star point, which in Fig. 15.8 makes $Z_2 = 0$. Then voltage oscillations cannot occur at the star point. Instead, very large star-point current oscillations are produced, which can cause high voltages in the transformer windings near the star point and which are treated in more detail in Chapter 20.

CHAPTER 16

JOINT ACTION OF COILS AND CONDENSERS

We have seen that inductances and capacitances have the same effect on the pulses transmitted through a junction but dissimilar effects on the reflected pulses. In the latter case, inductance produces a voltage increase and capacitance a voltage decrease. Thus it is of interest to examine their joint action to see if there is a canceling or reinforcement of effects. Several connections are possible by which the pulses can reach either the inductance or the capacitance first, or both at the same time in parallel. These and even more complex connections occur in practice, since relay coils and current transformers act inductively, while bus bars and terminal bushings have capacitive effects.

(a) *Inductance ahead of capacitance.* Figure 16.1 shows a pulse arriving on line 1, of surge

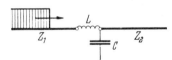

Fig. 16.1

impedance Z_1, that is connected in series with a lumped inductance L to a capacitance C and a line 2 of surge impedance Z_2. Thus the voltage at the ends of lines 1 and 2 differ by the voltage drop across the choke coil and

$$e_1 = e_2 + L \frac{di_1}{dt}. \qquad (16.1)$$

The currents at the ends of lines 1 and 2 differ by the condenser current and

$$i_1 = i_2 + C \frac{de_2}{dt}. \qquad (16.2)$$

Differentiation of Eq. (16.2) and substitution in Eq. (16.1) gives

$$e_1 = e_2 + L \frac{di_2}{dt} + LC \frac{d^2e_2}{dt^2}. \qquad (16.3)$$

The right-hand members of Eqs. (16.2) and (16.3) contain only the voltages and currents of the transmitted pulse in line 2, which is to be determined. As line 2 has only this forward-traveling pulse, while line 1 has pulses traveling in both directions, a resolution into partial pulses of Eq. (16.3) and a change from current to voltage gives

$$e_{v1} + e_{r1} = e_{v2} + \frac{L}{Z_2} \frac{de_{v2}}{dt} + LC \frac{d^2e_{v2}}{dt^2}. \qquad (16.4)$$

Similarly, Eq. (16.2) after multiplication by Z_1 gives

$$e_{v1} - e_{r1} = \frac{Z_1}{Z_2} e_{v2} + Z_1 C \frac{de_{v2}}{dt}. \qquad (16.5)$$

Addition of the last two equations cancels e_{r1} and we obtain

$$LC \frac{d^2e_{v2}}{dt^2} + \left(\frac{L}{Z_2} + Z_1 C \right) \frac{de_{v2}}{dt}$$
$$+ \left(1 + \frac{Z_1}{Z_2} \right) e_{v2} = 2e_{v1}. \qquad (16.6)$$

This is a linear differential equation of second order in e_{v2}, the transmitted voltage pulse, and shows that oscillations may arise. In fact, it is of the same form as that of Ref. 1, p. 47, Eq. (3) for the current in an oscillatory circuit.

Let us examine the effect of an incident voltage step pulse of amplitude

$$e_{v1} = E, \qquad (16.7)$$

that arrives at the end of line 1 at time $t = 0$ and then has the constant value E. After the transients have died out, only the time-invariant terms in Eq. (16.6) remain, and the equation becomes

$$e'_{v2} = \frac{2}{1 + \dfrac{Z_1}{Z_2}} e_{v1} = \frac{2Z_2}{Z_1 + Z_2} E = e_\infty. \qquad (16.8)$$

Thus the final value in the steady state is equal to that given in Eq. (6.6) with no inductance or capacitance at the junction. Thus again these devices even jointly can influence only the front-rise portion of the pulse upon arrival.

For the transient voltage e'', which effects the adjustment from initial conditions to the final steady state, we obtain by subtraction of Eq. (16.8) from Eq. (16.6) the homogeneous differential equation

$$\frac{d^2 e''_{v2}}{dt^2} + \left(\frac{1}{Z_2 C} + \frac{Z_1}{L}\right) \frac{de''_{v2}}{dt} + \frac{Z_1 + Z_2}{Z_2 LC} e''_{v2} = 0,$$
$$(16.9)$$

which has the solution

$$e''_{v2} = K \varepsilon^{\alpha t}, \qquad (16.10)$$

with the constant of integration K; by substitution in Eq. (16.9) and solution of the resulting quadratic equation, the value of α is found to be

$$\alpha = -\frac{1}{2}\left(\frac{1}{Z_2 C} + \frac{Z_1}{L}\right)$$
$$\pm \sqrt{\frac{1}{4}\left(\frac{1}{Z_2 C} - \frac{Z_1}{L}\right)^2 - \frac{1}{LC}}, \quad (16.11)$$

where the quantity under the radical has been put into its simplest form. For positive values of this quantity, α has two real values and thus the course of the voltage pulse is aperiodic, with two constants K corresponding to the two values of α in Eq. (16.11). For negative values of the quantity under the radical, α is a complex number and the pulse is periodic. The condition for oscillatory behavior is

$$\frac{4}{LC} > \left(\frac{1}{Z_2 C}\right)^2 - 2\frac{Z_1}{Z_2 CL} + \left(\frac{Z_1}{L}\right)^2 \quad (16.12)$$

or after rearrangement,

$$\frac{L}{Z_1 Z_2 C} + \frac{Z_1 Z_2 C}{L} < 2 + 4\frac{Z_2}{Z_1}. \quad (16.13)$$

Thus the shape of the transmitted voltage pulse depends strongly on the equivalence ratio of inductance and capacitance defined earlier in Eq. (12.27). If either is overwhelmingly effective and thus the ratio either very large or very small, the aperiodic course occurs. However,

if the two devices have comparable effectiveness then oscillations will occur because the left-hand member of Eq. (16.13) is close to 2. Equal effectiveness, signified by the condition

$$\frac{L}{Z_1 Z_2 C} = 1, \qquad (16.14)$$

always produces an oscillatory voltage pulse course.

The two members of Eq. (16.13) are shown graphically in Fig. 16.2. The straight line is the right-hand member and the hyperbolic

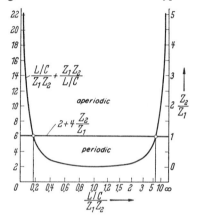

Fig. 16.2

curve is the left-hand member. The intersections of the two lines give the limits between periodic and aperiodic behavior of the circuit arrangement. We see that, as the straight line $2 + 4Z_2/Z_1$ is moved up or down with Z_2/Z_1, the region of periodic behavior changes in width about its center at $L/C = Z_1 Z_2$, the equal-effectiveness point. At this point no oscillation is possible since $Z_2 = 0$, that is, line 2 is short-circuited. As Z_2 increases, the periodic region around this central point increases. To obtain the limits we rewrite Eq. (16.13) in the form

$$\left(\frac{L/C}{Z_1 Z_2}\right)^2 - 2\left(1 + 2\frac{Z_2}{Z_1}\right)\frac{L/C}{Z_1 Z_2} + 1 = 0.$$
$$(16.15)$$

The solution of this quadratic equation is

$$\frac{L/C}{Z_1 Z_2} = \left(1 + 2\frac{Z_2}{Z_1}\right) \pm 2\frac{Z_2}{Z_1}\sqrt{1 + \frac{Z_1}{Z_2}}.$$
$$(16.16)$$

For equal lines with $Z_1 = Z_2 = Z$, which is the case shown in Fig. 16.2, this gives $\sqrt{L/C} = 0.42Z$ and $2.42Z$ at the two intersections shown. The region is already quite wide. For the case of main interest, with Z_2 very large, we can approximate the roots of Eq. (16.16) as $L/C = Z_1^2/4$ and $4Z_2^2$. Thus the arrangement is oscillatory if the impedance of the oscillatory LC circuit is within the range given by

$$\frac{1}{2}Z_1 < \sqrt{\frac{L}{C}} < 2Z_2. \qquad (16.17)$$

By comparison with Eq. (15.19) we see that the impedance of the oscillatory circuit here corresponds to the surge impedance of the protective line or device in Chapter 15. The range of oscillatory impedance over which the circuit will oscillate has, however, increased considerably in this case of lumped parameters over the former case of distributed parameters.

Because oscillations frequently produce excessive voltages, let us examine the most unfavorable case given by Eq. (16.14) at the center of the periodic range as shown in Fig. 16.2. Equation (16.10) can then be written, for a complex α,

$$e''_{v2} = K\varepsilon^{-\rho t}\cos(\nu t - \delta), \qquad (16.18)$$

where the damping factor is, according to Eq. (16.11),

$$\rho = \frac{1}{2}\left(\frac{1}{Z_2 C} + \frac{Z_1}{L}\right) \qquad (16.19)$$

and the natural frequency of the circuit is

$$\nu = \sqrt{\frac{1}{LC} - \frac{1}{4}\left(\frac{1}{Z_2 C} - \frac{Z_1}{L}\right)^2}. \qquad (16.20)$$

Here K and δ are constants of integration, which can be determined from the boundary conditions given. For $t = 0$ the total voltage on line 2, as the sum of the final and transient term, is

$$e_{v2} = e'_{v2} + e''_{v2}, \qquad (16.21)$$

and must be zero.

Thus from Eqs. (16.8) and (16.18)

$$K\cos\delta = -\frac{2Z_2}{Z_1 + Z_2}E. \qquad (16.22)$$

Further, at $t = 0$, $i_1 = 0$, since the current in the choke coil only starts then. Since $i_2 = 0$ also, Eq. (16.2) shows that at the start

$$\frac{de_2}{dt} = 0. \qquad (16.23)$$

Since e'_{v2} in Eq. (16.21) has no time-differential component, we need simply differentiate Eq. (16.18) with respect to time,

$$\frac{de''_{v2}}{dt} = K\varepsilon^{-\rho t}[-\rho\cos(\nu t - \delta)$$
$$- \nu\sin(\nu t - \delta)], \qquad (16.24)$$

and then at $t = 0$ Eq. (16.23) gives

$$\tan\delta = \rho/\nu. \qquad (16.25)$$

Thus the phase angle δ and the amplitude K of the transient oscillation are now determined and Eq. (16.21) becomes, by insertion of Eqs. (16.8), (16.18), and (16.22),

$$e_{v2} = \frac{2Z_2}{Z_1 + Z_2}\left[1 - \frac{\varepsilon^{-\rho t}}{\cos\delta}\cos(\nu t - \delta)\right]E \qquad (16.26)$$

for the transmitted voltage pulse on line 2. This is equivalent in form to the equation for

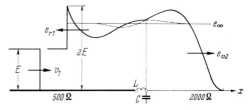

Fig. 16.3

a series RLC circuit connected to a direct-current source, as developed in Ref. 1, p. 60, Eq. (9). Thus the voltage rises gradually at first, then more rapidly in oscillatory fashion. Figure 16.3 shows the case of a traveling pulse under the specified conditions as represented by Eq. (16.26) in line 2. We see that the voltage exceeds the final value considerably, even though the oscillation is quite heavily damped.

The use of a series inductance and parallel capacitance thus forms an oscillatory circuit that is excited by a step pulse. It has a natural frequency given by Eq. (16.20) essentially as the square root of the reciprocal of LC. For the case of exact equivalence, according to Eq. (16.14) only L and C determine this frequency, but for deviations from this condition a slight reduction in frequency occurs.

Equation (16.19) shows the influence of the line impedances on the damping of these oscillations, which is of great importance and equivalent to ohmic resistances inserted into oscillatory circuits for this purpose. Here we have omitted such damping resistances, since the pulses propagated into both lines act already to dissipate energy and thus produce damping.

The highest voltage due to oscillation excited by the step pulse occurs after one-half period, that is, for

$$vt = \pi, \qquad (16.27)$$

and thus, according to Eq. (16.26), is related to the final value, given by Eq. (16.8), by the equation

$$E_{v2} = e_\infty(1 + \varepsilon^{-p\pi/v}). \qquad (16.28)$$

To evaluate the exponent we introduce

$$\sqrt{\frac{Z_2}{Z_1}LC} \qquad (16.29)$$

into the quotient of Eqs. (16.19) and (16.20) to obtain

$$\frac{p}{v} = \frac{\sqrt{\dfrac{L}{Z_1 Z_2 C}} + \sqrt{\dfrac{Z_1 Z_2 C}{L}}}{2\sqrt{\dfrac{Z_2}{Z_1} - \dfrac{1}{4}\left(\sqrt{\dfrac{L}{Z_1 Z_2 C}} - \sqrt{\dfrac{Z_1 Z_2 C}{L}}\right)^2}}. \qquad (16.30)$$

This is smallest and thus the voltage is highest by Eq. (16.28) if the equivalence condition of Eq. (16.14) for inductance and capacitance is met and is equal to 1. As it is used directly in Eq. (16.30), this now reduces for this least favorable case to the simple form

$$\tan \delta = \frac{p}{v} = \sqrt{\frac{Z_1}{Z_2}}, \qquad (16.31)$$

which at the same time gives the phase angle according to Eq. (16.25).

Thus the highest excess voltage that can be caused beyond the oscillatory circuit by the arrival of a step pulse is, from Eq. (16.28),

$$E_e = \frac{2}{1 + \dfrac{Z_1}{Z_2}}(1 + \varepsilon^{-\pi\sqrt{Z_1/Z_2}})\,E. \qquad (16.32)$$

Thus it is seen to depend only on the ratio of the surge impedances; it is shown graphically in Fig. 16.4.

For the case of equal lines, $Z_1 = Z_2$ and

$$\frac{E_e}{E} = 1 + \varepsilon^{-\pi} = 1.045,$$

which is an insignificant excess. For $Z_1 > Z_2$ the excess becomes even smaller. However, when the transition occurs from a smaller to a larger surge impedance, that is, when $Z_1 < Z_2$, the excess voltage can become considerable. For example, a transformer of

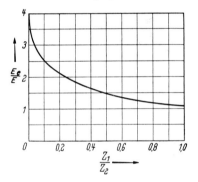

Fig. 16.4

$Z_2 = 5000\ \Omega$, fed by a cable of $Z_1 = 50\ \Omega$ and protected by an inductance and capacitance connected as in Fig. 16.1 and proportioned to meet the equivalence condition (16.14), gives rise to an excess voltage that exceeds the step-pulse voltage in the cable by the factor

$$\frac{E_e}{E} = \frac{2}{1 + \dfrac{50}{5000}}(1 + \varepsilon^{-\pi\sqrt{50/5000}}) = 3.46.$$

Thus this is not a protective arrangement. Of course, these oscillations decay rapidly, so that after 10 half periods only

$$\varepsilon^{-10\pi\sqrt{50/5000}} = \varepsilon^{-\pi} = 4.5 \text{ percent}$$

of the initial amplitude remains.

Thus this connection of choke coil and bypass condenser may be used only for transitions from lines or devices of higher to those of lower surge impedance as shown in Fig. 16.5. The choke coil is adjacent to the line of higher and the bypass condenser to the line of lower surge impedance. If this order is reversed, as

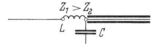

Fig. 16.5

shown in Fig. 16.6, serious excess voltages are possible that cannot arise in the arrangement shown in Fig. 16.5. The reason is that in Fig. 16.6 the low-impedance line and device are separated by the high-impedance choke coil, while the high-impedance line and device are separated by the low-impedance bypass

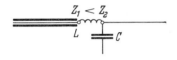

Fig. 16.6

condenser. We see that impedance transition is involved.

The unfavorable case of Fig. 16.6 arises if, for example, a relay or current-transformer coil is connected between a line and the transformer terminal bushings, which have a large capacitance to the transformer case, as shown in Fig. 16.7. The low values of L and

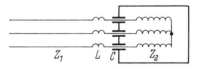

Fig. 16.7

C lead to high frequency and strong damping, by Eqs. (16.19) and (16.20). However, the high excess voltage given by Eq. (16.32) will occur briefly if L and C are close to equivalence as given by Eq. (16.14).

Even for choke coils of unsuitable size, such oscillations frequently occur. If, for example, in Fig. 16.7 the transformer winding has $Z_2 = 5000\ \Omega$, the terminal bushing has $C = 2 \cdot 10^{-10}$ F, the feeding line has $Z_1 = 500\ \Omega$,

and the choke coil has $L = 0.5 \cdot 10^{-3}$ H, the oscillatory impedance is

$$\sqrt{\frac{L}{C}} = \sqrt{\frac{0.5 \cdot 10^{-3}}{2 \cdot 10^{-10}}} = 1580\ \Omega,$$

which happens to be equal to the geometric mean surge impedance

$$\sqrt{Z_1 Z_2} = \sqrt{500 \cdot 5000} = 1580\ \Omega.$$

This, of course, implies exact equivalence according to Eq. (16.14), which leads to the oscillation of highest amplitude. The frequency in this case is

$$\frac{v}{2\pi} = \frac{1}{2\pi \sqrt{0.5 \cdot 10^{-3} \cdot 2 \cdot 10^{-10}}}$$

$$= 504{,}000 \text{ cy/sec.}$$

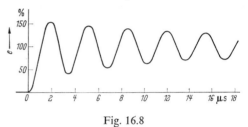

Fig. 16.8

Figure 16.8 shows a typical oscillogram for the oscillatory shaped pulse produced in a transformer by such an arrangement. We see that a small choke coil here gives rise to more danger than protection.

These oscillations can be avoided if L or C is increased to reach one of the aperiodic regions in Fig. 16.2. Frequently one can also increase the damping by the introduction of resistances, which absorb energy in addition to that propagated into the lines.

The reflected voltage pulse in line 1 can be easily obtained from Eq. (16.5):

$$e_{r1} = e_{v1} - \frac{Z_1}{Z_2} e_{v2} - Z_1 C \frac{de_{v2}}{dt}. \quad (16.33)$$

Here we can introduce the voltages from Eqs. (16.7) and (16.26) and the time differential from Eqs. (16.24) and (16.22) for amplitude to obtain

$$e_{r1} = \frac{Z_2 - Z_1}{Z_2 + Z_1} E + \frac{2Z_1}{Z_1 + Z_2} E \frac{\varepsilon^{-\rho t}}{\cos \delta}$$

$$\times \{\cos (vt - \delta) - Z_2 C[\rho \cos (vt - \delta)$$

$$+ v \sin (vt + \delta)]\}. \quad (16.34)$$

The front of the incident step pulse is thus reflected in full amplitude because at $t = 0$ the second and third terms in the right-hand member of Eq. (16.33) disappear. At later times Eq. (16.34) governs the reflected voltage pulse and is shown also in Fig. 16.3 for a given case. After some time the second term in the right-hand member of Eq. (16.34) has been damped to zero and the final value is the same in both lines, as given by Eq. (6.8) for two lines joined without L or C.

We see that choke coils and bypass condensers can act together to flatten the front of the transmitted pulse quite effectively. However, at the same time large excess voltages can occur if the arrangement between lines of differing surge impedances is not properly designed.

Still larger excess voltages can arise if an oscillatory circuit of L and C is present at

$$\sqrt{\frac{L}{C}} > \frac{1}{2} Z. \qquad (16.37)$$

For small capacitance C this is often the case. If we then neglect the small second correction factor in the radical of Eq. (16.36), the highest excess voltage across the bus bar C is given by Eqs. (16.8) and (16.28) a half period after arrival of the step pulse:

$$E_C = 2E(1 + \varepsilon^{-(\pi/2)Z/\sqrt{L/C}}). \qquad (16.38)$$

This shows that it is governed by the ratio of the surge impedance of the line to that of the oscillatory circuit. If the latter is very large, the voltage approaches four times the incident step value and gives rise to strong oscillations, which are also reflected into the line and are damped in accordance with Eq. (16.35). The shape of the reflected sinusoid can be obtained from Eq. (16.34) if Z_2 is made infinitely large.

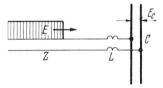

Fig. 16.9

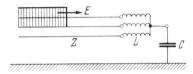

Fig. 16.10

the end of a single line. Figure 16.9 shows as an example how the capacitance C of a bus bar and the inductance L of a choke or other coil can form such an osillatory circuit. If the second line is absent, we can let $Z_2 = \infty$ and $Z_1 = Z$, and then Eq. (16.19) gives the damping factor

$$\rho = \frac{1}{2} \frac{Z}{L}, \qquad (16.35)$$

and Eq. (16.20) gives the natural frequency

$$\nu = \sqrt{\frac{1}{LC} - \left(\frac{Z}{2L}\right)^2}. \qquad (16.36)$$

The damping is smaller because the second line cannot remove energy. The frequency is also closer to that determined purely by L and C and is reduced slightly only by the surge impedance of the line. The voltage is oscillatory if in Eq. (16.36)

Figure 16.10 shows the star point of a three-phase transformer grounded by a capacitance, which may be due to a connecting bus bar or a condenser connected intentionally for this purpose. The oscillatory circuit formed by this capacitance C and the leakage inductance L of the transformer windings can also produce voltages, up to four times the incident ones, at the star point. Only if C is very large can this voltage remain significantly below this value.

For a 30,000-V transformer with a 10,000-kVA power rating, leakage inductance $L = 15$ mH, and a star-point bus bar 100 m long of capacitance $C = 7 \cdot 10^{-4} \mu F$ to ground, the oscillatory impedance is

$$\sqrt{\frac{L}{C}} = \sqrt{\frac{15 \cdot 10^{-3}}{7 \cdot 10^{-10}}} = 4640 \ \Omega.$$

Thus if it is connected to a three-phase line of combined surge impedance $Z = 250 \ \Omega$

then according to Eq. (16.38) a ratio of excess voltage to incident voltage of

$$\frac{E_C}{E} = 2(1 + \varepsilon^{-(\pi/2)250/4640})$$

$$= 2(1 + 0.92) = 3.84$$

is produced. The frequency of the oscillations is

$$\nu = \frac{1}{\sqrt{15 \cdot 10^{-3} \cdot 7 \cdot 10^{-10}}}$$

$$= 3.08 \cdot 10^5 \text{ in } 2\pi \text{ sec}$$

or $f = 49{,}000$ cy/sec.

Superimposed on these sinusoidally oscillating voltages will be the rectangular step-pulse oscillations examined in Chapter 15, Sec. (c), due to the impedance of the transformer winding. In combination they give rise to pulse shapes of the type shown in Fig. 13.9.

Still larger voltages can be produced by an oscillatory circuit at the junction of two lines or the end of one line if it is excited by an incident pulse train of a frequency close to its natural frequency instead of by a simple step pulse. These voltages are limited in amplitude only by the surge impedances of the lines and the resistances of the inductive windings. Thus in transformers and other machines high voltage differences between adjacent coils can occur. Apart from resonance due to its natural frequency, the capacitance-grounded star point of a transformer, as shown in Fig. 16.10, behaves as if it were solidly grounded for fast transients and isolated for very slow changes.

(b) *Inductance behind capacitance*. If a traveling pulse impinges on a choke coil and

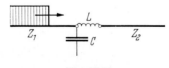

Fig. 16.11

bypass condenser connected as in Fig. 16.11, the voltage balance is

$$e_1 = e_2 + L\frac{di_2}{dt} \qquad (16.39)$$

and the current balance is

$$i_1 = i_2 + C\frac{de_1}{dt}. \qquad (16.40)$$

Combining these suitably we obtain

$$i_1 = i_2 + C\frac{de_2}{dt} + LC\frac{d^2i_2}{dt^2}. \qquad (16.41)$$

Introduction of partial pulses into Eqs. (16.39) and (16.41) then gives

$$e_{v1} + e_{r1} = e_{v2} + \frac{L}{Z_2}\frac{de_{v2}}{dt},$$

$$e_{v1} - e_{r1} = \frac{Z_1}{Z_2}e_{v2} + Z_1C\frac{de_{v2}}{dt} + \frac{Z_1}{Z_2}LC\frac{d^2e_{v2}}{dt^2}.$$

$$(16.42)$$

By addition of these two we obtain

$$\frac{Z_1}{Z_2}LC\frac{d^2e_{v2}}{dt^2} + \left(\frac{L}{Z_2} + Z_1C\right)\frac{de_{v2}}{dt}$$

$$+ \left(1 + \frac{Z_1}{Z_2}\right)e_{v2} = 2e_{v1}. \qquad (16.43)$$

This is a linear differential equation of second order in the transmitted voltage pulse e_{v2} which has the same form as Eq. (16.6).

Just as in Eq. (16.7) the arrival of a step pulse leads to a final steady-state term that consists of the third term of the left-hand member of Eq. (16.43) and thus

$$e'_{v2} = \frac{2}{1 + Z_1/Z_2}e_{v1} = \frac{2Z_2}{Z_1 + Z_2}E = e_\infty.$$

$$(16.44)$$

The remainder forms the transient term of the solution when equated to zero, which gives the homogeneous differential equation

$$\frac{d^2e''_{v2}}{dt^2} + \left(\frac{1}{Z_1C} + \frac{Z_2}{L}\right)\frac{de''_{v2}}{dt} + \frac{Z_1 + Z_2}{Z_1LC}e''_{v2} = 0.$$

$$(16.45)$$

By interchange of Z_1 and Z_2 this becomes the same as Eq. (16.9), because, as shown in Fig. 16.11, the inductance is now in series with Z_2 (instead of Z_1 as before) and the capacitance is now across Z_1 (instead of Z_2 as before). Thus the form of the solution is the same as before:

$$e''_{v2} = K\varepsilon^{-\rho t}\cos(\nu t - \delta) \qquad (16.46)$$

for the periodic case. With the surge impedances interchanged in Eqs. (16.19) and
(16.20), we now have a damping factor

$$\rho = \frac{1}{2}\left(\frac{1}{Z_1 C} + \frac{Z_2}{L}\right) \qquad (16.47)$$

and a natural frequency

$$\nu = \sqrt{\frac{1}{CL} - \frac{1}{4}\left(\frac{1}{Z_1 C} - \frac{Z_2}{L}\right)^2}. \qquad (16.48)$$

The condition for periodic behavior is now,
corresponding to Eq. (16.13),

$$\frac{L}{Z_1 Z_2 C} + \frac{Z_1 Z_2 C}{L} < 2 + 4\frac{Z_1}{Z_2}. \qquad (16.49)$$

Thus here also approximate equivalence of
L and C according to Eq. (16.14) leads to
periodic oscillations. In Fig. 16.2 we must
merely replace the right-hand ordinate Z_2/Z_1
by Z_1/Z_2 in this case to represent Eq. (16.49).
For the constants of integration K and δ in
Eq. (16.46) the boundary conditions give the
same relations as Eqs. (16.22) and (16.25),
so that the total transmitted voltage pulse
becomes, according to Eqs. (16.44) and (16.46),

$$e_{v2} = \frac{2Z_2}{Z_1 + Z_2}\left[1 - \frac{\varepsilon^{-\rho t}}{\cos \delta}\cos{(\nu t - \delta)}\right]E,$$
$$(16.50)$$

which agrees completely with the form of
Eq. (16.26). The highest voltage after one
half period is again

$$E_{v2} = e_\infty(1 + \varepsilon^{-\rho\pi/\nu}), \qquad (16.51)$$

and the exponent here, from Eqs. (16.47)
and (16.48), is

$$\frac{\rho}{\nu} = \frac{\sqrt{\dfrac{L}{Z_1 Z_2 C}} + \sqrt{\dfrac{Z_1 Z_2 C}{L}}}{2\sqrt{\dfrac{Z_1}{Z_2} - \dfrac{1}{4}\left(\sqrt{\dfrac{L}{Z_1 Z_2 C}} - \sqrt{\dfrac{Z_1 Z_2 C}{L}}\right)}}. \qquad (16.52)$$

The damping is again least when the
equivalence condition for L and C of Eq. (16.14)
is met,

$$\tan \delta = \frac{\rho}{\nu} = \sqrt{\frac{Z_2}{Z_1}}. \qquad (16.53)$$

This is the reciprocal of the earlier Eq. (16.31).
The largest excess voltage here is then, by
Eq. (16.51),

$$E_e = \frac{2}{1 + Z_1/Z_2}(1 + \varepsilon^{-\pi\sqrt{Z_2/Z_1}})E. \qquad (16.54)$$

For $Z_1 = Z_2$, that is, for lines of equal
impedance, the voltage excess as before is
only 4.5 percent above the final value and less
for the transition from low surge impedances
Z_1 to high surge impedances Z_2. However,

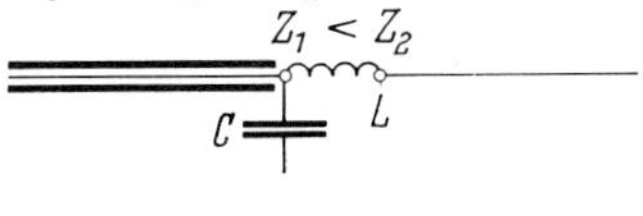

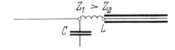

Fig. 16.12

transition from a high Z_1 to a low Z_2 can lead
to large voltage excesses. The mere transition
from an open line of $Z_1 = 500\ \Omega$ to a cable
of $Z_2 = 50\ \Omega$ produces an excess voltage of

$$1 + \varepsilon^{-\pi\sqrt{50/500}} = 1.37$$

times the final voltage e_∞. However, since e_∞
is reduced to

$$\frac{1}{1 + 500/50} = 0.182$$

times the incident step pulse by transition
from the line to the cable, we see that the

Fig. 16.13

largest possible excess voltage is only $1.37 \cdot$
$0.182 = 0.25$ times the incident voltage step.

We thus see that the connection of the series
inductance behind the bypass capacitance, as
shown in Fig. 16.12, produces only small
excess voltages for the transition from a line
of low surge impedance to a line of high
surge impedance. Hence this mode of connection is not dangerous. The opposite
mode of connection, shown in Fig. 16.13,
gives rise to somewhat higher oscillatory
excess voltages. However, the actual peak

voltage of the pulse propagated into the line of low surge impedance is so small that it is not capable of causing serious damage. As before, the development of larger oscillations in the mode of connection shown in Fig. 16.13 is due to the fact that the circuit elements of high capacitance are separated by a choke inductance and the circuit elements of high inductance are separated by a shunt capacitance. In this manner the choke inductance

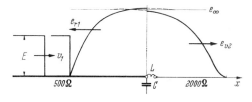

Fig. 16.14

and shunt capacitance form a more independent and undamped oscillatory circuit just by themselves.

The voltage pulse reflected at the junction back into line 1 can next be obtained from Eq. (16.42). This is not here done in detail. However, Fig. 16.14 shows the transmitted and reflected pulse shapes for a given case, connected as in Fig. 16.12, with the values of

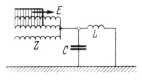

Fig. 16.15

line impedances chosen to give unfavorable conditions.

It is seen that the voltage pulses behave in almost aperiodic fashion and that the transmitted pulse has a gradually rising front. However, the incident voltage pulse is reflected along line 1 with a completely rectangular discharge drop due to the bypass capacitance. After this sharp drop it then gradually rises to its final value.

Large excess voltages can be produced if the oscillatory circuit consisting of L and C is directly closed by a ground connection, as shown in Fig. 16.15 for an inductive choke

grounding the star point of a three-phase winding.

If we put $Z_1 = Z$ and $Z_2 = 0$, Eq. (16.47) gives the damping factor

$$\rho = \frac{1}{2ZC}, \qquad (16.55)$$

and Eq. (16.48) gives the natural frequency

$$\nu = \sqrt{\frac{1}{CL} - \left(\frac{1}{2ZC}\right)^2}. \qquad (16.56)$$

The lower damping factor renders this arrangement more oscillatory and it is periodic if in Eq. (16.56)

$$\sqrt{\frac{L}{C}} < 2Z. \qquad (16.57)$$

This implies either large capacitances or low inductances, which will be encountered mostly in medium- to low-voltage installations of high power rating.

The voltage pulse across the capacitance can be shown to be

$$e_C = \frac{2E}{\cos \delta} \frac{\sqrt{L/C}}{Z} \varepsilon^{-\rho t} \sin \nu t. \qquad (16.58)$$

Thus the bus bar at the star point and the windings terminated at it are subject to a large amplitude oscillation. The peak value after a quarter period is

$$E_C = 2E \frac{\sqrt{L/C}}{Z} \varepsilon^{-(\pi/4)\sqrt{L/C}/Z}. \qquad (16.59)$$

Because Eq. (16.57) limits the oscillatory impedance to twice the surge impedance, the excess voltage stays below

$$E_C = 2 \cdot 2 \cdot \varepsilon^{-\pi/2} E = 0.84E.$$

While this is considerably lower than the case of the bus bar treated in Fig. 16.10, it still is large enough to require correspondingly adequate insulation for the bus bar and other connections to inductive grounding devices. Furthermore, there will be a superposition of oscillatory pulses of different modes at the star point, which may add up to higher peak excess voltages.

All these considerations lead to the following rule for the combined use of inductance and capacitance as a protection against step-front pulses: the choke inductance should always be connected to the line of higher surge impedance and the shunt (or bypass) capacitance to the line of lower surge impedance, as in Figs. 16.5 and 16.12. The advantage of this rule is that the self-oscillations of the circuit are always strongly damped by the pulses traveling away from the junction. Thus appreciable excess voltages are avoided irrespective of the line along which the exciting pulse first arrives at the junction.

Actual choke coils and bypass condensers are not always properly represented by lumped circuit parameters of inductance and capacitance, as has been assumed so far in the derivations. Instead it may sometimes be necessary to consider their dimensional extent and their surge impedance in distributed form, as shown in Fig. 16.16. This can be done by repeated use at each of the junctions of the laws of transmission and refraction developed in Chapter 6. The transmitted pulse can then have a very small initial front, because of the many transmissions, which at the same time then will lead to a reflected pulse with a large front. By multiple internal reflections the interaction of the distributed elements will lead to oscillatory or periodic changes in voltages and currents. However, as distinct from the continuous smooth changes in Figs. 16.3 and 16.14, they will occur in staircase form, as treated in detail in Chapter 15 for lumped circuit elements. It is possible to

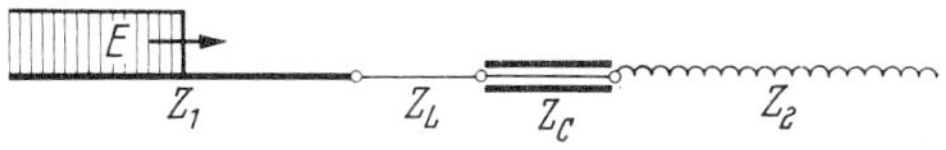

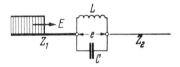

Fig. 16.16

use the stepped interaction of the internal reflections at the junctions of coils and condensers of selected surge impedances to disperse large pulses transmitted by one element, say the choke coil, by the addition of another suitable element.

(c) *Inductance and capacitance in parallel.* A parallel circuit between two lines is shown in Fig. 16.17. The voltage of the incident pulse arriving from the left as shown is given by

$$e_1 = e_2 + e, \qquad (16.60)$$

where e is the voltage drop across the parallel circuit. The currents in both lines are equal to the sum of the current in the capacitance i_C and the current in the inductance i_L; expressing the currents in terms of the common voltage drop e we obtain

$$i_1 = i_2 = i_C + i_L = C\frac{de}{dt} + \frac{1}{L}\int e\,dt. \qquad (16.61)$$

Fig. 16.17

The partial voltage pulses then sum as

$$e_{v1} + e_{r1} = e_{v2} + e, \qquad (16.62)$$

and using the current Eq. (16.61) we get

$$e_{v1} - e_{r1} = \frac{Z_1}{Z_2}e_{v2} = Z_1\left(C\frac{de}{dt} + \frac{1}{L}\int e\,dt\right). \qquad (16.63)$$

Addition of Eq. (16.62) and the first two members of Eq. (16.63) gives the transmitted voltage pulse,

$$e_{v2} = \frac{Z_2}{Z_1 + Z_2}(2e_{v1} - e), \qquad (16.64)$$

and, from the last member of Eq. (16.63),

$$(Z_1 + Z_2)\left(C\frac{de}{dt} + \frac{1}{L}\int e\,dt\right) = 2e_{v1} - e. \qquad (16.65)$$

Next let us designate the capacitive and inductive time constants as

$$T_L = \frac{L}{Z_1 + Z_2}, \quad T_C = C(Z_1 + Z_2), \qquad (16.66)$$

using simply the sum of the line surge impedances. Thus we obtain the differential equation

$$T_C\frac{de}{dt} + e + \frac{1}{T_L}\int e\,dt = 2e_{v1} = 2E \qquad (16.67)$$

for the voltage. Again the oscillatory circuit is driven by a voltage twice that of the given incident voltage step pulse E.

The solution is again the superposition of a forced and a free voltage component. For a step pulse with a sustained peak, the steady-state voltage across the oscillatory circuit must be zero because the integral term in Eq. (16.67) cannot increase indefinitely. Thus the complete solution must be of the form

$$e = K\varepsilon^{\alpha t}. \qquad (16.68)$$

The exponent α is obtained by substitution in Eq. (16.67) without the right-hand member and solution of the resulting quadratic:

$$\alpha = -\frac{1}{2T_C} \pm \frac{1}{2T_C}\sqrt{1 - 4\frac{T_C}{T_L}}. \qquad (16.69)$$

Periodic oscillations will occur if

$$T_L < 4T_C \qquad (16.70)$$

or, after insertion of the defining Eq. (16.66), if

$$\sqrt{\frac{L}{C}} < 2(Z_1 + Z_2). \qquad (16.71)$$

This limiting condition is similar to that of Ref. 1, p. 50, Eq. (16); however, the value of the damping resistance is here increased by a factor of 4 because the surge impedances Z_1 and Z_2 are here in parallel with the oscillatory circuit, whereas in Ref. 1 the resistance R was inserted in series.

For relatively small values of capacitance C the solution becomes aperiodic and from Eq. (16.69) one obtains, to a good approximation,

$$\alpha_1 = -\frac{1}{T_L},$$
$$\alpha_2 = -\frac{1}{T_C}. \qquad (16.72)$$

Thus the general solution for the voltage across the oscillatory circuit becomes

$$e = K\varepsilon^{\alpha_1 t} - K\varepsilon^{\alpha_2 t}. \qquad (16.73)$$

The fact that at time $t = 0$ the voltage e must also be zero, because of the uncharged state of the capacitance, is indicated by the minus sign in Eq. (16.73). At time $t = 0$ the integral term of Eq. (16.67) must also be zero,

so that for this boundary condition only the first differential term can be used to obtain the constant K. Thus differentiation of Eq. (16.73) and substitution at $t = 0$ in Eq. (16.67) gives

$$\alpha_1 T_C K - \alpha_2 T_C K = 2E. \qquad (16.74)$$

Using this now in Eq. (16.73) we get the solution for e,

$$e = \frac{2E(\varepsilon^{\alpha_1 t} - \varepsilon^{\alpha_2 t})}{T_C(\alpha_1 - \alpha_2)} = \frac{2E(\varepsilon^{\alpha_1 t} - \varepsilon^{\alpha_2 t})}{\sqrt{1 - 4T_C/T_L}}, \qquad (16.75)$$

where the difference between the two values of α in the denominator has been replaced by its exact form from Eq. (16.69).

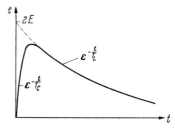

Fig. 16.18

For small capacitances then, according to Eq. (16.72), the voltage across the parallel circuit is given by the simple expression

$$e = 2E(\varepsilon^{-t/T_L} - \varepsilon^{-t/T_C}), \qquad (16.76)$$

and the transmitted voltage pulse, from Eq. (16.64), becomes

$$e_{v2} = \frac{2Z_2}{Z_1 + Z_2}E(1 + \varepsilon^{-t/T_C} - \varepsilon^{-t/T_L}). \qquad (16.77)$$

Figure 16.18 shows the voltage e across the parallel circuit as given as a function of time by Eq. (16.76). For a very small capacitance the peak voltage e becomes almost twice the value of the incident step voltage. The double value would occur instantaneously if only an inductance were connected between the lines. The capacitance has the effect of producing a more gradual rise to a value approaching this double limit. The gradual rise is governed by the capacitive time constant T_C given in Eq. (16.66).

A practical example may be an instrument, relay, or other intermediate transformer of leakage inductance $L = 75\,\text{mH}$ per phase connected between a transmission line of $Z_1 = 500\,\Omega$ and a power transformer of $Z_2 = 5000\,\Omega$. A small capacitance $C = 5 \cdot 10^{-4}$ μF is connected across it as a protection against step pulses. The time constants are

$$T_L = \frac{75 \cdot 10^{-3}}{500 + 5000} = 13.6\ \mu s;$$

$$T_C = 5 \cdot 10^{-10}(500 + 5000) = 2.8\ \mu s,$$

and thus, according to Eq. (16.70), the case is still just aperiodic. Thus each incident step-pulse front is expanded over a distance constant, equivalent to the time constant, of

$$X = vT_C = 3 \cdot 10^8 \cdot 2.8 \cdot 10^{-6} = 830\ \text{m},$$

which protects the windings of the intermediate transformer against insulation breakdown. The capacitance must, however, be able to withstand twice the peak value of the incident step pulse for a short time.

The transmitted pulse is shown in Fig. 16.19. Equation (16.77) gave its composition in the form of three terms. The voltage step pulse is transmitted instantaneously with full steepness of front. However, it then immediately decays almost to zero according to the time constant T_C. Later it rises again exponentially

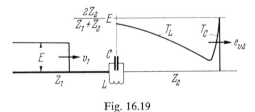

Fig. 16.19

to the final steady-state value according to the larger time constant T_L. Thus we obtain in effect a superposition of a capacitively transmitted impulse function, as in Fig. 12.9, and an inductively transmitted exponential rise, as in Fig. 12.1.

If the capacitance is made larger, so that, according to Eq. (16.71), the case becomes periodic, the solution of Eq. (16.67) takes the form

$$e = K\varepsilon^{-\rho t} \sin vt, \qquad (16.78)$$

and from Eq. (16.69) the damping exponent is

$$\rho = \frac{1}{2T_C} = \frac{1}{2C(Z_1 + Z_2)} \qquad (16.79)$$

and the natural frequency of the oscillatory circuit in radians per second is

$$v = \sqrt{\frac{1}{T_C T_L} - \frac{1}{4T_C^2}}$$

$$= \sqrt{\frac{1}{LC} - \frac{1}{4C^2(Z_1 + Z_2)^2}} \cong \frac{1}{\sqrt{LC}}. \qquad (16.80)$$

Owing to the large values of C and Z the final simple approximation in Eq. (16.80) is very close to the exact value.

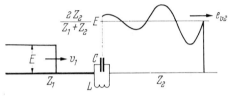

Fig. 16.20

The constants in Eq. (16.78) again are adjusted so that the voltage e becomes zero at $t = 0$. For $t = 0$ the first term of Eq. (16.67) gives by differentiation of Eq. (16.78) for moderate damping

$$vT_C K = 2E, \qquad (16.81)$$

and thus an oscillatory amplitude constant

$$K = \frac{2E}{vT_C} = 4\frac{\rho}{v}E = \frac{2\sqrt{LC}}{C(Z_1 + Z_2)}E, \qquad (16.82)$$

in which the good approximate frequency expression derived in Eq. (16.80) is used. Thus the voltage across the oscillatory circuit as a function of time is

$$e = 2\frac{\sqrt{L/C}}{Z_1 + Z_2}E\varepsilon^{-\rho t} \sin vt. \qquad (16.83)$$

The amplitude is governed by the ratio of the oscillatory impedance to the sum of the surge impedances.

The transmitted voltage pulse is, by Eq. (16.64),

$$e_{v2} = \frac{2Z_2}{Z_1 + Z_2}E\left(1 - \frac{\sqrt{L/C}}{Z_1 + Z_2}\varepsilon^{-\rho t}\sin vt\right), \qquad (16.84)$$

and is shown in Fig. 16.20. The steep front rise of the pulse is not changed by the periodic condition of the parallel circuit. However, the subsequent oscillations produce a sinusoidal train of voltage rises and depressions about the mean final and initial values. Thus large excess voltages can be produced.

The highest voltage E_e across the parallel circuit occurs at $vt = \pi/2$ and is, according to Eqs. (16.82) and (16.83),

$$E_e = 4\frac{\rho}{v} E\varepsilon^{-(\pi/2)\rho/v} \qquad (16.85)$$

This has a maximum for

$$\frac{\rho}{v} = \frac{2}{\pi} \quad \text{or} \quad \sqrt{\frac{L}{C}} = \frac{4}{\pi}(Z_1 + Z_2), \quad (16.86)$$

which is

$$E_e = 4\frac{2}{\pi} \varepsilon^{-1}E = 0.94E. \qquad (16.87)$$

We see that the maximum possible voltage across the intermediate parallel circuit is almost equal to the amplitude of the incident voltage step pulse.

Thus the voltage across the choke inductance, even in this most unfavorable condition, is reduced to less than half by the capacitance connected across it. For larger capacitances Eq. (16.83) shows that this voltage will become smaller owing to decreased oscillatory impedance, while for smaller capacitances it will also become smaller owing to more pronounced damping.

For the previous practical numerical example, a change in capacitance to $C = 5 \cdot 10^{-2}$ μF gives a natural frequency of 2600 cy/sec and an oscillatory impedance of

$$\sqrt{\frac{L}{C}} = \sqrt{\frac{75 \cdot 10^{-3}}{5 \cdot 10^{-8}}} = 1220 \ \Omega.$$

This is substantially smaller than the sum of the surge impedances and thus by Eq. (16.83) gives an amplitude only

$$2\frac{1220}{500 + 5000} = 45 \text{ percent}$$

of that of the incident step pulse, even without consideration of the damping exponent.

We see on the basis of the equations and the numerical examples that the periodic condition is more favorable toward the generation of voltages across the intermediate

circuit. However, it requires large capacitances. The benefit is further enhanced, according to Eq. (16.83), if the intermediate circuit precedes a winding of high surge impedance Z_2 instead of a connection between similar lines.

Figure 16.21 shows an inductance and a capacitance connected in series across the two terminals at the junction of two dissimilar lines. This case is essentially similar to that treated above if we interchange voltages and currents and inductances and capacitances.

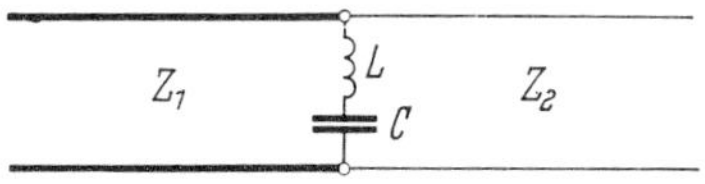

Fig. 16.21

The time constants of the oscillatory circuit then become

$$T_l = L\left(\frac{1}{Z_1} + \frac{1}{Z_2}\right), \quad T_c = \frac{C}{\dfrac{1}{Z_1} + \dfrac{1}{Z_2}}, \quad (16.88)$$

which correspond to Eqs. (12.23) and (12.51) for connection of single elements across a line junction. In this case also the steep pulse front is transmitted across the junction without change and then drops owing to the charging current of the capacitor, gradually returning to the initial value when the condenser is fully charged. Whether this final value is approached aperiodically or by periodic oscillations is again determined by the ratio of the two time constants.

(d) Protective connections using resistances. Protective choke coils and bypass condensers, singly or in combination, can protect machinery and transformers against all rapid voltage changes and in particular against voltage step pulses. This does not interfere with the regular performance of the systems in the low-frequency operating region. The step pulses are reflected fully back into the feeder lines and the protective devices modify only the transmitted pulses. The disturbances then travel to and fro along the lines in the system until their energy is dissipated in the line resistances.

To dissipate the energy of disturbances in addition to reflecting them away from protected devices it is thus necessary to introduce resistance elements. Figure 16.22 shows an ohmic resistance R in parallel with an inductive choke L connected in series between two lines. Figure 16.23 shows an ohmic resistance P in series with a bypass condenser C connected across the two terminals of the junction between two lines. Neither arrangement has an appreciable influence on the

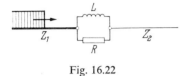

Fig. 16.22

regular low-frequency operation of the system, because the operating currents can flow through L without losses and the condenser C prevents losses in P at the low frequencies.

To examine the effect of the resistances on step-pulse disturbances, we can regard the choke coil as initially an open circuit that reflects the incident pulse and the bypass condenser as a short circuit that reflects a canceling pulse initially. Thus the resistances can influence only the amplitude development

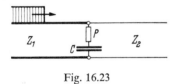

Fig. 16.23

of the pulse following the initial instantaneous behavior, which is analogous to that of the purely resistive elements between Figs. 16.22 and 6.10 and Figs. 16.23 and 6.11 respectively.

Thus the voltage step-pulse amplitude is reduced according to Eqs. (6.24) and (6.35) for the series resistance R or the parallel resistance P respectively, but is transmitted with full steepness into the second line. If the value of R or P is chosen to dissipate as much as possible of the energy of the traveling disturbance, here a step pulse, then the step pulse transmitted into line 2 will be reduced to only one-half the value that would be transmitted without any protective device. Thus while the resistances in dissipating energy reduce the reflected pulses, they at the

same time permit an appreciable portion of the incident pulse to be transmitted in step form to the second line. This limits the usefulness of such a dissipative protective arrangement.

Figure 16.24 shows the case of two equal lines with a purely reactive element at the

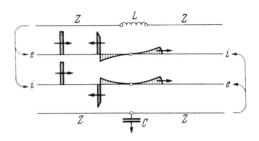

Fig. 16.24

junction when a short rectangular impulse is incident. The reflected pulse is of full amplitude, while the transmitted pulse has a small amplitude and time-stretched form. Figure 16.25 shows the pulse forms for the case of resistances chosen to dissipate the most energy. The reflected pulses are considerably reduced

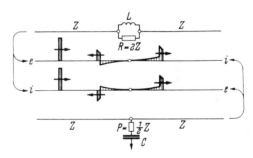

Fig. 16.25

in amplitude but the transmitted pulses now contain also a sharp frontal rectangular impulse.

The energy content of step-pulse fronts can be completely dissipated if circuit elements of inductance, capacitance and resistance are used, properly related in size and arranged in connection. Two connective arrangements are possible. Figure 16.26 shows one of these, in which the parallel resistance R and the choke L are in line 1 and the bypass condenser C

is connected across the terminals of the junction between the lines. The low-frequency operating currents are confined to L, while the high-frequency component voltage pulses cause currents in R. Similarly, C represents an open circuit for the operating low-frequency voltages and a short circuit for the high-frequency component voltage pulses. Thus all voltage pulses with steep fronts can cause currents to flow only in the directions of the

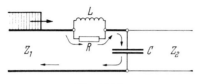

Fig. 16.26

arrows shown in the figure. They cannot enter into line 2 and their energy will be completely dissipated in R if it is made equal in value to Z_1.

Figure 16.27 shows the second arrangement, in which a choke L prevents the transmission of high-frequency components into line 2. Instead they are forced to cause current flow in the resistance P via the bypass condenser C. Again complete dissipation is possible if P is made equal to Z_1. In both arrangements complete dissipation is made possible because

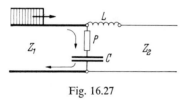

Fig. 16.27

the incoming line 1 is artificially terminated in a resistance equal to the surge impedance. We saw in Chapter 6, Sec. (c), that this was the only way to absorb traveling pulses without reflection or refraction. Of course these arrangements are completely effective only against the first steep rise of traveling pulses. More slowly changing portions of pulses, particularly their later time behavior, cannot be fully intercepted by the reactive elements. Instead, the gradual transmission of these portions of the pulses is governed by the energy storage in the inductances and

capacitances. However, because the insulation of machinery and transformer windings is stressed most dangerously by the steep portions of the voltage pulses, these arrangements offer very useful protection and dissipation.

Figure 16.28 shows both arrangements for lines of equal surge impedance and an incident step impulse. The steep step portions of the

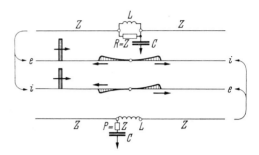

Fig. 16.28

impulse are completely dissipated. Both the reflected and the transmitted pulses rise gently to a moderate amplitude and then decay to zero at a somewhat slower rate. Oscillations do not occur in the arrangements of Figs. 16.26 and 16.27 because the surge impedances of the lines, as well as the resistances, introduce sufficient damping.

For lines in which impulses can arrive from either side, a double arrangement such

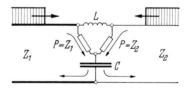

Fig. 16.29

as is shown in Fig. 16.29 can be used to dissipate them if the resistances are matched to their respective lines as shown. Such dissipative protection against steep pulses caused by switching or similar processes can be very useful at places of danger in systems or along lines, particularly at the junction of dissimilar lines or elements.

The most comprehensive presentation of the behavior of traveling pulses is given by a time sequence of oscillograms in film-strip

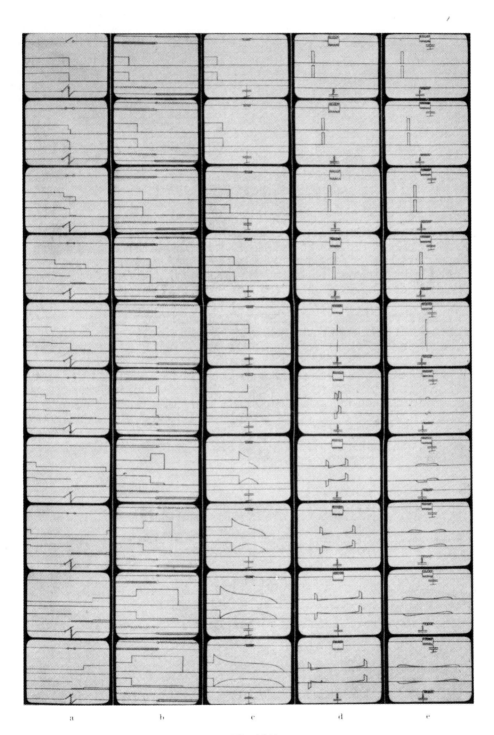

Fig. 16.30

form. Figure 16.30 shows several such strips for typical processes.

Strip (*a*) shows the switching on of the right-hand line with and without a protective resistance. The pulses travel in both directions and are reflected at the right-hand open-ended termination of the line switched on. Strip (*b*) shows the reflection and refraction of a long-back rectangular step pulse at the junction of two dissimilar lines. Strip (*c*) shows the change in pulse form when a long-back rectangular step pulse impinges on a winding protected by inductance or capacitance. Strip (*d*) shows what happens to a short impulse that encounters resistance in combination with inductance or capacitance in or across a line, respectively. Strip (*e*) shows the combined effect of *L*, *C*, and *R* properly proportioned and arranged. In each frame of each strip the circuit arrangement is shown at the top and bottom. Figures 16.24 to 16.28 make it easy to recognize which of the two oscillogram traces represents current and which represents voltage.

All power networks and switching, distributing, and generating stations contain numerous oscillatory circuit elements among their relays, bushings, connectors, insulators, instrument transformers, bus bars, and similar deviations from homogeneous line structures. Many of these oscillatory circuit elements can be excited into natural modes of oscillation by incident traveling pulses. These modes can result in overvoltages that are rather hard to predict. To reduce such hazards and at the same time to dissipate the energy in the pulses, it is advisable to insert suitable resistances in parallel with all inductances and in series with all capacitances. For example, one should shunt current transformers with resistances and insert resistances in series with capacitive voltage dividers. We should note that bus bars with many connected lines are far less endangered by such oscillations than the end of a single branch line, because each line contributes damping by means of its surge impedance and at a junction these are added effectively in parallel.

PART IV

TRANSFORMER AND MACHINE WINDINGS

CHAPTER 17

PROPAGATION VELOCITY AND SURGE IMPEDANCE IN SIMPLE WINDINGS

When a pulse travels from an open line into a transformer or machine winding, the inhomogeneous structure of the winding causes a considerable change in the shape of the pulse. The more discontinuous, lumped rather than continuous, and open distributed structure of these windings makes it necessary to modify the simple assumptions used in Chapter 1 to examine propagation in homogeneous lines. The general observation that such windings can be driven into oscillatory behavior and have natural frequencies supports the hypothesis that currents and voltages can travel in them in the form of pulses. The mathematical equivalence of standing and traveling wave modes or functions developed in Chapter 1, Sec. (d), shows that there is a basic conceptual relation between oscillatory modes and traveling-pulse or wave phenomena. We shall develop numerical values characteristic of the oscillatory and traveling-pulse or surge phenomena in such windings.

(a) Transformer windings. Figure 17.1 shows a simple single-layer tubular winding as used in many transformers. It is of great length and its diameter is large compared with the distance b between the winding and the concentric inner iron core. The distance d to the outer iron shell can have any value. The thickness of the winding is small. Assume that the inner core and outer shell have surfaces that are concentric with the cylindrical winding, smooth, and perfectly conducting. Then no rapidly changing magnetic flux can penetrate these surfaces and the lines of force of the magnetic field must be parallel and those of the electric field perpendicular to the surfaces, as indicated by the broken lines. First we derive the steady-state values of capacitance and self-inductance of the winding (or coil), since these are basic for the surge impedance.

The electromagnetic system of units is used, with conversion factors added for practical units where these are convenient.

A uniform voltage E in the winding produces over the internal insulating distance b and the external insulating distance d electric field gradients E_i and E_a, as shown in Fig. 17.1

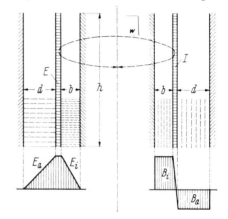

Fig. 17.1

on the left. If w is the length of one turn of the winding, equal to its circumference, and h is the axial length of the winding, then the total electric charge on the winding due to the voltage E stored in the electric field in the insulating spaces of dielectric constant ε is

$$Q = \frac{wh}{4\pi v_0^2}\left(\varepsilon\frac{E}{b} + \varepsilon\frac{E}{d}\right) = \frac{\varepsilon wh}{4\pi v_0^2}E\left(\frac{1}{b} + \frac{1}{d}\right),$$

(17.1)

where v_0 is the velocity of light in vacuum. The capacitance of the coil to ground is thus

$$C = \frac{Q}{E} = \frac{\varepsilon wh}{4\pi v_0^2}\frac{b+d}{bd}\ (\cdot\ 10^9 \text{ in farads}).$$

(17.2)

149

A steady current I in the winding produces a magnetic flux that is entirely in the axial direction and is equal in magnitude and opposite in direction within the insulating spaces on the two sides of the coil. The flux densities B_i and B_a in the internal and external spaces are shown in Fig. 17.1 in inverse proportion to the distances b and d. If N is the total number of turns of the winding, the flux in either gap is

$$\Phi = \frac{4\pi\mu wNI}{\dfrac{h}{b} + \dfrac{h}{d}}, \qquad (17.3)$$

where μ is the magnetic permeability of the insulating spaces. Thus the self-inductance of the winding is

$$L = \frac{N\Phi}{I} = 4\pi\mu\frac{w}{h}N^2\frac{bd}{b+d} (\cdot\ 10^{-9} \text{ in henrys}). \qquad (17.4)$$

Equations (17.2) and (17.4) contain the physical dimensions w, d, b, and h of the transformer.

If we denote the total length of wire in the winding by

$$a = wN, \qquad (17.5)$$

the specific self-inductance l and capacitance c per unit length of the winding are simply obtained by dividing Eqs. (17.2) and (17.4) by a. The velocity of traveling current or voltage waves, surges, or pulses within the winding is then derived according to Chapter 1:

$$v = \frac{1}{\sqrt{lc}} = \frac{a}{\sqrt{LC}}. \qquad (17.6)$$

If we now insert the values of L and C from Eqs. (17.2) and (17.4) we see that all physical dimensions cancel and we obtain

$$v = \frac{v_0}{\sqrt{\varepsilon\mu}}. \qquad (17.7)$$

This shows that the velocity of surges along the wire of such transformer windings is always equal to the universal velocity of light in a vacuum, $v_0 = 300$ m/μsec, divided by the square root of the product of dielectric constant (Table 17.1) and magnetic permeability of the insulating spaces.

This fundamental law was also found empirically as a result of measurements on many actual transformers as a limiting case for windings having small internal capacitance. We thus see that electromagnetic waves travel on conductors wound into coils of negligible internal capacitance with the same velocity as on a straight line conductor. The increase in self-inductance resulting from winding the wire in a cylindrical coil is exactly compensated by the decrease in capacitance to ground due to the close arrangement of the turns.

The surge impedance of the winding, which determines the ratio of voltage to current of traveling pulses, is also given by Eqs. (17.2) and (17.4):

$$Z = \sqrt{\frac{l}{c}} = \sqrt{\frac{L}{C}} = 4\pi v_0\sqrt{\frac{\mu}{\varepsilon}}\cdot\frac{N}{h}\frac{bd}{b+d}$$

$$(\cdot\ 10^{-9} \text{ in ohms}). \qquad (17.8)$$

TABLE 17.1. Dielectric constant ε of various insulating materials.

Material	ε	Material	ε
Bakelite	4.5–8	Pyranol	4–5
Impregnated fiber	3–5	Polyethylene	2.3
Glass	3.5–6–20	Polystyrenes	3–20
Mica	5–7–8	Porcelain	5.5–6.5–9
Natural rubber	2.4–4	Quartz	3.8–4
Synthetic rubber	3–5	Shellac	3.8
Cardboard	4.8	Silicones	2.5–3.5–5.5
Micanite	4.5–6–12	Steatites	5.5–6.5
Oil	2–2.5	Titanium compounds	100–1000–8000
Oil-impregnated paper	4–4.5	Barium titanate	1000–3000
Paper	3.3	Ceramic titanates	30–150
		Distilled water	80

Because v_0 is a universal constant, we can simply obtain in practical units

$$Z = 120\pi \sqrt{\frac{\mu}{\varepsilon} \frac{N}{h} \frac{b}{1 + b/d}} \, \Omega. \qquad (17.9)$$

Thus we see that in addition to ε and μ the surge impedance of transformer windings is dependent mainly on the number of turns per unit length N/h and on the internal insulating distance b while the larger external distance d enters only in a correction term.

A model oil transformer with grounded core consisting of a cylindrical steel sheet had a single-layer winding of flat copper wire 1.2×7 mm in cross section and of length $a = 59$ m. The wire was wound with $N = 52$ turns on an insulating tube of diameter 36.3 cm and axial length $h = 65$ cm. The inner distance from coil to core was $b = 5.4$ cm and there was no outer metal tank. From the shape of the switching-on voltage curves, measured with a cathode-ray oscillograph, in comparison with those of a known homogeneous transmission line, the velocity of propagation through the turns of the winding was found to be $v = 155$ m/μs and the surge impedance $Z = 920 \, \Omega$ approximately. With an average $\varepsilon = 3.5$ and $\mu = 1$, Eqs. (17.7) and (17.9) give

$$v = \frac{300}{\sqrt{3.5}} = 160 \text{ m}/\mu s$$

and

$$Z = 120\pi \sqrt{\frac{1}{3.5} \frac{52}{65}} \, 5.4 = 870 \, \Omega.$$

The results of similar experiments are listed in Table 17.2. We see that the simple assumptions have produced good relations to determine the desired characteristic properties.

Average transformer windings in oil with $\varepsilon = 3.5$ and $\mu = 1$, having a winding density of $N/h = 10$ turns/cm and insulating distances $b = 3$ cm to the core and $d = 30$ cm to the tank will have a surge impedance

$$Z = 120\pi \sqrt{\frac{1}{3.5}} \cdot 10 \cdot \frac{3}{1 + 3/30} = 5500 \, \Omega.$$

This is about ten times the value of the surge impedance of ordinary overhead lines.

A study of many power transformers of various types shows that for rated sizes of 5 to 500 to 50,000 kVA the winding density varies from 100 to 10 to 1 turn/cm. These values refer to the high-voltage side; for the low-voltage side the values of winding densities are about $\frac{1}{5}$ to $\frac{1}{10}$ of the foregoing figures. Since the insulating distance b varies only from 1 cm for low voltages to 5 cm for high voltages, while the tank distance d is always relatively large, we can expect that the windings of oil transformers in the above power rating range will have the surge impedances shown in Table 17.3, which agree

TABLE 17.3. Surge impedance of average transformer windings.

Power rating (kVA)	5	500	50,000
$Z_{\text{high voltage}}$ (Ω)	20,000	5,000	1,000
$Z_{\text{low voltage}}$ (Ω)	4,000	600	100

with sporadic measurements published in the literature.

If a pulse of any form travels with high velocity v along the wire of a helical winding, the axial velocity of propagation of the electromagnetic state will be much less than v. The shape of the circumferential pulse is preserved in the axial direction; however, it is condensed

TABLE 17.2. Velocities of propagation and surge impedances from a number of experiments on transformer windings.

Experimenter	Winding	N/h (cm^{-1})	$\dfrac{b}{1 + b/d}$ (cm)	ε	Calculated		Observed	
					v (m/μs)	Z (Ω)	v (m/μs)	Z (Ω)
Strigel	Model in oil	0.79	5.4	3.5	160	870	155	920
Norinder	Transformer	23.1	1.4	3.5	160	6530	148	6600
Lerstrup	Air cored	35.6	0.81	1.34	259	9350	260	9020

in the ratio of wire length w to pitch s of a turn. Thus the axial velocity of the pulse is only

$$v_y = \frac{s}{w}v = \frac{h}{a}v, \qquad (17.10)$$

expressed also in terms of the ratio of axial coil length h to total wire length a.

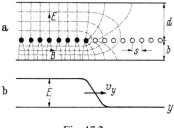

Fig. 17.2

If a voltage pulse E of the form shown in Fig. 17.2b travels through the coil, the turns ahead of the pulse front are still without voltage and current while the turns behind the front have full voltage and current. Hence the electric and magnetic lines of force are as shown in Fig. 17.2a. The pulse front travels circumferentially along the helical wire with the velocity of light v in the dielectric medium. The entire electromagnetic excitation of the coil, however, propagates in the axial direction of the coil with the much lower velocity v_y given by Eq. (17.10). Because of this slow axial propagation, the electromagnetic field in the longitudinal cross section shown in Fig. 17.2a is nearly static, so that it is easy to trace the lines of force. Thus the structure of the electromagnetic field of the transformer coil is pulselike or wavelike in the circumferential direction but quasistationary in the axial direction.

So far we have assumed that the winding is infinitely long. In order to consider the effect of a finite end, as shown in Fig. 17.3a, the electromagnetic field of this end may be conceived as being composed of the field of an infinite winding, as in Fig. 17.3b, and the field of a finite winding, as in Fig. 17.3c, extending to the other side and having the opposite sign from the infinite winding. Superposition of the two windings of Fig. 17.3b, c gives exactly the winding of Fig. 17.3a

with a finite end. If F is the well-known electromagnetic field of the infinite winding, and f the unknown field at the end of the finite winding in Fig. 17.3a, the field at the end of the auxiliary coil in Fig. 17.3c is $-f$. Superposition gives

$$F - f = f \quad \text{or} \quad f = \frac{F}{2}. \qquad (17.11)$$

Hence the electromagnetic field and therefore the specific properties, such as electromagnetic inductance and electrostatic influence, both per unit length, decrease toward the end of a winding to exactly half the value that they have in an infinitely long coil or at the center of a finite but still long coil. This is represented in Fig. 17.3d, while Fig. 17.3e shows that the specific capacitance c, the reciprocal of the static influence, is doubled at the end.

The velocity v of the pulses, determined according to Eq. (17.6) as the product of l and c, remains spatially constant up to the end of the coil. The surge impedance, however, determined by the quotient of l and c, as in

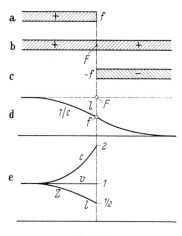

Fig. 17.3

Eq. (17.8), decreases toward the end to one-half of its original value. Figure 17.3e shows the distribution of all four characteristic values at the finite end of the winding.

The axial distance to which these changes in l and c extend into the winding depends entirely on the breadth b of the insulating gap and actually is only a small multiple of the distance b. As a whole, a finite end of the winding acts as if it were loaded with some additional capacitance and some inductance were removed. Those properties of the coil that are determined by the wave velocity, such as the natural frequencies, are not changed by the alterations near a finite end. The surge impedance, on the other hand, comes into effect mainly with transmission of surges coming from or going to adjacent lines, which for the most part have much smaller surge impedance than the winding. Halving of the surge impedance near the winding end therefore produces only moderate alterations in the transmitted voltage.

(b) *Minor influences.* In actual transformers, as shown in Fig. 17.4, neither the core nor the tank has a smooth surface as in our idealized earlier model, but the insulating distances vary around the circumference. However, since we shall not determine the variation of voltage around a single turn, we may replace, without actual loss of accuracy, the irregular surfaces of both core and tank by concentric cylinders that give an equivalent interior and exterior insulating distance. Since the distance d to the tank is comparatively

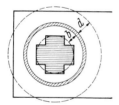

Fig. 17.4

large and its influence, according to Eq. (17.9), is small, it does not matter if the tank is eccentric to the winding in order to enclose the second leg of the transformer.

Very often the dielectric between the winding and the core is not uniform but is as shown on the left-hand side of Fig. 17.5a. If we denote the wall thickness of the insulating tube by b_i and its dielectric constant by ε_i, while the remainder of the distance to the core is b_0 with the dielectric constant ε_0, we have instead of Eq. (17.1) for the charge of the coil under the voltage E the equation

$$Q = \frac{wh}{4\pi v_0^2} E \left[\frac{\varepsilon_0}{d} + \left(\frac{b_i}{\varepsilon_i} + \frac{b_0}{\varepsilon_0} \right)^{-1} \right]. \quad (17.12)$$

By comparison with Eq. (17.1) we obtain therefore for the mean dielectric constant for the whole transformer

$$\varepsilon = \varepsilon_0 \frac{\left(\frac{\varepsilon_0}{\varepsilon_i} \frac{b_i}{b} + \frac{b_0}{b} \right)^{-1} + \frac{b}{d}}{1 + b/d}. \quad (17.13)$$

With representative figures this gives for oil-filled transformers resultant values of

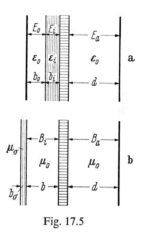

Fig. 17.5

$\varepsilon = 3.5$, which may be used for the velocity of propagation in Eq. (17.7) as well as for the surge impedances in Eqs. (17.8) and (17.9).

The interior and exterior boundaries of the electromagnetic fields of actual transformers do not consist, as in our idealized model, of material with perfect electrical conductivity but are made of iron with finite electrical conductivity and high magnetic permeability. Figure 17.4 shows that the surface consists partly of rolled steel sheets, partly of magnetic or nonmagnetic structural steel plates, and partly of the edges of the steel sheets that form the active magnetic core. High-frequency magnetic oscillations, of which every traveling pulse is composed, penetrate only into a thin layer of these materials, the effective depth of which is

$$b_\sigma = \frac{1}{4\pi} \sqrt{\frac{\sigma}{\mu_\sigma f}}, \quad (17.14)$$

where f is the equivalent frequency, σ the

specific resistance of the steel, and μ_σ its permeability.

If we put, for example, $f = 10^4$ cy/sec, $\sigma = 1.5 \cdot 10^4$ cm²/sec, and $\mu_\sigma = 200$, we obtain a depth of penetration

$$b_\sigma = \frac{1}{4\pi} \sqrt{\frac{1.5 \cdot 10^4}{200 \cdot 10^4}} = 6.9 \cdot 10^{-3} \text{ cm.}$$

The magnetic flux within this skin supplements the air flux and, since the flux density in steel is μ_σ times that in air, we have an additional equivalent air breadth for the internal magnetic flux of $\mu_\sigma \cdot b_\sigma = 1.38$ cm in our example.

For several reasons it is difficult to decide without further experiments what value the permeability actually has. The edges of the highly magnetic steel sheets are damaged magnetically by the cutting process, the magnetization by traveling pulses is superposed on the steady-state magnetization, the magnetic circuit for the pulse flux is not entirely in iron, and the hysteresis in steel parts of poorer magnetic quality gives an additional screening effect.

If the effective permeability μ_σ of the steel surfaces were known, we could compute the additional effect on the permeability μ_0 of the insulating space as in Fig. 17.5b. In confining ourselves to the effect of the more important core surface, we can derive the resultant surge permeability for a steel-core winding,

$$\mu = \mu_0 \frac{1 + b/d}{\left(\dfrac{\mu_\sigma b_\sigma}{\mu_0 b} + 1\right)^{-1} + \dfrac{b}{d}}, \quad (17.15)$$

an expression that is similar in structure to Eq. (17.13). If the value of μ_σ is considerably greater than that of μ_0, it decreases the velocity of propagation given by Eq. (17.7) and increases the surge impedance given by Eqs. (17.8) and (17.9). Since b_σ, according to Eq. (17.14), is dependent on frequency, we then obtain a variation of the characteristics of the winding with frequency.

With the values of the last example and insulating distances $b = 3$ cm and $d = 30$ cm, we obtain a resultant permeability

$$\mu = 1 \frac{1 + 3/30}{\left(\dfrac{200}{1} \dfrac{6.9 \cdot 10^{-3}}{3} + 1\right)^{-1} + \dfrac{3}{30}} = 1.43.$$

For lower frequencies the value may be considerably larger and for higher frequencies considerably smaller.

The copper winding also permits some penetration of the magnetic flux, since it is not perfectly conducting. Since its permeability is always $\mu = 1$, the depth of penetration, or skin depth, for $\sigma = 2 \cdot 10^3$ cm²/sec and $f = 10^4$ cy/sec is from Eq. (17.14)

$$b_c = \frac{1}{4\pi} \sqrt{\frac{2 \cdot 10^3}{1 \cdot 10^4}} = 3.56 \cdot 10^{-2} \text{ cm.}$$

This value is so small compared with the insulating distances that one is almost always justified in neglecting it.

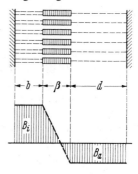

Fig. 17.6

We have assumed that the winding thickness in Fig. 17.1 is very small. Actual transformers, however, have a considerable winding thickness β, as shown in Fig. 17.6, where the variation of the flux density between the interior and exterior insulating spaces is indicated by an inclined broken line. The intermediate flux within the cross section of the winding is only partially linked with the turns and therefore gives, as is well known, only an effective thickness $\beta/3$. However, as transformer windings are usually made of solid copper wire with large skin effect, as mentioned above, the transient flux penetrates only the insulation of the wires and not the copper. With β' as the total radial insulating distance within the winding of thickness β, and perhaps several layers as shown in Fig. 17.6, we thus obtain a correction factor

$$\mu' = 1 + \frac{1}{3}\frac{\beta'}{b}, \quad (17.16)$$

which may be used as an additional permeability for the electromagnetic field.

As Fig. 17.6 indicates, the capacitance of the winding to ground is not noticeably altered by finite thickness of the winding, even if it consists of single spaced sections as shown. Since neither the total self-inductance nor the total ground-capacitance is materially influenced by subdivision of the winding into single disks, sectionalized cylindrical windings, as shown in Fig. 17.6, can also be treated by the method already outlined. However, the last disks at the ends of the winding stack will have some extra ground capacitance due to the flat outer faces of the stack. This can be considered as lumped end capacitance of the coil in addition to the other end effects already mentioned.

If the transformer is built up of pancake coils with interleaved primary and secondary sections, as in Fig. 17.7, it is still possible to compute the total self-inductance and ground capacitance in a manner similar to that already described, when the secondary winding is effectively short-circuited to ground, thus forming a part of the entire ground. We have only to consider, according to familar rules,

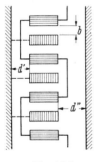

Fig. 17.7

the formation of the electric and magnetic fields within all interspaces of the primary and secondary coils, with distances b, d', and d'' as shown in Fig. 17.7. The multiple concentric layer winding, shown in Fig. 17.8, can be treated similarly, except that in this case the internal capacitance between adjacent layers far outweighs the ground capacitance, so that velocity of propagation and surge impedance may become dependent on fre-

quency; this will be treated in more detail later.

The last simplifying assumption to be reviewed here is the cylindrical shape of the insulating space between the winding and the core and tank. For the usual transformer construction it is not difficult to arrive at an approximately equivalent cylindrical arrangement such as that shown in Fig. 17.4, even

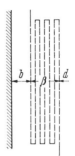

Fig. 17.8

when the winding itself does not have a straight axis, as would be the case, for example, of a toroidal winding or some other deviation from a true cylinder. Equations (17.7) and (17.8) will still give us v and Z to a good approximation as long as at least one of the ground electrodes, either the inner core or the outer tank cylinder, is near the winding.

If the inner electrode is taken away and the outer one is moved to a great distance, matters are different. The surge impedance increases considerably and the velocity of propagation along the wire becomes greater than the velocity of light, owing to the interturn capacitance. However, the much smaller axial velocity of propagation of the electromagnetic field always is less than that of light. Experiments with long windings, even of considerable thickness, confirm such results when only the distant earth is used as a ground. Such coils are often used as delay lines for pulses or surges.

In most stray-field calculations on transformers, the deviation from true straight cylindrical shape of the winding is usually neglected in any case, even though the air spaces for the stray magnetic fields are the same as those that entered into our considerations above.

(c) Magnetic-core flux. Up to now we have neglected the magnetic flux in the core of the transformer because it cannot in its totality change with the speed required to follow traveling pulses. However, measurements made on transformer windings, over a wide frequency range starting at $v = 0$, show that there are several pronounced resonances of the winding that are often caused by the core flux. The mechanism is now entirely different because the core flux is closely linked to all portions of the winding, while the stray flux lines in the air spaces, as shown in Fig. 17.2, can more easily close between the inner and outer spaces around portions of the winding. Thus the voltage induced in each turn of the winding by the time variation of core flux

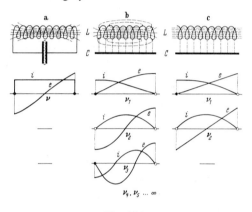

Fig. 17.9

is the same and hence the voltage along the length of the winding must increase linearly, as was demonstrated by R. Willheim and J. Pirenne.

In Fig. 17.9 are shown three different distributions of L and C in interaction. Figure 17.9a gives the distribution of current and voltage in a lumped oscillatory circuit; i is constant and e increases almost linearly along the length of the winding. The only natural frequency is

$$v = \frac{1}{\sqrt{LC}}, \qquad (17.17)$$

as derived in Ref. 1, p. 49, Eq. (15), neglecting saturation influences on the value of L.

Figure 17.9b shows the degrees of freedom for voltage and current distributions in a

circuit involving interaction between distributed L and C in the air space of importance to traveling pulses. Adjacent portions of the winding can interact on account of their localized l and c. Thus a number of resonance oscillations can arise whose frequencies belong to the harmonic series

$$v = \frac{n\pi}{2\sqrt{LC}}, \qquad n = 1, 2, \ldots \quad (17.18)$$

This was developed in detail earlier for uniform lines in Eqs. (1.50) and (4.17). The current i and voltage e are distributed in sinusoidal form over the winding, which increases the lowest natural frequency by a factor $\pi/2 = 1.57$ over that of Fig. 17.9a.

Figure 17.9c shows L completely lumped but C distributed. This leads to a linear voltage and parabolic current distribution along the length of the winding, as shown. Because the average value of a linear distribution is $\frac{1}{2}$ and that of a parabolic distribution is $\frac{2}{3}$, the capacitive current and induced voltage are related by

$$I = \tfrac{1}{2} v C E, \qquad E = \tfrac{2}{3} v L I. \quad (17.19)$$

Elimination of E and I gives the two natural frequencies as

$$v = n\sqrt{\frac{3}{LC}}, \qquad n = 1, 2, \quad (17.20)$$

corresponding to the two modes shown in Fig. 17.9c. Here also saturation influences on the value of L are not considered.

These equations give the natural frequencies for the three cases shown in Fig. 17.9. For the core flux the sole higher harmonic that can exist must satisfy the boundary condition that i or e is zero at the open or short-circuited ends of the winding. For the same reason, a core-flux resonance can occur only if at least one end of the winding is open-circuited.

The two lower core-flux natural frequencies do not in general influence traveling surges. Observations by means of oscillograms show that they vary in time with changing harmonic current owing to saturation phenomena, which lead to peaked current and flattened voltage waveforms even for moderate flux densities. This is treated in detail in Ref. 1, Chapter 50. Figure 17.10 shows such an oscillogram of the voltage at the high-tension terminals and the current at the center of the winding of a 100-kV transformer that was rapidly disconnected. The voltage decreases in slow, almost rectangular, form, with a fundamental frequency lower than the operating frequency.

The consideration of eddy currents in the core is another criterion that upon examination shows the small influence of the core flux

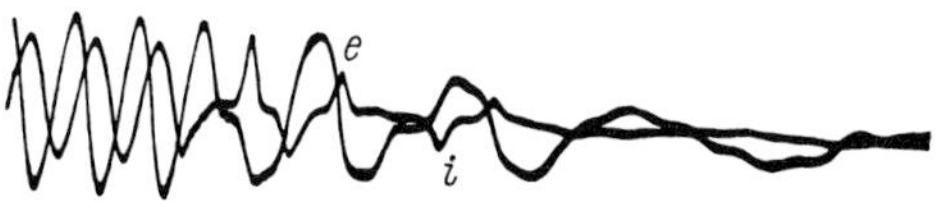

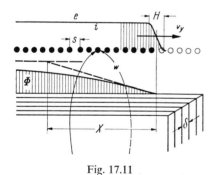

Fig. 17.10

upon rapid traveling phenomena. The time constant for the penetration of the magnetic flux into sheet steels of thickness $\delta = 0.035$ cm, specific resistance $\sigma = 1.5 \cdot 10^4$ cm²/sec and permeability $\mu_\sigma \geq 200$ is

$$T \geq \frac{4}{\pi} \frac{\mu_\sigma}{\sigma} \delta^2 = \frac{4}{\pi} \cdot \frac{200}{1.5 \cdot 10^4} \cdot 0.035^2 = 20 \ \mu s.$$

(17.21)

Fig. 17.11

This was developed in Ref. 1, p. 111, Eq. (24), and the effective permeability μ_σ was derived there on p. 108, Eqs. (8) and (9), in a form given here by Eq. (17.14).

Figure 17.11 shows a winding of turn length w and pitch s around a magnetic core. The axial velocity of propagation of pulses can be only relatively small; for example, if $s/w = 0.001$ and $v = 150$ m/μs, the axial velocity v_y is

$$v_y = \frac{s}{w} v \geq \frac{1}{1000} \cdot 150 \text{ m/}\mu s = 15 \text{ cm/}\mu s.$$

(17.22)

The magnetic flux in the core can develop only slowly after arrival of the head of the pulse and is delayed behind it by a space constant

$$X = v_y T \geq 15 \cdot 20 = 300 \text{ cm}, \quad (17.23)$$

which is shown in Fig. 17.11. This is usually larger than the actual core length and thus shows that from this point of view also the core flux cannot have a significant influence on the effect of a steep pulse front upon the winding. The small remaining effect has already been considered in Eq. (17.14) as a small addition to the air-space flux.

(d) *Rotating-machinery windings.* The windings of rotating machinery do not have a simple symmetric structure like the transformer windings considered previously. Instead they contain elements such as single turns, coils, and layers that are interconnected to form a winding. The detailed specific effects in such nonuniformly structured windings will be treated in Chapter 21. The different structure, however, still permits the propagation of pulses in the windings. We may even use Eqs. (17.6) and (17.8) to derive the characteristic values of v and Z on the basis of the partial inductance L and capacitance C of each of the elements composing the winding. Thus let us first develop these values for a number of turns in a slot and then extend this to complete windings.

Figure 17.12 shows the usual way in which adjacent turns are wound in the slots of generators or motors. The conductor has most of its capacitance to ground in the narrow core slots, of length k, and most of its self-inductance in the overhang portion, of length s. Of course the slot portion also has some self-inductance, but this is much less than its usual stray-field inductance because the rapidly changing stray-field flux lines due to traveling pulses cannot penetrate the boundary steel or copper surfaces and are thus confined to the narrow insulating space around the conductors in the slot. On the other hand, the overhang portions also have some capacitance to ground but here the usually large distance to ground again makes this a small effect. For the sake of simplicity in our initial review, we shall at first neglect these small additional effects.

Then for the capacitance C we can use the simple relation for a parallel-plate capacitor,

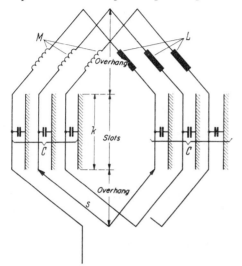

Fig. 17.12

using b_i as the thickness of the insulation in the slot, with the dielectric constant ε, u as the mean circumference of the insulation, and k as the slot length of the active magnetic core, as shown in Fig. 17.13. Because the

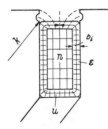

Fig. 17.13

capacitance is determined by the quotient of area and separation, we obtain for one slot

$$C = \frac{\varepsilon}{4\pi v_0^2} \frac{ku}{b_i} \ (\cdot 10^9 \text{ in farads}). \quad (17.24)$$

Here we have considered the slot closed by a conducting insert (not shown in Fig. 17.13) to avoid a more complex calculation due to

stray-field considerations; some machines actually use such inserts.

While this slot capacitance is not influenced by any surrounding slots, the self-inductance of the overhang portion cannot be isolated in this simple way. Instead we must deduce the specific self-inductance of a coil by considering the total magnetic field surrounding the overhang conductors. This includes then the mutual inductance M between conductors, as shown in Fig. 17.12.

Let A be the current density along the active magnetic surface due to the current I in the n conductors per slot with slot spacing ϑ; then

$$A = \frac{nI}{\vartheta}, \quad (17.25)$$

and this produces a magnetic-flux density B_z perpendicular to the twofold current grid shown in Fig. 17.14, where

$$B_z = 2\pi\mu \cdot \frac{s}{\lambda} A \cdot \cos\frac{\pi x}{\tau} \cdot \cos\frac{\pi y}{\lambda}. \quad (17.26)$$

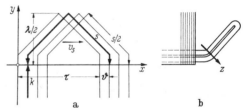

Fig. 17.14

Here x, y, and z are the space coordinates in the transformed overhang space as shown in Fig. 17.14, τ is the distance between centers of poles, λ is the axially projected extension of the overhang on both sides of the core, s is the total overhang length of a half turn, and μ is the permeability of the overhang space, which is introduced merely for the sake of rigor, as was done earlier in the case of transformer windings. This field has the form of a cosine-shaped hill, as shown in Fig. 17.15, with the peak at $x = y = 0$, which decreases to zero amplitude along the x-axis over a distance $\tau/2$ and along the y-axis over a distance $\lambda/2$. In the direction of the z-axis the field decreases exponentially; however, this does not enter our considerations here.

This overhang stray field progresses along the circumference of the core x with a velocity

$$v_s = 2\tau f = \frac{\omega\tau}{\pi}, \qquad (17.27)$$

if the winding is excited by a multiphase current of frequency f. In the y-direction, perpendicular to this velocity, voltage is induced in the overhang conductors over a length $\lambda/2$ perpendicular to v_s on both sides of the core, that is, over a total length λ. Thus the induced voltage in the n overhang conductors in series per slot is

$$E_s = n\lambda v_s \frac{B_z}{2}. \qquad (17.28)$$

Because, as shown in Fig. 17.15, B_z decreases over the overhang region from its

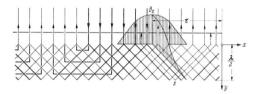

Fig. 17.15

peak value at $x = y = 0$ in both the x- and y-directions, it decreases, by Eq. (17.26), as the square of a cosine function, the mean of which is $\frac{1}{2}$, and this is included in Eq. (17.28).

If we now substitute the peak value from Eq. (17.26) and then introduce Eqs. (17.25) and (17.27), we obtain

$$E_s = \pi\mu n s v_s A = \mu n^2 \omega \frac{s\tau}{\vartheta} I. \quad (17.29)$$

This voltage, however, is also given by the self-inductance due to the overhang stray field,

$$E_s = \omega L I, \qquad (17.30)$$

and comparison with Eq. (17.29) leads to the overhang stray-field self-inductance

$$L = \mu n^2 \frac{\tau}{\vartheta} s \, (\cdot \, 10^{-9} \text{ in henrys}). \quad (17.31)$$

This very simple expression gives the contribu-

tion to the self-inductance L due to a slot-conductor bundle of overhang length s, as shown in Figs. 17.14 and 17.15. It can be used to develop the total overhang self-inductance of a field winding.

Because the self-inductance given by Eq. (17.31) includes all mutual effects between conductors in the overhang and because Eq. (17.24) fully represents capacitive effects, as discussed earlier, the propagation velocity and surge impedance can be directly derived from these equations for a half-turn length equal to $k + s$ of a conductor bunch.

Thus we obtain for the surge impedance

$$Z = \sqrt{\frac{L}{C}} = \sqrt{4\pi \frac{\mu}{\varepsilon}} \, v_0 \cdot n \sqrt{\frac{\tau}{\vartheta} \frac{b_i}{u} \frac{s}{k}}$$
$$(\cdot \, 10^{-9} \text{ in ohms}), \quad (17.32)$$

and we see that, in addition to the known universal constants, it is determined also by the ratio of overhang wire length s to core length k, by the ratio of insulation thickness b_i to circumference u, and by the ratio of pole separation τ to slot separation ϑ, which is also the number of slots per pole. All these enter the expression only to the power one-half, while n, the number of conductors per slot, enters directly. In practical units, substituting for the velocity of light in vacuum v_0, we obtain

$$Z = 30n \sqrt{4\pi \frac{\mu}{\varepsilon}} \sqrt{\frac{\tau}{\vartheta} \frac{b_i}{u} \frac{s}{k}} \, \Omega. \quad (17.33)$$

As an example, take a high-voltage motor with $n = 20$ conductors per slot, $\mu = 1$, $\varepsilon = 5$ (see Table 17.1), $\tau/\vartheta = 9$ slots per pole, slot insulation of thickness $b_i = 3$ mm and circumference $u = 8$ cm, and $s/k = 1.5$ as ratio of overhang to core length. Then

$$Z = 30 \cdot 20 \sqrt{\frac{4\pi}{5}} \sqrt{9 \cdot \frac{0.3}{8} \cdot 1.5} = 677 \, \Omega,$$

which agrees with sporadic measurements on machines in this category. Depending on the number of conductors and slots and on the interconnection, values of $Z = 30$ to $3000 \, \Omega$ may be encountered. Thus, depending on circumstances, the surge impedance of machinery windings may be larger than, equal to, or smaller than that of the connecting lines or cables. This can lead to reflections of incident surges that are useful or dangerous with respect to their effect on the windings.

On the other hand, the velocity of propagation in the conductors of each element of total length $n(k + s)$ is

$$v = \frac{n(k + s)}{\sqrt{LC}} = \sqrt{\frac{4\pi}{\varepsilon\mu}} \cdot v_0 \sqrt{\frac{(k + s)^2 \, \vartheta \, b_i}{ks \quad \tau \; u}}.$$
(17.34)

This contains the same constants and parameters in a different relation. However, n, the number of conductors per slot, has no influence. We can rewrite this expression as a ratio to v_0, the velocity of light in a vacuum:

$$\frac{v}{v_0} = \sqrt{\frac{4\pi}{\varepsilon\mu}} \left(\sqrt{\frac{k}{s}} + \sqrt{\frac{s}{k}} \right) \sqrt{\frac{\vartheta \, b_i}{\tau \; u}}, \quad (17.35)$$

and we see that if k and s are not too different the central term in parentheses will always be close to 2.

A large turbogenerator had $n = 2$ conductors per slot, an overhang-to-core ratio $s/k = 1$, $\tau/\vartheta = 24$ slots per pole, insulation of thickness $b_i = 8.25$ mm and circumference $u = 20$ cm consisting of micanite tubes with $\varepsilon = 5$. For $\mu = 1$ the calculated velocity of propagation was

$$v = 300 \sqrt{\frac{4\pi}{5}} (1 + 1) \sqrt{\frac{1}{24} \frac{0.825}{20}}$$

$$= 39.6 \text{ m}/\mu\text{s},$$

while a measurement gave 37.5 m/μs, which is in good agreement.

We see that this velocity is not only much less than that of 300 m/μs in the air around the overhang conductors, but even less than that of about 130 m/μs which would prevail in the slot insulating material. This is because the effective velocity in a machinery winding is determined by the reciprocal of the product of the large overhang self-inductance and the large slot capacitance. This combination of two large factors reduces the velocity of propagation very considerably. Because the factors of Eq. (17.35) do not vary very much between even quite different machines, experimental results usually give values for v of 5–15 percent of the velocity v_0 of light in a vacuum.

We can deduce a correction factor for the self-inductance L' of the conductors in the slot and the capacitance C' of the conductors to ground in the overhang, which we have so far neglected. We note that in each of the spaces concerned the lines of force of the magnetic and electric fields are orthogonal. This orthogonality of the fields leads to mutual relations between the neglected values C' and L' and the previously calculated values of L and C,

$$C' = \frac{s^2}{v_s^2 L} \quad \text{and} \quad L' = \frac{k^2}{v_k^2 C}, \quad (17.36)$$

where v_s is the velocity of propagation in the overhang and v_k that in the slots. If we add these terms to Eqs. (17.32) and (17.34), we obtain a very small increase in Z and decrease in v.

Because a rotating machine always contains a secondary winding in the rotor that is either completely or partially short-circuited, a core flux cannot arise unless these secondary windings are open-circuited for a measurement. However, if measurements are made at much lower frequencies than those obtained by dividing the velocity of propagation given by Eq. (17.35) by the length of winding, as in Chapter 8, we must be aware that the effect of close steel surfaces, for example in the slots, can raise the self-inductance to the value of stray inductance that governs the normal operating calculations and behavior of such machines.

CHAPTER 18

ANALYSIS OF PULSE PROPAGATION IN TRANSFORMER WINDINGS BY MAXWELL'S EQUATIONS

In Chapter 17 good agreement was found between the analytic results, based on an arrangement of lumped circuit subelements, and the experimental results, particularly for the numerical values of surge impedance and velocity of propagation. However, the approach used in Chapter 17 does not give us a true physical picture of the propagation of the pulses in the winding and surrounding

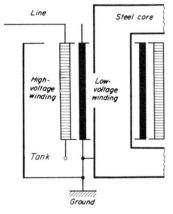

Fig. 18.1

structures. We shall therefore here develop a rigorous solution, using Maxwell's equations, for the propagation of pulses in helical conductors such as transformer windings. This will serve to explain several interesting details of the propagation process. Some assumptions are necessary to obtain tractable boundary conditions. These are, however, in good agreement with many actual transformers.

(a) *Winding geometry.* Figure 18.1 shows the cross section of a usual transformer with concentric primary and secondary windings, the resistances of which may be neglected. Let us examine the primary or high-voltage winding, which is exposed to pulses impinging

from the incoming supply line and is constructed as a long single-layer coil. As in Chapter 17, the windings are disposed between concentric cylindrical electrodes formed by the steel core and the enclosing tank. Even the secondary winding can be regarded as an inner cylindrical electrode if it is grounded and short-circuited for rapid pulses by the low external load impedance.

Thus the rapid pulses cannot produce a magnetic flux that penetrates into the core, being prevented from doing so either by the conductive cladding or surface of the core or by the conductive shield formed by the secondary winding.

Further, let us assume that the radial thickness of the uniformly wound primary coil and also, but to a lesser extent, the insulating distance to the inner electrode or secondary winding are small compared with the diameter of the coil. If the interspace is relatively small, as it is in most actual designs, the problem becomes one that can be expressed in rectangular instead of cylindrical coordinates, which allows simpler analytical treatment. Finally, we assume that the winding is long compared with the insulating distances.

All these assumptions have made the capacitance between adjacent turns of the winding negligible. The effects of this self-capacitance of the winding will be examined in the next chapter. All other external and mutual effects between windings, capacitive as well as inductive, and between neighboring as well as widely separated portions, are fully covered by the analysis to be developed.

The coil in Fig. 18.2a is shown cut longitudinally and flattened out into the xy-plane in Fig. 18.2b. Figure 18.2c shows the insulating spaces around the winding, which now consists of a linear sloping grid, as well as the coordinate system and field-vector directions.

The *x*-coordinate is in the direction of the former coil circumference *u*, the *y*-coordinate along the axis of the coil, and the *z*-coordinate perpendicular to the thin grid of windings in which currents flow in parallel. The separation

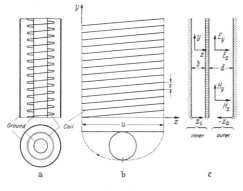

Fig. 18.2

between adjacent wires is *s*, which is also the pitch of the sloping grid.

Because the two ends of the coil as split in Fig. 18.2*b* are identical, the electromagnetic-field conditions are identical at these two

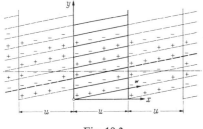

Fig. 18.3

boundaries, and we may repeat the presentation of the developed coil periodically in the +*x* and −*x* directions, as shown by dashed lines in Fig. 18.3. Thus we cover the entire *xy*-plane with a grid of close parallel sloping conductors. We can now solve our problem mathematically by a statement valid for any values of *x* and *y* within the infinite plane.

In Fig. 18.3 let us represent a pulse of arbitrary shape by a sequence of + and − signs along the helical wire developed in the central *x*-axis sector, which is shown by solid lines since it represents the actual coil. We can also enter this pulse shape in the dotted

periodic extensions as shown. In this case we see that the pulse shape can be developed along a single sloping conductor in the *xy*-plane. Further we see by inspection that, given a value of *y*, the pulse shape is the same at any of the periodic wire intersections representing a series of values of *x*, for example along the dot-dash horizontal line in Fig. 18.3. Only when the value of *y* is changed does the pulse shape at the intersections vary. This periodic symmetry along a line of constant *y* will simplify our analysis.

In the interspace between the inclined conductors there may, of course, be some change of the electromagnetic state in the *x*-direction, but certainly only a small one. We see this immediately if we either think of our conductors as being wide enough to fill the entire *xy*-plane or think of the space between the thin wires as being filled up by interlined wires so as to make a multiple winding.

Thus for our mathematical development we shall consider a coil consisting of a thin layer that is perfectly conducting in the direction of the wires and perfectly insulating perpendicular to the wires. With this idealization, we can solve rigorously the problem of electric oscillations in such a coil. We neglect here only the change of electromagnetic field in the immediate neighborhood of each single wire, which is the same as neglecting the capacitance between adjacent turns. However, by our assumption of constancy in the *x*-direction we do not restrict any variation of the electromagnetic field in the *y*-direction, and therefore also in the direction of the inclined wires. Hence we include in our development even those oscillations that may be shorter in space than a single turn.

(*b*) *Fundamental equations.* If the electromagnetic field varies only to a negligible extent in the *x*-direction, it is even more reasonable to assume that it remains constant at some distance from the wire grid. Thus with the condition

$$\frac{\partial}{\partial x} = 0 \qquad (18.1)$$

for every field component within the entire space, we have the following set of Maxwell equations in a rectangular coordinate system defined in Figs. 18.2 and 18.3 and in electromagnetic units: the variation of the electric intensity E with time is

$$\frac{\varepsilon}{v_0^2} \frac{\partial E}{\partial t} = \text{curl } H \qquad (18.2a)$$

or

$$\frac{\varepsilon}{v_0^2} \frac{\partial E_x}{\partial t} = \frac{\partial H_z}{\partial y} - \frac{\partial H_y}{\partial z},$$

$$\frac{\varepsilon}{v_0^2} \frac{\partial E_y}{\partial t} = \frac{\partial H_x}{\partial z}, \qquad (18.2)$$

$$\frac{\varepsilon}{v_0^2} \frac{\partial E_z}{\partial t} = -\frac{\partial H_x}{\partial y},$$

while the time variation of magnetic intensity H is

$$-\mu \frac{\partial H}{\partial t} = \text{curl } E, \qquad (18.3a)$$

or

$$-\mu \frac{\partial H_x}{\partial t} = \frac{\partial E_z}{\partial y} - \frac{\partial E_y}{\partial z},$$

$$-\mu \frac{\partial H_y}{\partial t} = \frac{\partial E_x}{\partial z}, \qquad (18.3)$$

$$-\mu \frac{\partial H_z}{\partial t} = -\frac{\partial E_x}{\partial y},$$

where ε and μ are the dielectric constant and magnetic permeability, respectively, within the space considered and v_0 is the velocity of propagation of light in a vacuum. Furthermore, the spatial variation of the magnetic intensity is

$$\text{div } H = 0,$$

or

$$\frac{\partial H_y}{\partial y} + \frac{\partial H_z}{\partial z} = 0, \qquad (18.4)$$

and that of the electric intensity is

$$\text{div } E = 0,$$

or

$$\frac{\partial E_y}{\partial y} + \frac{\partial E_z}{\partial z} = 0. \qquad (18.5)$$

These relations are valid throughout the insulating spaces shown in Fig. 18.2c. At the surfaces of the unidirectionally conducting coil layer, however, the electric intensity must be zero in the conductive direction. Figure 18.4 shows that the electric intensity E_w in the direction of the turn length w, which in terms of the circumference u and the pitch s is

$$w = \sqrt{u^2 + s^2}, \qquad (18.6)$$

disappears only if the components in the x- and y-directions have the ratio

$$\frac{E_x}{E_y} = -\frac{s}{u}. \qquad (18.7)$$

Further, since all currents in Fig. 18.3 flow in the w-direction, they cannot produce a

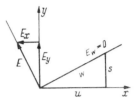

Fig. 18.4

magnetic intensity in that direction. Therefore the same geometric considerations of Fig. 18.4 applied to H give the ratio

$$\frac{H_x}{H_y} = -\frac{s}{u}. \qquad (18.8)$$

These relations also are valid throughout the insulating space considered.

Equations (18.1), (18.7), and (18.8) give the structural relations of the electromagnetic field of the coil and determine an important part of the solution of the Maxwell equations. The boundary conditions at all conducting surfaces enclosing the insulating spaces determine another part and will be considered later.

If by means of Eqs. (18.7) and (18.8) we eliminate the x-components of the electric and magnetic intensities in Eqs. (18.2) and (18.3), we obtain from the first group

$$-\frac{s}{u} \frac{\varepsilon}{v_0^2} \frac{\partial E_y}{\partial t} = \frac{\partial H_z}{\partial y} - \frac{\partial H_y}{\partial z},$$

$$\frac{\varepsilon}{v_0^2} \frac{\partial E_y}{\partial t} = -\frac{s}{u} \frac{\partial H_y}{\partial z}, \qquad (18.9)$$

$$\frac{\varepsilon}{v_0^2} \frac{\partial E_z}{\partial t} = \frac{s}{u} \frac{\partial H_y}{\partial y},$$

and from the second group

$$\frac{s}{u}\mu\frac{\partial H_y}{\partial t} = \frac{\partial E_z}{\partial y} - \frac{\partial E_y}{\partial z},$$

$$-\mu\frac{\partial H_y}{\partial t} = -\frac{s}{u}\frac{\partial E_y}{\partial z}, \quad (18.10)$$

$$-\mu\frac{\partial H_z}{\partial t} = \frac{s}{u}\frac{\partial E_y}{\partial y}.$$

Before proceeding further we note from Eqs. (18.7) and (18.8) that the x-components, corresponding to the field in the circumferential direction of the coil, are very weak if the pitch s is only a small fraction of the circumference u, as is true for most actual windings. Furthermore, the second and third of Eqs. (18.9) and (18.10) show that the time variations of both the electric and the magnetic fields in the y- and z-directions are very small compared with the variations of the field components in space, determined by the ratio s/u. This means that the field in the y- and z-directions is almost static, although rapid time variations in the x-direction may occur. The smaller the pitch or the greater the number of turns per unit of coil length, the more closely the field approaches the static state.

(c) *Solution for the field components.* We eliminate $\partial/\partial z$ from the first two of Eqs. (18.9) and (18.10) to obtain

$$\frac{\varepsilon}{v_0^2}\left(\frac{u}{s} + \frac{s}{u}\right)\frac{\partial E_y}{\partial t} = -\frac{\partial H_z}{\partial y},$$

$$\frac{\varepsilon}{v_0^2}\left(\frac{u}{s} + \frac{s}{u}\right)\frac{\partial H_y}{\partial t} = \frac{\partial E_z}{\partial y}. \quad (18.11)$$

If we combine the first of Eqs. (18.11) with the last of Eqs. (18.10), we obtain

$$\frac{\varepsilon\mu}{v_0^2}\left(1 + \frac{u^2}{s^2}\right)\frac{\partial^2 E_y}{\partial t^2} = \frac{\partial^2 E_y}{\partial y^2}, \quad (18.12)$$

and exactly the same relation can be derived for H_z. If we combine the second of Eqs. (18.11) with the last of Eqs. (18.9) we find that E_z and H_y also are given in the form of Eq. (18.12).

This well-known wave equation (18.12) can be solved by an arbitrary function of the form

$$E_y, E_z, H_y, H_z = f(y \pm v_y t), \quad (18.13)$$

which shows that the entire electromagnetic field is propagated undistorted in the y-direction, that is, along the axis of the coil, with a velocity given in terms of the constant coefficient in the left-hand member of Eq. (18.12) by

$$v_y = \frac{v_0}{\sqrt{\varepsilon\mu(1 + u^2/s^2)}} = \frac{s}{w}\frac{v_0}{\sqrt{\varepsilon\mu}}. \quad (18.14)$$

Since this axial velocity depends only on the fundamental electromagnetic velocity $v_0/\sqrt{\varepsilon\mu}$ and on the relative pitch or slope of the turns, s/w, we see that it is a natural velocity inherent in every coil, dependent only on the slope and independent of all other dimensions of the coil. It is a small fraction of the velocity of light, in conformity with the fact that the change of the phenomena in the axial direction is nearly static. The velocity is independent of the variation of the field in time or of the exciting frequency. Thus pulses of any shape, including those with very steep fronts, are propagated with the same axial velocity along the coil.

For coils of small pitch, the velocity may be very low. However, if the pitch s approaches the turn length w, the velocity of propagation approaches that of light in the insulating medium. For stranded cables with finite pitch, which constitute the limiting case of a multiconductor coil, the velocity is still some-

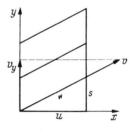

Fig. 18.5

what less than that of light. The velocity of light is fully attained only in solid wires.

We can now also determine the phase velocity of the pulses along the wires. Figure 18.5 shows that this is larger in the ratio w/s and thus

$$v = \frac{w}{s}v_y = \frac{v_0}{\sqrt{\varepsilon\mu}}. \quad (18.15)$$

Thus every electromagnetically excited component travels along the wire itself with the velocity of light pertaining to the insulating space surrounding the wire. The structure of the Maxwell equations shows that this results from the fact that the electric and magnetic intensities vanish in the direction of the wires and that the lines of force are therefore perpendicular to that direction. Since $\varepsilon\mu$ is about 4 for oil-filled transformers with conductors embedded in solid insulation, the velocity of propagation along the wire is about one-half of that in a vacuum. Thus the natural frequencies, fundamental as well as higher harmonic, can be determined easily for any given length of wire. The experimental results given in Chapter 17 are in good agreement with this.

As we now have expressed the dependence of the field on the x-, y-, and t-coordinates in a form of general validity, it remains to determine the variation in the z-direction. We eliminate the time variation $\partial/\partial t$ from the first two of Eqs. (18.9) and (18.10) to obtain

$$\left(1 + \frac{s^2}{u^2}\right)\frac{\partial H_y}{\partial z} - \frac{\partial H_z}{\partial y} = 0,$$

$$\left(1 + \frac{s^2}{u^2}\right)\frac{\partial E_y}{\partial z} - \frac{\partial E_z}{\partial y} = 0. \qquad (18.16)$$

If we combine these relations with the second of Eqs. (18.4) and (18.5), we obtain for the variation of the component H_y in z and y

$$\left(\frac{w}{u}\right)^2 \frac{\partial^2 H_y}{\partial z^2} + \frac{\partial^2 H_y}{\partial y^2} = 0 \qquad (18.17)$$

and exactly the same equation for each of the other three field components, H_z, E_y, and E_z.

Equation (18.17) is similar to the well-known potential equation and thus we can use any suitable solution given by potential theory, except that the z-dimension here is extended by the ratio of the turn length w to the circumference u of the coil. For tightly wound coils this ratio is very close to 1; for loosely wound coils it is considerably larger, and it increases to infinity for the limiting case of a straight conductor.

A simple solution of Eq. (18.17) has the form of sinusoidal variations along the y-axis of the coil. This imposes an exponential distribution in the z-direction perpendicular to the coil surface. We see at once that the magnetic intensities

$$H_y = H_1 \cosh \beta z \sin \alpha\,(y \pm v_y t),$$
$$H_z = H_2 \sinh \beta z \cos \alpha\,(y \pm v_y t) \qquad (18.18)$$

and the electric intensities

$$E_y = E_1 \sinh \beta z \cos \alpha\,(y \pm v_y t),$$
$$E_z = E_2 \cosh \beta z \sin \alpha\,(y \pm v_y t) \qquad (18.19)$$

satisfy the differential equation (18.17). One such group of field components holds within the internal insulating space of the coil, and another one within the external insulating space. Let us distinguish the two groups by the subscripts i and a respectively, as shown in Fig. 18.2c.

We have expressed Eqs. (18.18) and (18.19) in terms of hyperbolic functions, rather than exponential functions, in order to make the equations symmetric in $+z$ and $-z$. Moreover, in this manner we can immediately satisfy the boundary conditions $H_z = E_y = 0$ at the perfectly conducting ground electrodes in Fig. 18.2c, if we fix the two origins $z = 0$ for the internal and the external spaces at the respective positions of these electrodes, as shown.

By inserting the field components, Eqs. (18.18) and (18.19), in Eq. (18.17), we obtain an equation between the attenuation constant β in the z-direction and the wave density α in the y-direction,

$$\beta = \frac{u}{w}\alpha, \qquad (18.20)$$

which holds for both the internal and the external insulating space.

The field intensities of Eqs. (18.18) and (18.19) vary in time, with an angular frequency given by the coefficient of t,

$$\omega = \alpha v_y. \qquad (18.21)$$

If we excite the winding with this frequency ω, the wave density in the y-direction becomes, with the use of Eqs. (18.14) and (18.15),

$$\alpha = \frac{\omega}{v_y} = \omega\frac{w}{s}\frac{\sqrt{\varepsilon\mu}}{v_0} = \frac{w}{s}\frac{\omega}{v}. \qquad (18.22)$$

From this we can easily determine the wavelength of the field intensities along the y-axis:

$$\lambda = \frac{2\pi}{\alpha} = \frac{2\pi}{\omega}\frac{s}{w}v. \qquad (18.23)$$

Along the wire of the coil the wavelength is w/s times greater, namely

$$\Lambda = \frac{2\pi}{\omega}v. \qquad (18.24)$$

These wavelengths are given, as usual, by the quotient of velocity and frequency.

The spatial attenuation of Eq. (18.20) transverse to the wires now becomes, from Eqs. (18.22) and (18.23),

$$\beta = \frac{u}{s}\frac{\omega}{v} = \frac{u}{w}\frac{2\pi}{\lambda}. \qquad (18.25)$$

Thus we see that the extension of the field in the z-direction, given by $1/\beta$, is directly proportional to the relative pitch s/u of the coil, and inversely proportional to the exciting frequency ω. Thus for high frequency and small slope of the turns the electromagnetic field concentrates heavily in the close neighborhood of the coil.

If, for example, as in many power transformers, the slope of the winding $s/u = 0.001$ and the velocity of propagation of the pulses $v = 1.5 \cdot 10^{10}$ cm/sec, the attenuation space constant with an exciting frequency of 10^4 cy/sec is

$$\frac{1}{\beta} = \frac{1}{1000}\frac{1.5 \cdot 10^{10}}{2\pi \cdot 10^4} = 240 \text{ cm};$$

for 10^5 cy/sec this reduces to 24 cm and for 10^6 cy/sec to 2.4 cm.

Figure 18.6 shows the general configuration of the electric and magnetic lines of force according to Eqs. (18.18) and (18.19). In addition to these components in the y- and z-directions, there exist also circumferential components in the x-direction, given by Eqs. (18.7) and (18.8); these are, however, negligible as long as the pitch s is only a small fraction of the circumference u.

The dashed sinusoidal curve in the lower portion of Fig. 18.6 indicates at which points along the winding the voltage to ground is a maximum or zero. If the interaction of pulse trains produces standing waves, then at such points the winding may be opened or shorted to ground, respectively. This allows us to visualize the voltage and field distribution in

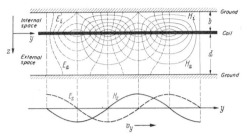

Fig. 18.6

coils that have open- or short-circuited terminations and an arbitrary multiple of quarter wavelengths.

To derive the constants H and E in Eqs. (18.18) and (18.19), we insert the field intensities in Eq. (18.16). Then we obtain for the magnetic amplitudes, using Eqs. (18.6) and (18.20),

$$H_2 = -\frac{w}{u}H_1, \qquad (18.26)$$

and for the electric amplitudes,

$$E_2 = \frac{w}{u}E_1. \qquad (18.27)$$

Furthermore, an interdependence of the magnetic and electric amplitudes can be derived by inserting the field components in Eqs. (18.9) and (18.10). If we take E_z from Eq. (18.19) and H_y from Eq. (18.18) and introduce them into the last of Eqs. (18.9), we obtain

$$\frac{\varepsilon}{v_0^2}v_yE_2 = \frac{s}{u}H_1, \qquad (18.28)$$

and if then we use Eq. (18.15) this simplifies to

$$H_1 = \frac{1}{v_0}\frac{u}{w}\sqrt{\frac{\varepsilon}{\mu}}E_2. \qquad (18.29)$$

All these relations refer to the internal as well as to the external insulating space.

(d) Voltage and current in the winding. To establish continuity of the field solutions obtained with the voltage and current in the winding, we must consider the boundary conditions at the surface of the coil. The line integral of the electric field between coil and ground electrodes in Fig. 18.2c gives the voltage E at the point considered. Thus we have over the internal distance b, using the second of Eqs. (18.19) and confining ourselves to the amplitude of the spatial sine wave,

$$E = \int_0^b E_z \, d_z = \frac{E_2}{\beta} \sinh \beta b. \quad (18.30)$$

Since a similar expression yields the voltage E for the external distance d, only with opposite sign, the amplitudes in both spaces are determined by

$$E_{2i} = \frac{\beta}{\sinh \beta b} E,$$
$$E_{2a} = \frac{-\beta}{\sinh \beta d} E. \quad (18.31)$$

These maximum electric intensities can be easily computed, with β as given by Eq. (18.25), if the amplitude E and frequency ω of the impressed voltage are given.

The continuity conditions for the magnetic intensities across the boundary are that the components H_z, perpendicular to the coil surface, must be equal, while the difference of the axial components H_y is determined by the current within the layer. From the first condition, using the last of Eqs. (18.18), we obtain for $z_i = b$ and $-z_a = d$

$$H_{2i} \sinh \beta b = -H_{2a} \sinh \beta d = H_0, \quad (18.32)$$

where H_0 is merely an abbreviation for these maximum values. From the second condition, using the first of Eqs. (18.18), we obtain

$$H_{1i} \cosh \beta b - H_{1a} \cosh \beta d = 4\pi A_x, \quad (18.33)$$

where A_x is the amplitude of the linear current density in the circumferential direction of the coil, measured perpendicular to the field intensities H_y. Since the real current I within the turns flows at an angle to the x-direction, we see from Fig. 18.7 that the x-component of the linear current density A is

$$A_x = \frac{u}{w} A = \frac{u}{w} \frac{NI}{k} = \frac{NI}{h}, \quad (18.34)$$

where N is the number of turns in a coil of axial length h and k is the projection of this length perpendicular to the wires.

Considering Eq. (18.32), we can express both amplitudes H_2 by the maximum value H_0 of the transverse magnetic field through the

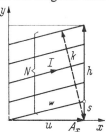

Fig. 18.7

coil layer, and using Eq. (18.26) obtain the components H_1 as

$$H_{1i} = -\frac{u}{w} \frac{H_0}{\sinh \beta b},$$
$$H_{1a} = \frac{u}{w} \frac{H_0}{\sinh \beta d}. \quad (18.35)$$

If we introduce these values into the left-hand member of Eq. (18.33) and the value of Eq. (18.34) into its right-hand member, there results

$$4\pi \frac{N}{h} I = -\frac{u}{w} H_0 (\operatorname{ctnh} \beta b + \operatorname{ctnh} \beta d), \quad (18.36)$$

from which H_0 can be computed if the amplitude of the current is I and its frequency ω.

We can now consider the interdependence of magnetic intensity H and electric intensity E, given by Eq. (18.29), in which we insert the values given by Eqs. (18.31) and (18.35). The internal as well as external fields give the same relation,

$$-H_0 = \frac{\beta}{v_0} \sqrt{\frac{\varepsilon}{\mu}} E \quad (18.37)$$

between the maximum magnetic intensity H_0 and the maximum coil voltage E, both in the z-direction and thus perpendicular to the wire layer. Inserting Eq. (18.37) in Eq. (18.36) to eliminate H_0, we obtain a definite interdependence between current I in the turns and voltage E to ground at any point of the coil. The surge impedance Z of the coil, as the ratio of voltage to current, is therefore given by

$$Z = \frac{E}{I} = 4\pi v_0 \sqrt{\frac{\mu}{\varepsilon}} \cdot \frac{w}{u} \frac{N}{h}$$

$$\times \frac{1/\beta}{\operatorname{ctnh} \beta b + \operatorname{ctnh} \beta d} (\cdot 10^{-9} \text{ in ohms}). \quad (18.38)$$

To obtain Z in ohms, the dimensional factor 10^{-9} is introduced as shown.

(e) Surge impedance of transformer windings. In an examination of the surge impedance, as given by Eq. (18.38), we can separate the first group of factors and the last fractional term in the right-hand member. The first group simply consists of the numerical term 4π, the velocity of light in a vacuum v_0, the quotient μ/ε of the insulating space, and a few parameters of the winding. These are the ratio w/u of turn length to circumference, which for transformers is always very close to 1, and the number of turns N per length h of the coil, which is a characteristic of the winding that varies from about 100 to 1 turn/cm for small-size to large-size power transformers respectively. Thus into this group of factors there enter only very few data of the transformer, namely, the electromagnetic constants of the insulating space and the number of turns per unit axial length.

The second part of Eq. (18.38), that is, the fraction

$$g = \frac{1/\beta}{\operatorname{ctnh} \beta b + \operatorname{ctnh} \beta d}, \quad (18.39)$$

determines the equivalent ground distance of the winding. We see this immediately if at first small values of the arguments βb and βd are specified, which means either relatively slow oscillations, by Eq. (18.25), or small distances to ground. Then the hyperbolic cotangents are equal to the reciprocals of their arguments, and hence

$$g_0 = \frac{1/\beta}{1/\beta b + 1/\beta d} = \frac{1}{1/b + 1/d}. \quad (18.40)$$

In this case the equivalent ground distance g_0 is given by the actual interior and exterior ground distances in parallel, and thus the surge impedance for low frequency or for narrow insulating spaces becomes

$$Z_0 = 4\pi v_0 \sqrt{\frac{\mu}{\varepsilon}} \cdot \frac{N}{h} \frac{b}{1 + b/d} \cdot 10^{-9} \; \Omega. \quad (18.41)$$

We see that this value is determined by only two more numerical data of the transformer than those mentioned above, namely the insulating distances b and d, and that in the range of small βb and βd the surge impedance is independent of the frequency and of the many other design parameters of the transformer.

If only one insulating distance, perhaps the internal one b, is small and the external one d is large, measured in terms of βb and βd as mentioned above, $\operatorname{ctnh} \beta b$ approaches $1/\beta b$ and $\operatorname{ctnh} \beta d$ approaches 1, so that the equivalent ground distance becomes

$$g = \frac{1/\beta}{(1/\beta b) + 1} = \frac{b}{1 + \beta b} \cong b. \quad (18.42)$$

Thus the influence of one remote ground electrode disappears, as already indicated in Eq. (18.40).

We should note that for low frequencies the approximation, Eq. (18.41), here given by a rigorous treatment of distributed parameters, is exactly the same as Eq. (17.8), which was derived by the use of elementary concepts of lumped capacitance and inductance. We note also that the velocity of propagation along the wire is in both treatments equal to the velocity of propagation of light in the insulating spaces. However, the more rigorous treatment in this chapter shows that only the velocity of propagation of the pulses is independent of frequency, while the surge impedance decreases for higher frequencies. This is caused by a gradual contraction of the field lines shown in Fig. 18.6 with increasing frequency, which in turn leads effectively to a lower self-inductance and increased capacitance between neighboring winding elements.

Note also that in Chapter 17 experimental results were found to give good agreement with the analytic approach used.

For relatively high frequencies and large values of β, or for large distances b and d, the values of the hyperbolic cotangents in

Eq. (18.39) approach 1 and thus we obtain as equivalent ground distance the asymptotic value

$$g_\infty = \frac{1}{2\beta} = \frac{s}{u}\frac{v}{2\omega}. \qquad (18.43)$$

In this range the surge impedance, by Eq. (18.15) and since $s = h/N$, is thus

$$Z_\infty = \frac{2\pi}{\omega} v_0^2 \sqrt{\frac{\mu}{\varepsilon}}\frac{w}{u^2}. \qquad (18.44)$$

It varies inversely as the frequency, and for tightly wound coils ($w \simeq u$) also inversely as the circumference of the coil.

If we draw a curve for the equivalent ground distance g, dependent on β, we obtain a shape

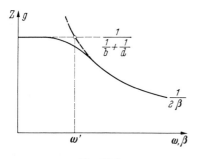

Fig. 18.8

like that shown in Fig. 18.8, which also shows the dependence of the surge impedance Z on frequency ω. The lower and the upper asymptotic ranges are separated by a boundary frequency ω', which is determined by the intersection of the asymptotes of the two ranges just mentioned. The boundary frequency, therefore, according to Eqs. (18.40) and (18.43), is

$$\omega' = \frac{s}{u}\frac{v}{2}\left(\frac{1}{b} + \frac{1}{d}\right). \qquad (18.45)$$

For a power transformer of average size let us take $v = 150$ m/μs, $s/u = 0.001$, $b = 2$ cm, and $d = 20$ cm. Then we obtain

$$\omega' = \frac{1}{1000}\frac{1.5 \cdot 10^{10}}{2}\left(\frac{1}{2} + \frac{1}{20}\right)$$
$$= 4.12 \cdot 10^5 \text{ cy/sec.}$$

This is much higher than the fundamental natural frequency of such a transformer,

which is of the order of 10^4 cy/sec. The surge impedance of transformers of usual design thus remains constant for frequencies of the order of the natural frequency up to a first group of higher harmonics and decreases only for frequencies of the order of very high harmonics.

Another way of expressing the boundary between the two ranges in Fig. 18.8 is to compute the corresponding wavelength λ'. By Eq. (18.23) this boundary wavelength is

$$\lambda' = 4\pi\frac{u}{w}\frac{b}{1 + b/d}. \qquad (18.46)$$

Thus we see that the surge impedance decreases with frequency only for waves that are shorter in the axial direction than 4π times the resultant insulating distance. For $b = 2$ cm and $d = 20$ cm,

$$\lambda' = 4\pi \cdot 1 \cdot \frac{2}{1.1} = 22.8 \text{ cm,}$$

which is small as compared with a coil length h of about 100 cm for such a transformer, which would have natural oscillations of wavelength λ about 200 or 400 cm in axial length, depending on the end conditions. In this example the boundary between the two ranges of constant and hyperbolically decreasing surge impedance is near the 9th or 17th harmonic, respectively. For most transformers, with coils that are long compared with the insulating distance, the use of Eq. (18.41) for the lower frequency range is therefore sufficiently accurate.

However, if we remove the ground electrodes on both sides of the coil to greater distances, the boundary frequency decreases and therefore the upper frequency range approximation of the surge impedance in Eq. (18.44) extends to actual lower frequencies. The reason for the variation of surge impedance with frequency now occurring is, as seen in Fig. 18.6, that not all the electric lines of force end at the ground electrode, but that many of them pass directly between the sections, with alternating polarity of voltage and current along the winding. Thus we have here solved rigorously the problem of the mutual influence of neighboring parts of the winding in the case of harmonic waves of voltage and current.

We realize that this effect has no influence on the velocity of propagation, which, according to Eqs. (18.14) and (18.15), is independent of the frequency.

There still remains a discussion of the factor w/u in the surge impedance, Eq. (18.38), which is the ratio of turn length to circumference. For tightly wound coils w/u is always near 1, but for loosely wound coils we can express it, together with the turn density N/h, and using Eq. (18.6), as

$$\frac{w}{u}\frac{N}{h} = \frac{w}{su} = \sqrt{\frac{1}{u^2} + \frac{1}{s^2}}. \qquad (18.47)$$

Thus both pitch and circumference of the coil influence the surge impedance. The deviation from the tight-coil formula becomes greatest when we consider a stranded concentric cable in which the wires are wound in helices. Since such cables always have lengths that are very large as compared with the insulating distances, the boundary wavelength λ' is comparatively very small, and so for the lower range of frequencies the surge impedance becomes, by Eq. (18.40),

$$Z = 4\pi v_0 \sqrt{\frac{\mu}{\varepsilon}} \cdot \frac{\sqrt{1/u^2 + 1/s^2}}{1/b + 1/d} \cdot 10^{-9}\,\Omega. \qquad (18.48)$$

This expression agrees with the usual relation for cables with small insulating distances, except for the influence of the twist of the conductors, which is accurately considered here.

If we gradually stretch out a helical winding in the axial direction, the ratio of turn length w to circumference u increases in the limit to infinity. Since Eq. (18.17) has shown that the z-dimension of any potential solution must then be lengthened increasingly up to infinity, we see from Fig. 18.6 that the lines of force straighten out radially more and more, until for a linear conductor they become straight lines perpendicular to the axis. This gives the transition to the well-known picture of lines of force for waves on smooth wires, where the electric and magnetic forces lie strictly in radial planes.

For the exact treatment of windings of small diameter or helical conductors far removed from ground electrodes, the Maxwell equations should be formulated in cylindrical coordinates. The solution then contains Bessel function terms and is of a very complex form. However, it provides no novel features of significant value to the problem considered.

For many applications it is useful to derive rigorous expressions for self-inductance l and ground capacitance c of the coil. Since velocity of propagation and surge impedance are, in terms of these two quantities,

$$v = \frac{1}{\sqrt{lc}}, \quad Z = \sqrt{l/c}, \qquad (18.49)$$

we obtain for the self-inductance

$$l = \frac{Z}{v} = 4\pi\mu \frac{w}{u}\frac{N}{h}g, \qquad (18.50)$$

and for the capacitance to ground,

$$c = \frac{1}{Zv} = \frac{\varepsilon}{4\pi v_0^2}\frac{u}{w}\frac{h}{N}\frac{1}{g}, \qquad (18.51)$$

both per centimeter wire length of the coil. For low frequencies and constant equivalent ground distance g_0, as in Eq. (18.40), these two equations are in agreement with those derived in Chapter 17 from elementary considerations of transformer-coil dimensions in lumped form.

For very high frequencies, however, by use of Eq. (18.43), we obtain

$$l = \frac{2\pi}{\omega}\mu v_0 \frac{w}{u^2},$$

$$c = \frac{\omega}{2\pi}\frac{\varepsilon}{v_0^3}\frac{u^2}{w}. \qquad (18.52)$$

The self-inductance now decreases with frequency, owing to the rapid spatial attenuation of the magnetic field perpendicular to the winding area. The capacitance, however, increases with frequency, owing to the diminishing distances between the alternating-polarity sections of voltage along the coil. For this reason, more and more electric lines of force close between adjacent parts of the coil, without touching the ground electrodes, as shown in the external space of Fig. 18.6. Finally, for very high frequencies or for remote ground electrodes, far above the criterion of

Eq. (18.45), the ground plays no part whatever. Thus we have found in Eqs. (18.52) the electromagnetic characteristics for a long single coil, not too small in diameter, situated in empty space without ground.

As an important result of the rigorous theory we have found that, under the stated conditions, tightly wound coils behave, up to a boundary frequency, with respect to pulses traveling along their wires exactly like smooth lines. In the direction of their axis, however, their behavior is nearly static or quasi-stationary. Perpendicular to this axis the spatial field of harmonic oscillations appears exponentially attenuated above this boundary frequency.

CHAPTER 19

LINKAGES IN COIL WINDINGS

Breakdowns of insulation between turns and layers occur in practice far more frequently in the terminal sections of windings than in the main body. One might be tempted to explain this simply on the hypothesis that the incident pulse impinges first on these terminal sections and dissipates some energy when breakdown occurs, and that this dissipation in turn leads to some protection for the main body of the winding. However, experiments with very sharply rising pulses incident on tightly wound coils show that even without any insulation failure the terminal turns or sections are subjected to far higher voltage differences and hence insulation stresses than the main body. None of our previous treatments of windings can explain these phenomena. In these treatments we regarded the windings as either lumped self-inductances or homogeneous lines with distributed self-inductance and capacitance, and even in the rigorous analysis in Chapter 18 by means of Maxwell's equations all portions of a winding were weighted equally. Thus we must take a further step in our treatment and assumptions to obtain analytic insight into these phenomena, which are of great importance to high-voltage technology.

(a) *Propagation of pulses.* Figure 19.1 shows a tightly wound coil consisting of individual pancake turns (or layers), which are parallel and closely spaced. In Chapter 18 we quite rigorously considered the capacitances between distant portions of the winding and between portions of the winding and the iron core and other ground electrodes. Here we see that we must also consider the electric field in the narrow spaces between adjacent pancake layers due to the voltage difference between these layers.

Since steep-front pulses tend to produce large voltage differences along the conductors,

they will give rise to considerable charging or displacement currents in the capacitances between layers. If these capacitances are large owing to the area and close spacing of the pancake structure, they may exceed capacitances to ground and thus give rise to currents

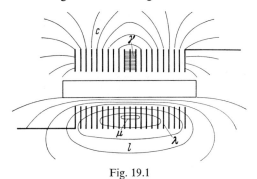

Fig. 19.1

so large that they tend to dominate the behavior of the winding.

A rigorous treatment using Maxwell's equations and including this internal winding capacitance would become exceedingly complex. Thus we shall employ a somewhat simpler treatment, using differential equations for long coils. The end effects of shorter coils can be approximated as in Fig. 17.3.

The charging current for each element of conductor length Δx of the nth turn of the winding, which is due to the external capacitance and thus flows mainly to ground, is

$$i_{cn} = c\Delta x \frac{\partial e_n}{\partial t}, \qquad (19.1)$$

where c is the external capacitance per unit length of conductor. Let γ be the internal capacitance per unit length between adjacent pancake layers. Then the charging current that flows from the nth turn to the $(n-1)$th turn is

$$i'_{\gamma n} = \gamma\Delta x \frac{\partial(e_n - e_{n-1})}{\partial t} = \gamma\Delta x \frac{\partial\Delta' e_n}{\partial t}, \quad (19.2)$$

172

and that to the $(n + 1)$th turn is

$$i''_{\gamma n} = \gamma \Delta x \frac{\partial(e_n - e_{n+1})}{\partial t} = -\gamma \Delta x \frac{\partial \Delta'' e_n}{\partial t}. \tag{19.3}$$

The sum of these two currents,

$$i_{\gamma n} = \gamma \Delta x \frac{\partial}{\partial t} (\Delta'' e_n - \Delta' e_n) = -\gamma \Delta x \frac{\partial \Delta^2 e_n}{\partial t}, \tag{19.4}$$

is the total internal capacitive current per unit conductor length. Here Δe is the voltage difference between adjacent turns and $\Delta^2 e$ is the difference of this difference between successive turns. The voltage difference is thus that due to a length of conductor

$$\Delta x = w \tag{19.5}$$

of one complete turn of length w. Thus the second difference of voltage can be rewritten in differential rather than difference form

$$\Delta^2 e = w^2 \frac{\Delta^2 e}{\Delta x^2} = w^2 \frac{\partial^2 e}{\partial x^2}. \tag{19.6}$$

This is valid if the coil contains many turns and if the detailed distribution of the phenomena along a single turn length are not of interest. The total external and internal capacitive charging current of each turn, as given by Eqs. (19.1) and (19.4), is then equal to the decrease $-\Delta i_n$ in the current in the nth turn. Thus, using Eq. (19.6), we obtain

$$-\Delta i_n = c\Delta x \frac{\partial e_n}{\partial t} - \gamma w^2 \Delta x \frac{\partial^3 e_n}{\partial t \partial x^2}, \tag{19.7}$$

and hence the space derivative of current is

$$\frac{\partial i}{\partial x} = -c \frac{\partial e}{\partial t} + \gamma w^2 \frac{\partial^3 e}{\partial t \partial x^2}, \tag{19.8}$$

which shows that it depends on the change of voltage not only with time but also with space. For fast traveling pulses with steep portions, and thus pronounced changes in space along the conductor, the second term of Eq. (19.8) can become very significant, particularly if the internal winding capacitance γ is substantially larger than the external capacitance c, which is to ground. In practice this is frequently the case.

The current as well as the voltage in a long coil varies from turn to turn. If the nth turn,

of self-inductance λ per unit length, were separated from the rest of the coil, the voltage induced by the current in it would be

$$e_{\lambda n} = \lambda \Delta x \frac{\partial i_n}{\partial t}. \tag{19.9}$$

The two adjacent turns coupled by the mutual inductance μ per unit length further add an induced voltage

$$e_{\mu n} = \mu \Delta x \frac{\partial(i_{n+1} + i_{n-1})}{\partial t}. \tag{19.10}$$

Here the sum of the currents differs in form from that for voltages in Eq. (19.4). To avoid the complex relation that would result from a summation of the positive terms, we shall substitute a form similar to that of Eq. (19.4). We obtain this by subtracting i_n from the current in each of the two adjacent turns. This leads, on rearrangement, to the following expression for the effect of the adjacent turns:

$$e_{\mu n} = \mu \Delta x \frac{\partial}{\partial t} [(i_{n+1} - i_n) - (i_n - i_{n-1}) + 2i_n]$$

$$= 2\mu \Delta x \frac{\partial i_n}{\partial t} + \mu \Delta x \frac{\partial \Delta^2 i_n}{\partial t}. \tag{19.11}$$

Thus the mutual inductance of the adjacent turns contributes an induced voltage that can be thought of as consisting of two parts, one of which is proportional to effects in the nth turn and the other due to second differences between the currents in the turns. Each succeeding turn will produce an induced voltage in the nth turn that will have the same form as Eq. (19.11), except that μ will become smaller with increasing distance between turns. All induced voltages proportional to the current i_n in the nth turn can be summed from Eqs. (19.9) and (19.11), to give

$$e_{ln} = (\lambda + \Sigma\mu)\Delta x \frac{\partial i_n}{\partial t} = l\Delta x \frac{\partial i_n}{\partial t}. \tag{19.12}$$

Here l is the self-inductance of the coil per unit length derived from the total inductive effects between all turns; thus it is the self-inductance of the entire coil divided by the length of wire. This is identical with the winding self-inductance used in the quasi-stationary treatment of Chapter 17 and with the effective self-inductance derived in Chapter 18 by the use of Maxwell's equations.

In the last term of Eq. (19.11) we can introduce Eq. (19.5) and write

$$\Delta^2 i = w^2 \frac{\Delta^2 i}{\Delta x^2} = w^2 \frac{\partial^2 i}{\partial x^2}, \quad (19.13)$$

when many turns are again used as a basis for transformation from the difference to the differential form. For this term we take into account only the influence of the immediately adjacent turns, because the induced effects due to the farther ones have already been included in Eq. (19.12) via the "external" magnetic field and thus should not be counted again. The additional inductive influence of the two adjacent turns is effective here, like that of the interturn capacitance treated earlier. The two effects always occur together because their force fields are orthogonal. By comparison of Eq. (19.11) with Eq. (19.4), we see that the additional inductive term occurs with the opposite sign to the capacitive term. The internal self-inductance will also be small if the internal capacitance is large. Thus the net result here will probably be much less affected by the internal inductive effect. The total voltage induced in the nth turn, in accordance with Eqs. (19.11) and (19.12), must be equal to the decrease $-\Delta e_n$ of the observable voltage difference. Using Eq. (19.13) we thus obtain

$$-\Delta e_n = l\Delta x \frac{\partial i_n}{\partial t} + \mu w^2 \Delta x \frac{\partial^3 i_n}{\partial t \partial x^2}, \quad (19.14)$$

and the voltage difference along the coil is

$$\frac{\partial e}{\partial x} = -l\frac{\partial i}{\partial t} - \mu w^2 \frac{\partial^3 i}{\partial t \partial x^2}. \quad (19.15)$$

Again we see that this depends on the variation of current with time as well as with space. For fast traveling pulses with steep portions the third-order derivative may attain appreciable magnitudes. However, its influence here is smaller than the corresponding capacitive one on voltage in Eq. (19.8) because the mutual inductance μ between adjacent turns is small compared with the self-inductance component l per unit turn of the entire coil.

In this chapter we shall not deduce numerical values for l and c, since this has already been done in the two preceding chapters, on the basis of typical coil dimensions. However, we should note that for such calculations it will be advisable to use values of l, c, v, and Z given by Eqs. (18.50), (18.51), (18.15), and (18.38) respectively, derived from Maxwell's equations, with due regard to the necessary approximations and the frequency range under consideration.

In principle it would be possible to take into account deviations from uniformity of structure within a coil, such as variable turn spacing, diameter, and so on. However, this would require rather elaborate and necessarily simultaneous consideration of the interacting electric and magnetic fields for the many now different subelements. Solutions could be obtained only for specifically chosen examples by means of modern computing facilities which can handle the complex differential-integral equations. This makes it hard to develop insight into generaly valid considerations.

The two differential equations (19.8) and (19.15) link the time and space dependence of current and voltage in the coil. Thus there is no simple solution for an arbitrary pulse shape such as was obtained in Chapter 1 for homogeneous lines. We shall here develop a solution valid only for a pulse of sinusoidal time variation that is impressed on the coil. Because any arbitrary pulse shape can be synthesized by a series of such sinusoidal components, we can still ultimately use a summation approach to obtain the response to a more complex pulse shape.

If we assume the voltage, and thereby also the current, in the coil to vary sinusoidally (or harmonically) in time, then the linear form of the differential equations (19.8) and (19.15) allows a similarly structured trial solution in the space domain. Thus we may try the solution

$$e = E\varepsilon^{j\omega t}\varepsilon^{j\alpha x}, \\ i = I\varepsilon^{j\omega t}\varepsilon^{j\alpha x}, \quad (19.16)$$

where ω is the angular frequency of the impressed oscillations and α is a reciprocal length, undetermined as yet. If we differentiate Eqs. (19.16) with respect to x and t, substitute the derivatives in Eqs. (19.8) and

(19.15), and eliminate superfluous terms, we obtain the two real relations

$$\alpha I = -c\omega E - \gamma w^2 \omega \alpha^2 E, \quad (19.17)$$

$$\alpha E = -l\omega I + \mu w^2 \omega \alpha^2 I. \quad (19.18)$$

Before we proceed to establish the relations between the constants E and I, we shall develop the conditional equation for the exponent α. We divide Eq. (19.18) by Eq. (19.17), which gives

$$-\frac{E}{I} = \frac{\alpha}{\omega(c + \gamma w^2 \alpha^2)} = \frac{\omega(l - \mu w^2 \alpha^2)}{\alpha}. \quad (19.19)$$

By cross multiplication we obtain the biquadratic equation

$$\alpha^4 w^4 \mu\gamma + \alpha^2 w^2 \left[\frac{1}{w^2 \omega^2} - (l\gamma - c\mu) \right] - lc = 0, \quad (19.20)$$

written in terms of the product αw, which represents a dimensionless number. The solution for the square of this number is

$$(\alpha w)^2 = -\frac{1}{2} \left[\frac{1}{\mu\gamma w^2 \omega^2} - \left(\frac{l}{\mu} - \frac{c}{\gamma} \right) \right]$$

$$\pm \sqrt{\frac{1}{4} \left[\frac{1}{\mu\gamma w^2 \omega^2} - \left(\frac{l}{\mu} - \frac{c}{\gamma} \right) \right]^2 + \frac{lc}{\mu\gamma}}. \quad (19.21)$$

This equation can be simplified by the following means. The mutual inductance μ will in general be small compared with the self-inductance l. However the interturn capacitance γ will be of the order of magnitude of the ground capacitance c. Thus c/γ may be neglected in comparison with l/μ. Further let

$$v_1 = \frac{1}{w\sqrt{\mu\gamma}} \quad (19.22)$$

designate an ideal natural frequency corresponding to the capacitance between two adjacent turns and the mutual inductance between them. Thus it is very high and of the order of magnitude of the natural frequency of just two turns. This then reduces Eq. (19.21) to

$$(\alpha w)^2 = -\frac{1}{2} \left[\left(\frac{v_1}{\omega} \right)^2 - \frac{l}{\mu} \right]$$

$$\pm \sqrt{\frac{1}{4} \left[\left(\frac{v_1}{\omega} \right)^2 - \frac{l}{\mu} \right]^2 + \frac{lc}{\mu\gamma}}. \quad (19.23)$$

This quadratic equation has two real and two imaginary roots, owing to the plus-and-minus sign on the square-root term. These roots are

$$\alpha^I = +\alpha_1, \quad \alpha^{III} = +j\alpha_2, \quad (19.24)$$
$$\alpha^{II} = -\alpha_1, \quad \alpha^{IV} = -j\alpha_2,$$

where α_1 and α_2 are positive real numbers given by

$$w\alpha_1 = $$
$$\sqrt{\sqrt{\frac{1}{4} \left[\left(\frac{v_1}{\omega} \right)^2 - \frac{l}{\mu} \right]^2 + \frac{lc}{\mu\gamma}} - \frac{1}{2}\left[\left(\frac{v_1}{\omega} \right)^2 - \frac{l}{\mu} \right]},$$

$$w\alpha_2 = $$
$$\sqrt{\sqrt{\frac{1}{4} \left[\left(\frac{v_1}{\omega} \right)^2 - \frac{l}{\mu} \right]^2 + \frac{lc}{\mu\gamma}} + \frac{1}{2}\left[\left(\frac{v_1}{\omega} \right)^2 - \frac{l}{\mu} \right]}. \quad (19.25)$$

The actual value depends only on the frequency ω, once the coil parameters are given.

Corresponding to the four possible values of α we must expand the solution of Eq. (19.16) by introducing the four values of Eq. (19.24). This gives the voltage,

$$e = \varepsilon^{j\omega t}(E_1' \varepsilon^{+j\alpha_1 x} + E_1 \varepsilon^{-j\alpha_1 x} + E_2 \varepsilon^{-\alpha_2 x} + E_2' \varepsilon^{+\alpha_2 x}). \quad (19.26)$$

Because our interest is concentrated on the effects at the beginning of the coil we may assume that the coil is very long in the x-direction and that no energy can be arriving from the far end of the coil. In Eq. (19.26) a large value of x makes the first and last terms in parentheses negligibly small, on account of the positive sign in the exponent. Thus we are left with

$$e = E_1 \varepsilon^{j(\omega t - \alpha_1 x)} + E_2 \varepsilon^{j\omega t} \varepsilon^{-\alpha_2 x}, \quad (19.27)$$

and we can derive a similar expression for the current,

$$i = I_1 \varepsilon^{j(\omega t - \alpha_1 x)} + I_2 \varepsilon^{j\omega t} \varepsilon^{-\alpha_2 x}. \quad (19.28)$$

These two equations show that two different voltage and current distributions are superposed in the coil. One has the amplitudes E_1

and I_1, is harmonic in time and space, and thus constitutes an undamped sinusoidal pulse progressing into the coil. The other, of amplitudes E_2 and I_2, does not vary sinusoidally but decreases exponentially with space. However, it too varies harmonically (or sinusoidally) with time. The exponential decrease with x of this component leads to the result that the inner portion of the coil experiences only a fraction of the voltage impressed at the terminal. Thus it represents a reflection in the initial portion of the coil. Let us examine in detail the distribution of voltage over a range of frequencies.

For low frequencies ω the ratio v_1/ω will dominate the factor in square brackets in Eqs. (19.25). Thus a close approximation for the two roots will be

$$\alpha_1 = \omega\sqrt{lc},$$
$$\alpha_2 = \frac{v_1}{w\omega}. \qquad (19.29)$$

The decaying voltage component E_2 of Eq. (19.27) is extinguished with a space constant that is the reciprocal of the exponent α_2 and thus

$$X = \frac{1}{\alpha_2} = \frac{\omega}{v_1}w. \qquad (19.30)$$

Because v_1 is high, according to Eq. (19.22), X is only a small fraction of the turn length w. Thus the decaying voltage distribution really plays no part at these low frequencies.

The sinusoidal component E_1 of Eq. (19.27) propagates with the velocity

$$v = \frac{\omega}{\alpha_1} = \frac{1}{\sqrt{lc}} \qquad (19.31)$$

along the wire of the coil. Thus the coil behaves as an ordinary homogeneous line. However, the total self-inductance of the coil per unit length is much larger than that of the unit length owing to the flux linkages between turns. Thus if the capacitance to ground is about equal to that of single wire, the velocity of propagation will be much smaller along the wire of the coil than along an open line. However, if the capacitance to ground is reduced owing to the winding and disposition with respect to ground of the coil, the velocity of propagation along the wire may become

equal to or even higher than that along the open line.

If we let the idealized natural frequency of each winding depend on its capacitance to ground and self-inductance per unit length, that is,

$$v_2 = \frac{1}{w\sqrt{lc}}, \qquad (19.32)$$

this is much lower than the natural frequency v_1, which, as given in Eq. (19.22), depends on the mutual capacitance and inductance. Thus, from Eq. (19.31),

$$v = v_2w, \qquad (19.33)$$

and v_2 is also the velocity of propagation in turns per second; this, like the absolute velocity of propagation as given by Eq.(19.31), is independent of the impressed low frequency.

The spatial wavelength of the harmonic oscillations in the coil is, by Eqs. (19.29) and (19.32),

$$\Lambda = \frac{2\pi}{\alpha_1} = 2\pi\frac{v_2}{\omega}w. \qquad (19.34)$$

Thus it is a large multiple of the turn length w and decreases with increasing frequency.

If the sinusoidal or other pulses are excited in a coil open at both ends, they are reflected at both ends just as in a homogeneous line open at both ends. The natural period of the entire coil, which represents a complete traversal of the wire length a in both directions, is thus, according to Eq. (19.31),

$$\tau = \frac{2a}{v} = 2a\sqrt{lc}. \qquad (19.35)$$

If we replace the specific values per unit length with the self-inductance L and capacitance to ground C of the entire coil, we obtain

$$\tau = 2\sqrt{al \cdot ac} = 2\sqrt{LC}. \qquad (19.36)$$

Thus the natural frequency of the entire coil, which is the number of oscillations in 2π sec, is

$$v_0 = \frac{2\pi}{\tau} = \frac{\pi}{a\sqrt{lc}} = \frac{\pi}{\sqrt{LC}}. \qquad (19.37)$$

This can also be expressed in terms of the

number of turns N in the coil, which is always large; by Eq. (19.32),

$$v_0 = \pi \frac{w}{a} v_2 = \frac{\pi}{N} v_2. \qquad (19.38)$$

Thus the natural frequency of the coil is given entirely by the lumped constants of the coil and is only a small fraction of the natural frequency due to self-constants v_2 of Eq. (19.32) and an even smaller one of the natural frequency due to mutual constants v_1 of Eq. (19.22) for the distributed winding. This further justifies the approximations leading to Eqs. (19.29), on the basis of the first of which we developed the final form in Eqs. (19.37) and (19.38).

The wavelength of this natural oscillation of the coil is found by substitution of Eq. (19.38) in Eq. (19.34):

$$\Lambda = 2\pi \frac{v_2}{v_0} w = 2a. \qquad (19.39)$$

Thus the wavelength of the oscillation of a finite coil open at both ends is twice the length of the coil.

As given by Eq. (19.37), the natural frequency of a coil is π times that of an oscillatory circuit with the same value of lumped constants. This is because both capacitance and inductance are distributed along the winding and are thus always exposed to only a part of the full voltage and current. This formulation also agrees exactly, and for the same reason, with that for a homogeneous line open at both ends.

For extremely high frequencies ω the ratio v_1/ω becomes negligible in comparison with the other terms in square brackets in Eqs. (19.25). Close approximations to the roots then are

$$\alpha_1 = \frac{1}{w} \sqrt{\frac{l}{\mu}},$$
$$\alpha_2 = \frac{1}{w} \sqrt{\frac{c}{\gamma}}. \qquad (19.40)$$

The velocity of propagation of the first voltage component is

$$v = \frac{\omega}{\alpha_1} = \omega w \sqrt{\frac{\mu}{l}}. \qquad (19.41)$$

It increases with increasing frequency and develops a wavelength along the wire

$$\Lambda = \frac{2\pi}{\alpha_1} = 2\pi w \sqrt{\frac{\mu}{l}}. \qquad (19.42)$$

This no longer depends on frequency. Owing to the relatively small value of μ it has also become much smaller than w. We shall see later that this voltage component has lost its significance for our considerations at this frequency.

The space constant of the decaying voltage component is now

$$X = \frac{1}{\alpha_2} = w \sqrt{\frac{\gamma}{c}}. \qquad (19.43)$$

It depends solely on the ratio of interturn capacitance γ to ground capacitance c. Because in tightly wound coils this ratio is usually large, the space constant is a multiple

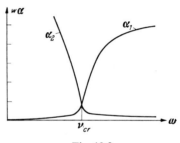

Fig. 19.2

of w. Thus again this voltage component cannot penetrate deeply into the central portion of the winding, which is protected by the capacitance to ground, somewhat bypassed, however, by the interturn capacitance. Incident pulses are largely reflected back into the line.

Thus the coil winding behaves quite differently for slow (low-frequency) and fast (high-frequency) oscillations, owing to the difference in magnitude of α_1 and α_2 as given by Eq. (19.25). For slow oscillations propagation into the central portion of the coil prevails, on account of the preponderance of α_1 in Eqs. (19.27) and (19.28). For fast oscillations the preponderance of α_2 leads to reflection at the terminals.

Figure 19.2 shows $w\alpha$ for both α_1 and α_2 as a function of frequency for a capacitance ratio $c/\gamma = 0.1$ and an inductance ratio $\mu/l = 0.01$.

(b) *Voltage distribution and critical frequency.* The two voltage distributions have the same significance when

$$\alpha_1 = \alpha_2. \tag{19.44}$$

This is the case when the last terms in Eqs. (19.25) are equal to zero and thus

$$\left(\frac{\nu_1}{\omega}\right)^2 = \frac{l}{\mu}. \tag{19.45}$$

Let us call this the critical frequency because it occurs at the point of transition from one dominant voltage distribution to the other. Equations (19.45) and (19.22) give the value as

$$\omega = \nu_{cr} = \frac{1}{w\sqrt{l\gamma}}. \tag{19.46}$$

Inspection of Eqs. (19.25) now shows that the two values of α reduce to the single value

$$\alpha_{cr} = \frac{1}{w}\sqrt[4]{\frac{lc}{\mu\gamma}}. \tag{19.47}$$

This gives the velocity of propagation as

$$v_{cr} = \frac{\nu_{cr}}{\alpha_{cr}} = \frac{1}{\sqrt{lc}}\sqrt[4]{\frac{\mu c}{l\gamma}}. \tag{19.48}$$

This is very much lower than that for slow oscillations as given by Eq. (19.31) because μ is usually much smaller than l and γ is usually larger than c.

The wavelength of the sinusoidal pulse is

$$\Lambda_{cr} = \frac{2\pi}{\alpha_{cr}} = 2\pi w \sqrt[4]{\frac{\mu\gamma}{lc}}, \tag{19.49}$$

and the space constant of decay is

$$X_{cr} = \frac{1}{\alpha_{cr}} = w \sqrt[4]{\frac{\mu\gamma}{lc}}. \tag{19.50}$$

Thus in the critical case both are larger than the turn length w if the capacitance ratio γ/c is larger than the inductance ratio l/μ. However, if the interturn capacitance γ is small compared with the capacitance to ground c, the critical wavelength and decay space constant are shorter than one turn. Thus the voltage within one turn will vary very significantly. Such variations were neglected for this analysis at the outset of this chapter. We

see that conditions in the vicinity of the critical frequency will be quite confused and complicated.

To establish a clearer view of the critical frequency as given by Eq. (19.46) we can use Eq. (19.32) to eliminate the capacitance ratio, and we obtain

$$\nu_{cr} = \sqrt{\frac{c}{\gamma}} \frac{1}{w\sqrt{lc}} = \sqrt{\frac{c}{\gamma}} \nu_2. \tag{19.51}$$

Next, using Eqs. (19.38) and (19.37), we get

$$\nu_{cr} = \frac{N}{\pi}\sqrt{\frac{c}{\gamma}} \nu_0 = \sqrt{\frac{c}{\gamma}} \frac{N}{\sqrt{LC}}. \tag{19.52}$$

Thus for coils with a large number of turns N the critical frequency is a large multiple of the natural frequency ν_0 of the entire coil.

If the interturn capacitance γ is large, the critical frequency is entirely within the range of frequencies of traveling pulses that are encountered in practical systems. Thus a portion of these pulses will be reflected by the end turns of the coil. If, however, the interturn capacitance is small, the critical frequency is quite high, so that even pulses with quite steep features can travel into the coil without undue distortion just as they travel along homogeneous lines.

The magnitude of the two voltage distributions that enter the winding is determined by the boundary conditions of a given case. Let us assume that at the terminal of the coil, that is, for $x = 0$, the total voltage is a sinusoidal alternating voltage of amplitude E supplied by a large source. Then, according to Eq. (19.27),

$$E_1 + E_2 = E. \tag{19.53}$$

At the beginning of the coil the first turn is linked electromagnetically only on one side, so that it is not as completely coupled as a turn deep inside the coil. However, in the earlier development of the differential equations we had used a fully coupled turn in the winding. Thus, if we wish our solution to apply at the start of the winding, we must take into account the reduced coupling by means of an appropriate boundary condition. Owing to this looser coupling, the changes in current along the turn will be less pronounced and as a first approximation for $x = 0$ let us assume

$$\frac{\partial i}{\partial x} = 0. \tag{19.54}$$

Equation (19.28) differentiated with respect to x then gives for the coil termination

$$-j\alpha_1 I_1 - \alpha_2 I_2 = 0. \tag{19.55}$$

The current amplitudes I_1 and I_2 are functions of the voltage amplitudes E_1 and E_2. The functional relation can be established from Eq. (19.17), the roots α being inserted according to Eqs. (19.24) and (19.26). Then

$$I_1 = -\frac{\omega}{\alpha^{II}} (c + \gamma w^2 \alpha^{II2}) E_1$$

$$= \frac{\omega}{\alpha_1} (c + \gamma w^2 \alpha_1^2) E_1,$$

$$I_2 = -\frac{\omega}{\alpha^{III}} (c + \gamma w^2 \alpha^{III2}) E_2 \tag{19.56}$$

$$= j\frac{\omega}{\alpha_2} (c - \gamma w^2 \alpha_2^2) E_2.$$

If these are substituted in Eq. (19.55), the partial voltages are given by

$$(c + \gamma w^2 \alpha_1^2) E_1 + (c - \gamma w^2 \alpha_2^2) E_2 = 0. \tag{19.57}$$

Next, using Eqs. (19.53) and (19.57), we obtain the two voltage-distribution amplitudes,

$$E_1 = \frac{w^2 \alpha_2^2 - c/\gamma}{w^2 \alpha_1^2 + w^2 \alpha_2^2} E,$$

$$E_2 = \frac{w^2 \alpha_1^2 + c/\gamma}{w^2 \alpha_1^2 + w^2 \alpha_2^2} E. \tag{19.58}$$

For low frequencies Fig. 19.2 and Eqs. (19.29) indicate that α_1 is small and α_2 is large. Thus, neglecting α_1^2 and using Eq. (19.29), we have as a good approximation in the region below the critical frequency

$$E_1 = \left[1 - \frac{c}{\gamma} \left(\frac{\omega}{\nu_1} \right)^2 \right] E,$$

$$E_2 = \frac{c}{\gamma} \left(\frac{\omega}{\nu_1} \right)^2 E. \tag{19.59}$$

The penetrating sinusoidal voltage component E_1 predominates, while the reflected damped component E_2 is much smaller. As frequency increases, however, the component E_1 becomes smaller and E_2 grows.

At the critical frequency, with $\alpha_1 = \alpha_2$ as in Eq. (19.44), we obtain, using Eq. (19.47), the two voltage components

$$E_{1cr} = \frac{1 - \sqrt{c\mu/\gamma l}}{2} E,$$

$$E_{2cr} = \frac{1 + \sqrt{c\mu/\gamma l}}{2} E. \tag{19.60}$$

Thus at the critical frequency the two voltage components will be about equal because the square-root term in the numerator is usually quite small. For even higher frequencies the sinusoidal penetrating component E_1 decreases and the decaying reflected component E_2 increases.

For very high frequencies α_1 is very much larger than α_2 according to Eqs. (19.40) and thus Eqs. (19.58) give the asymptotic approximations

$$E_1 = 0,$$

$$E_2 = E. \tag{19.61}$$

The sinusoidal penetrating voltage has disappeared completely and only the damped

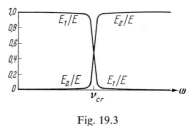

Fig. 19.3

reflected voltage distribution is present. Figure 19.3 shows the two voltage components normalized to E as a function of frequency. As in Fig. 19.2, the values $c/\gamma = 0.1$ and $\mu/l = 0.01$ are used here.

We can now determine the voltage difference between turns caused by the superposition of the two voltage components entering the coil. This voltage difference imposes a stress on the insulation between turns. By Eq. (19.5) it is

$$\mathfrak{e} = -\Delta e = -w \frac{\partial e}{\partial x}. \tag{19.62}$$

Its amplitude for the first turn is, from Eq. (19.27),

$$\mathfrak{E}_0 = w(j\alpha_1 E_1 + \alpha_2 E_2). \qquad (19.63)$$

Of most interest is the absolute magnitude of this voltage,

$$\mathfrak{E}_0 = w\sqrt{\alpha_1^2 E_1^2 + \alpha_2^2 E_2^2}, \qquad (19.64)$$

which can be expressed, by substitution of Eqs. (19.58) and neglect here of the small ratio c/γ, as

$$\mathfrak{E}_0 = wE \sqrt{\frac{\alpha_1^2 \alpha_2^2}{\alpha_1^2 + \alpha_2^2}}$$

$$= E \frac{\sqrt{lc/\mu\gamma}}{\sqrt[4]{[(\nu_1/\omega)^2 - l/\mu]^2 + 4lc/\mu\gamma}}. \qquad (19.65)$$

The product and the sum of squares of α are here obtained from Eq. (19.25).

The denominator of this expression shows that the absolute maximum voltage across the end-turn insulation varies as a function of frequency in the form of a resonance curve. For the critical frequency, as given by Eqs. (19.45) and (19.46), a very high resonance peak voltage can occur.

For low frequencies, when α_1 is very much smaller than α_2 according to Eq. (19.29), the first form of Eq. (19.65) together with Eqs. (19.29) and (19.31) leads to

$$\mathfrak{E}_0 = wE\alpha_1 = \omega w\sqrt{lc}E = \frac{\omega w}{v}E. \qquad (19.66)$$

Because the product of angular frequency and turn length will be small compared with the velocity of propagation, the voltage difference between end turns will be relatively small for low-frequency alternating-current components.

However, it increases with increasing frequency and at the critical frequency with $\alpha_1 = \alpha_2$ the first form of Eq. (19.65) together with Eq. (19.47) gives

$$\mathfrak{E}_{0cr} = wE\frac{\alpha_{cr}}{\sqrt{2}} = \frac{E}{\sqrt{2}}\sqrt[4]{\frac{lc}{\mu\gamma}}. \qquad (19.67)$$

If the inductance ratio l/μ is larger than the capacitance γ/c, the turn voltage may reach a multiple of the impressed voltage. However, strong coupling of the turn to the winding by means of a large γ and μ may keep this resonancelike peak within reasonable limits.

For still higher frequencies the voltage drops again. In the limit, when α_2 is very much smaller than α_1, the second of Eqs. (19.40) gives the asymptotic value

$$\mathfrak{E}_{0\infty} = wE\alpha_2 = E\sqrt{\frac{c}{\gamma}}. \qquad (19.68)$$

Thus at very high frequencies the turn voltage difference becomes independent of frequency and depends only on the capacitance ratio.

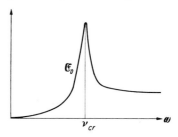

Fig. 19.4

For all frequencies significantly above the critical frequency this turn voltage difference is thus essentially the same. It can also be kept small by a high interturn capacitance γ. Figure 19.4 shows the voltage difference between the end turns of a coil as a function of frequency. The resonance maximum associated with the critical frequency is quite evident.

Pulses without steep fronts or other steep features that may impinge on the coil will in general contain only low-frequency components, which lie on the portion of the curve in Fig. 19.4 that is far below the critical frequency. Thus they can penetrate a coil just as they travel along an open line. They are not influenced in the main by the interturn coupling in the coil. However, very steep pulses such as step pulses or impulses have a very large content of high-frequency components. Equation (19.68) becomes applicable for the end-turn voltage difference caused by such incident steep pulses. In this case E will be the voltage across the feeder line at the junction, caused by reflection of the incident voltage pulse in the end turns of the coil; only for feeder lines of very low surge impedance will it be almost equal to the incident pulse voltage. If the front of the incident pulse has been flattened somewhat, there may be some frequency components near the critical frequency of the coil. These components can then cause, as shown in Fig. 19.4 and Eq. (19.67), larger end-turn voltage differences.

The asymptotic approximation of Eq. (19.68) for the voltage difference across the first turn could also be derived by means of a ladder network of capacitances. Such a ladder network, consisting of series condensers γ and parallel grounding condensers c, is shown in Fig. 19.5. The detailed analysis of

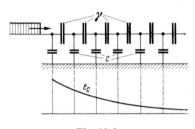

Fig. 19.5

such a ladder is treated in Chapter 27. Here, too, the voltage across the ladder decreases exponentially with distance and a space constant as given by Eq. (19.43). Thus tightly wound coils behave for high-frequency voltages and particularly steep pulses essentially as if the turns were coupled only capacitively to each other and ground and not conductively at all. The current development is governed not by the conductive paths but by the displacement currents in the insulation around the conductors and between them and the surrounding grounded structures.

The magnitude of the displacement current across the interturn capacitance γ can be derived from Eqs. (19.3) and (19.5); it is

$$\mathfrak{i} = -\gamma w^2 \frac{\partial^2 e}{\partial t \partial x}. \tag{19.69}$$

If we differentiate the voltages given in Eq. (19.27) with respect to time and space we obtain for this displacement current also sinusoidal and exponentially decaying components, of amplitudes

$$\begin{aligned} I_1 &= -\gamma w^2 \omega \alpha_1 E_1, \\ I_2 &= +j\gamma w^2 \omega \alpha_2 E_2. \end{aligned} \tag{19.70}$$

These are equal in magnitude, though opposite

in sign, to the second terms in parentheses in Eqs. (19.56), which gave the currents in the conductors. Thus if we add the currents in the conductors, given by Eqs. (19.56), and the displacement currents, given by Eqs. (19.70), only the first terms in Eqs. (19.56) remain, and thus at the beginning of the coil, at $x = 0$,

$$\Sigma I_0 = \omega c \left(\frac{1}{\alpha_1} E_1 + \frac{j}{\alpha_2} E_2 \right). \tag{19.71}$$

For low frequencies, large values of α_2 give the total current as a function of E_1 only. From Eqs. (19.29) it becomes

$$\Sigma I_0 = \frac{\omega c}{\alpha_1} E = \sqrt{\frac{c}{l}} E. \tag{19.72}$$

Thus the coil accepts a current due to the impressed voltage E that is in phase with it and whose amplitude is determined solely by the surge impedance

$$Z = \sqrt{\frac{l}{c}} = \sqrt{\frac{L}{C}}. \tag{19.73}$$

This can, as shown, be calculated from either the distributed or the lumped constants of the entire coil just as in the case of a homogeneous line. For low frequencies this surge impedance is real and constant.

At the critical frequency the two α's and, from Eqs. (19.60), the two voltage components are equal. Thus the total coil current, from Eq. (19.71) together with Eqs. (19.46) and (19.47), becomes

$$\Sigma I_{0cr} = (1+j) \frac{v_{cr} c}{\alpha_{cr}} \frac{E}{2} = \frac{1+j}{2} \sqrt{\frac{c}{l}} \sqrt[4]{\frac{\mu c}{\gamma l}} E. \tag{19.74}$$

The current has been reduced considerably owing to the small value of the ratio under the fourth-root sign; in addition, a phase shift of 45° with respect to the voltage E has been introduced.

For very high frequencies Eqs. (19.61) show that the voltage component E_1 becomes negligible and thus the current is, by Eqs. (19.40) and (19.71),

$$\Sigma I_{0\infty} = j \frac{\omega c}{\alpha_2} E = j\omega w \sqrt{\gamma c} E. \tag{19.75}$$

The current leads the voltage E by $90°$ and has an amplitude equal to that which would flow into a condenser whose capacitance

$$\bar{C} = w\sqrt{\gamma c} \qquad (19.76)$$

is the geometric mean of the interturn and ground capacitances of a single turn. This can be shown to be exactly the value of the capacitance looking into the ladder network shown in Fig. 19.5.

We thus see that coils behave under low-frequency alternating voltages exactly like homogeneous lines with a given surge impedance. For very high frequencies (well above the critical frequency), they behave like a capacitance ladder network with given capacitances. For low frequencies the current flows predominantly in the conductive portions of the winding, while at very high frequencies it flows predominantly as a displacement current across the insulating portions of the winding. At the critical frequency the coil accepts only a small current from the driving source. However, in this case both the conductive and the displacement currents in the coil are large owing essentially to a resonance condition among the internal subelements of the coil.

This can be seen most simply by a comparison between Eqs. (19.69) and (19.62), which gives for the displacement current

$$i = \gamma w \frac{\partial e}{\partial t} = j\omega w \gamma e. \qquad (19.77)$$

Thus the resonance-voltage equation (19.65) can be used directly for the displacement current after multiplication by $j\omega w \gamma$. This then makes it valid for use as the second term of the conductive current as given in (19.56). Adjacent turns within the coil now oscillate strongly with respect to each other over distances governed by the critical wavelength as given in Eq. (19.49). However, as seen in Eq. (19.74), only a small current is supplied to the coil to support these strong oscillations. The resistance and skin effects in the conductors of the coil, which have been neglected so far, will of course broaden out the resonance curve considerably and reduce its peak value.

If very short and, in particular, steep pulses, such as are caused by switching processes, impinge on a coil, then, as we have shown, they produce a very considerable, though exponentially damped, voltage-difference distribution. Thus predominantly they stress the insulation between the first few turns or layers of a coil and may give rise to failures if they are excessive. However, if periodic pulses with sinusoidal components of frequency below the critical value impinge on the coil, they can travel without decay into all portions. Such pulses may be due to oscillatory processes in other portions of the network having appropriate natural frequencies. Such pulses stress the insulation to ground of all turns in the coil equally. However, they do not severely stress the insulation between adjacent turns or even layers. The full voltage will occur only between turns that are separated by a half wavelength. According to Eqs. (19.34) and (19.38), this represents a number of turns given by

$$N_{\Lambda/2} = \frac{\Lambda}{2w} = \pi \frac{\nu_2}{\omega} = \frac{\nu_0}{\omega} N. \qquad (19.78)$$

Thus, for pulses with frequency components ω greater than the natural frequency ν_0 of the entire coil, high stresses occur between

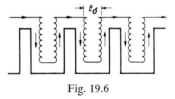

Fig. 19.6

points that are separated only by a fraction of the total number of turns N in the coil.

The windings of high-voltage transformers often consist of a number of coils connected as shown in Fig. 19.6. For the regular steady-state voltage distribution only moderate voltage differences exist between adjacent portions of the winding. However, the entire amplitude e_σ of an incident high-frequency voltage pulse may fall between adjacent coil terminals, as shown in Fig. 19.6, if the number of turns of the two connected coils are as given by Eq. (19.78). Such pulses then can easily cause insulation breakdown between coil sections or layers.

In the windings of machinery, transformers, and other devices, the design objective is to minimize insulation stresses due to impinging step pulses. Equations (19.67) and (19.68) show that this can be achieved if the interturn capacitance γ is made as large as possible compared with the ground capacitance c. This can be achieved by the use of flat-conductor helical coils arranged so that the wide sides of adjacent turns face each other while the narrow sides face the ground or the iron core. It is possible to carry this too far, in which case the space constant X as given by Eq. (19.43) becomes very large. Then the step pulses are not fully intercepted by the strongly insulated first few turns but can penetrate to the interior turns of the coil with weaker insulation.

Protective choke coils intended to protect against traveling pulses are expected to do so even for steep pulses with high-frequency components Thus it is necessary to construct such coils with low interturn capacitance γ. If this is done, the critical frequency, as given by Eq. (19.46), is so high that many of the incident frequency components are below it and are thus reduced by the effect of the surge impedance as given by Eq. (19.73). On the other hand, the few frequency components above the critical frequency that are contained in the dangerous pulses will be damped completely within the coil owing to the small space constant as given by Eq. (19.43). However, if this space constant should be larger than the length of wire $a = Nw$ in the entire coil, then the protective effect would be small because the most dangerous pulse components would be essentially transmitted through the coil by interturn capacitive coupling.

(c) *Measurement of voltage distribution in windings.* The distribution of voltage over turns, layers, and coils of alternating-current windings has frequently been measured. In early work spark gaps, and more recently cathode-ray oscillographs, have been used, with good agreement between the two methods and the analysis we have presented here.

Figure 19.7 shows the voltage on a line connected via a flat-wound choke coil to a

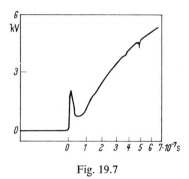

Fig. 19.7

voltage step pulse. We see that initially the voltage step is transmitted as a sharp spike owing to the large interturn capacitance. Only after this spike is over does the voltage on the line rise in accordance with the inductance of the coil. Thus for many analytic purposes the coil can be replaced by a capacitance and an inductance connected in parallel, as shown in Fig. 16.17.

If high-frequency oscillations are impressed upon flat-wound choke coils, then, as shown by R. Rücklin, moderate frequencies produce standing waves with pronounced nodes and antinodes. This is shown in Fig. 19.8a for a coil grounded at one end. The observations were made with a rectifier instrument and so did not include phase observation. With increasing frequency the spacing of the nodes and antinodes decreases and in Fig. 19.8b an exponential voltage distribution is superposed. At a still higher frequency, shown in

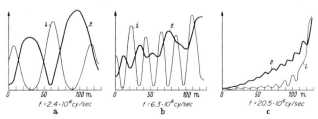

Fig. 19.8

Fig. 19.8*c*, this exponential voltage distribution completely dominates and the nodes and antinodes produce only very minor and closely spaced ripples.

Figure 19.9 shows the voltage e_λ between the coils of a large induction motor of 5000-V

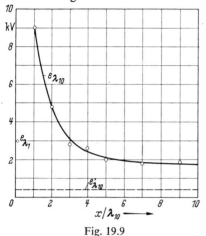

Fig. 19.9

rating during switching, both on and off. The supply line was a cable. The maximum voltage difference was measured by means of a spark gap connected across sections of the

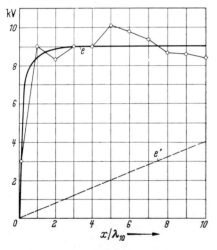

Fig. 19.10

winding. One reading, $e_{\lambda 1}$, was taken across the first turn and then the others, $e_{\lambda 10}$, were taken across successive groups of 10 turns. We see that a very large fraction of the total voltage difference is impressed upon the

first turn and that the succeeding turns are stressed less and less. Finally, the inner turns are under considerably less voltage stress. In Fig. 19.10 the voltage difference e between the terminal and the several tapping points is shown. Again we see that most of the change occurs over the first few turns. However, we

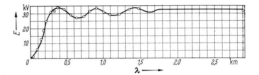

Fig. 19.11

note that the insulation to ground of the inner turns is highly stressed. In both Figs. 19.9 and 19.10 broken lines, labeled $e'_{\lambda 10}$ and e' respectively, indicate the steady-state operating voltages measured at the same locations.

Figure 19.11 shows a measurement made by O. Böhm to observe the penetration of a periodic high-frequency pulse into a transformer winding. Here, too, the voltage differences between the entrance terminal and various taps along the length of the winding were measured. We see here the superposition of

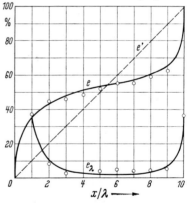

Fig. 19.12

the oscillatory sinusoidal and exponentially damped behavior in the form of successive nodes and antinodes about the final value.

Figure 19.12 shows the voltage distribution over the 10 coils of a single-phase transformer winding as measured with a cathode-ray oscillograph. The highest voltage upon arrival of a step pulse was determined both as *e* between the terminals and successive taps and as e_λ across each of the 10 coils. Again the broken line *e'* shows the steady-state operating voltage distribution but e'_λ as a line at 10 percent is not shown. We see that again the

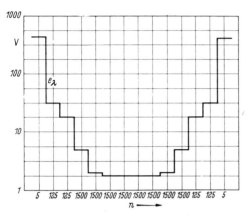

Fig. 19.13

and that 4 percent of the pulse occurs across just the first turn, whose insulation is thus stressed by 230 times the steady-state operating voltage.

High-voltage and voltage-measurement transformers usually have windings of a length much greater than the space constants,

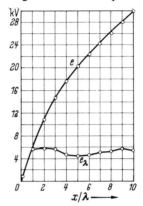

Fig. 19.14

voltage stresses are concentrated in the end portions of the winding. This is shown even more clearly in Fig. 19.13, for a voltage transformer. The end portions were divided more finely to permit the effects to be observed more closely by means of a spark gap. The voltage difference e_λ is shown as a function of the number of turns *n*. We see that 25 percent of the incident step-voltage pulse occurs across the first 2 percent of all of the turns,

so that step pulses as shown stress mainly the end windings. In contrast the windings of series-connected transformers and current transformers are often shorter than the space constant. Figure 19.14 shows measurements on a 10-kV current transformer. We see that the single coils have almost the same voltage difference e_λ across them, while the voltage *e* between the terminal and successive taps increases rather uniformly.

CHAPTER 20

ENTRY OF STEP PULSES INTO WINDINGS

It is extremely important for the reliable design of the insulation of machinery and transformers to know the stresses produced in the windings by traveling pulses with steep features, such as step pulses. Hence we must examine the entry into and penetration of windings by such pulses in some detail. Our objective will be to gain further insight into these phenomena on a basis simple enough to allow quantitative assessments for various types of winding.

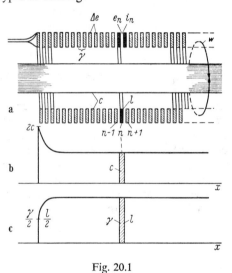

Fig. 20.1

Figure 20.1a shows a long tightly wound coil made of a flat conductor with the narrow edge facing the iron core. All elements or turns have equal self- and mutual-characteristics. Figures 20.1b, c show that the end effects of the coil can be taken into account by allowing the external or ground capacitance c to rise to twice its value at the end. Similarly the interturn capacitance γ and self-inductance l decrease at the end to half their value. This simple way of including the end effects, which is applicable to the long coils that are generally used in the construction of

transformers, avoids the rigorous treatment of the field interactions of each element, which would be extremely complex. The mutual inductance between turns is fully allowed for by the term l, derived from the total effective self-inductance of the coil, as has already been done in Chapters 17 and 18. We shall neglect damping due to ohmic-resistance losses, skin effect, or hysteresis of the core because they merely lead to a gradual decay of the initial conditions and have only a secondary effect on the magnitude of the initial conditions that produce the peak stresses of interest.

We can thus use here the differential equations derived in Chapter 19. However, to simplify the following considerations, we shall neglect the mutual inductance μ of adjacent turns. As we saw in Chapter 19, the effect of μ either disappears completely or plays only a minor role in the equations of practical importance, namely, Eq. (19.37) for the natural frequency, Eq. (19.46) for the critical frequency, Eq. (19.68) for the voltage difference across the end turn at very high frequencies, and Eq. (19.76) for the effective entrance capacitance at high frequencies, so that omission of this parameter appears justifiable. Thus we obtain from Eqs. (19.8) and (19.15) the differential equations effective for our consideration here,

$$-\frac{\partial i}{\partial x} = c\frac{\partial e}{\partial t} - \gamma w^2 \frac{\partial^3 e}{\partial t \partial x^2},$$
$$-\frac{\partial e}{\partial x} = l\frac{\partial i}{\partial t}. \tag{20.1}$$

The term with the interturn capacitance γ represents the only difference between Eq. (20.1) and Eqs. (1.2) and (1.4) for homogeneous lines. It is this term that changes the phenomena in machinery and transformer windings from those in long homogeneous lines.

Before we proceed it will be useful to extend the validity of Eq. (20.1) to other cases of importance. If the coil has not only single turns as shown in Fig. 20.1 but a number of turns per layer as shown in Fig. 20.2, it is possible to consider average values e_n and i_n for each layer n. Correspondingly, we now let c be the ground capacitance of each layer and

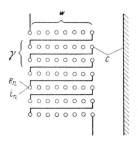

Fig. 20.2

γ the capacitance between adjacent layers, both per unit length. The definition of self-inductance l is unchanged. We can then use Eq. (20.1) provided only that w is now the length of conductor per layer.

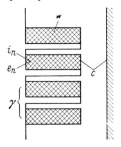

Fig. 20.3

If, as shown in Fig. 20.3, the winding is arranged by the interconnection of a number of subcoils, we can use the characteristic values as applicable to these subcoils and Eqs. (20.1) will give the average value of voltage e_n and current i_n for each of the n subcoils.

We can treat in this manner the windings of alternators and motors as well as those of transformers and chokes. In each case we must consider the conductor length w and corresponding capacitances c and γ over which we wish to obtain average values of voltage and current. For the coarsest distribution this may be the length over an entire coil, for a finer distribution it may be the length for a layer

in a coil, and for the highest resolution of distribution it may be a single turn within a coil or layer.

(a) *The development of standing and traveling waves.* If we differentiate the first of Eqs. (20.1) with respect to t and the second with respect to x, we can eliminate the current i by addition to obtain

$$\frac{\partial^2 e}{\partial x^2} - lc\frac{\partial^2 e}{\partial t^2} + l\gamma w^2 \frac{\partial^4 e}{\partial t^2 \partial x^2} = 0. \quad (20.2)$$

A similar fourth-order differential equation holds for the current i.

The simplest form of solution for Eq. (20.2) is

$$e = E\varepsilon^{j\omega t}\varepsilon^{j\alpha x}. \quad (20.3)$$

According to this the voltage, of amplitude E, varies sinusoidally in time and also in space along the conductor in the x-direction; ω is the angular frequency in 2π sec, and α is the wave density along the wire, expressed in wavelengths per 2π cm of wire. If we use Eq. (20.3) in Eq. (20.2) we obtain

$$-\alpha^2 + lc\omega^2 + l\gamma w^2\omega^2\alpha^2 = 0. \quad (20.4)$$

This gives the relation between frequency and wave density.

Instead of taking the standing-wave form of solution used in Eq. (20.3), we can also solve Eq. (20.2) by the trial form of solution

$$e = E\varepsilon^{j\omega(t - x/v)}, \quad (20.5)$$

which represents traveling waves within the winding. These oscillate in time with the frequency ω and propagate with the velocity v along the conductors in the coils. There is complete mathematical equivalence between these two forms of solution. However, the solution of the form in Eq. (20.3) gives us an entirely different physical understanding of the phenomena, as was seen in Chapter 1, Sec. (d). By introducing Eq. (20.5) into Eq. (20.2) we obtain

$$-\left(\frac{\omega}{v}\right)^2 + lc\omega^2 + l\gamma w^2\omega^2\left(\frac{\omega}{v}\right)^2 = 0. \quad (20.6)$$

This shows the relation between the frequency ω and the velocity of propagation v of traveling waves. Which of the two physical approaches, standing or traveling waves, is more useful depends entirely on the particular technical problem under analysis.

Let us first examine the development for standing waves. From Eq. (20.4) we obtain

$$\alpha = \sqrt{\frac{cl\omega^2}{1 - \gamma lw^2\omega^2}},$$

or   (20.7)

$$\omega = \frac{\alpha}{\sqrt{cl + \gamma lw^2\alpha^2}}.$$

For low frequencies the second term in both denominators is very small and thus frequency and wave density are essentially proportional to each other. The expression for ω shows that for higher wave densities α the effect of interturn capacitance γ becomes additive to that of the external or ground capacitance c. The expression for α shows that with increasing frequency ω the wave density α increases sharply as a frequency is approached for which the denominator becomes zero and thus α becomes infinite. This is the case for

$$\omega = \nu = \frac{1}{w\sqrt{\gamma l}}.$$ (20.8)

We again call this value ν the critical frequency of the coil because it plays a critical role in the behavior of the winding. Standing waves cannot arise at this or any higher frequency because for higher frequencies α will become imaginary, which would invalidate the solution in the form of Eq. (20.3).

Let us first examine the region below this critical frequency. On any finite line, with open or shorted ends, there can be only two series of well-defined standing waves. These are given by the boundary conditions at the ends. If m is the number of waves over the entire length of the conductor a, only wavelengths

$$\lambda = \frac{a}{m}, \text{ with } m = \begin{cases} \frac{1}{4} \times 1, 3, 5, 7, \ldots \\ \frac{1}{4} \times 2, 4, 6, 8, \ldots \end{cases}$$
                                                                      (20.9)

are compatible with the stated boundary conditions. Thus the wave density of these standing waves is given by

$$\alpha = \frac{2\pi}{\lambda} = \frac{2\pi}{a} m.$$ (20.10)

If we insert this into the second of Eqs. (20.7), we obtain the frequencies of natural oscillations in the winding,

$$\omega = \frac{2\pi m/a}{\sqrt{lc\left[1 + 4\pi^2 \frac{\gamma}{c}\left(\frac{w}{a}m\right)^2\right]}}.$$
                                                                      (20.11)

While in homogeneous lines the natural frequencies are directly proportional to the

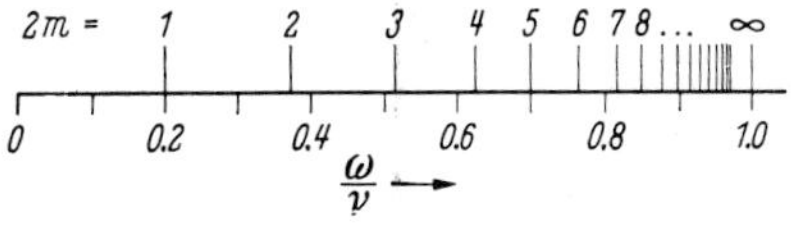

2m =   1   2   3   4 5 6 7 8... ∞

0   0.2   0.4   0.6   0.8   1.0
                    $\frac{\omega}{\nu}$ ⟶

Fig. 20.4

number m of wavelengths and inversely proportional to the length a of the line, this relation holds in windings only for small values of m and relatively large values of wire length a.

However, if the ratio m/a increases, the intervals between successive natural frequencies become smaller. Finally, all natural

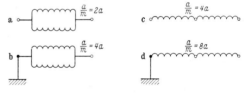

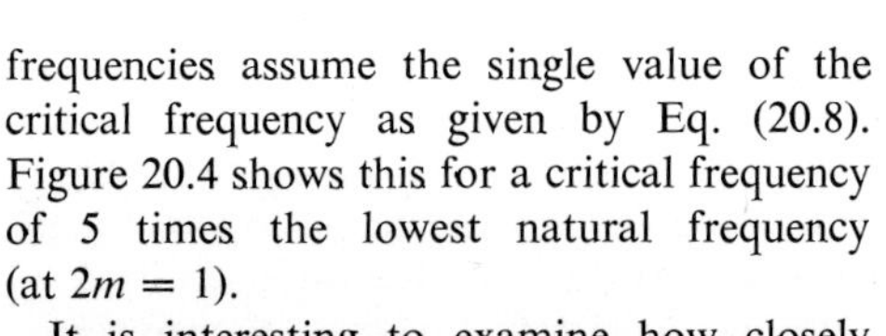

Fig. 20.5

frequencies assume the single value of the critical frequency as given by Eq. (20.8). Figure 20.4 shows this for a critical frequency of 5 times the lowest natural frequency (at $2m = 1$).

It is interesting to examine how closely observations on actual transformer windings agree with this analysis. Since it is hardly possible to observe the nodes of standing waves along a winding inside an oil-filled tank, an experiment was conducted in which the lower natural frequencies for different lengths of winding a could be measured. To do this, different interconnections of the winding sections of a large transformer were used, as shown in Fig. 20.5. J. S. Cliff used a single-phase transformer of 16,600-kVA rating that had pancake windings for 11 kV on the low-voltage side and for 19 kV on the high-voltage side. One entire winding, connected either in series or in parallel, ungrounded or grounded at one end, shown in Fig. 20.5, was excited to resonance while the other winding was grounded at both ends. Table 20.1 lists the results of the measurements.

We must note carefully that the grounding of one end of the winding did not, as expected, lead to a halving of the natural frequency in any of the cases of series or parallel connection for either the high- or the low-voltage windings. This is due to the effect of the denominator in Eq. (20.11). However, the natural frequency of the parallel connection grounded at one end is observed to be almost the same as that of the ungrounded series connection. This is because for these two cases m/a is the same, so that the denominator in Eq. (20.11) is also constant.

To compare the results more exactly with Eq. (20.11) we can use the abbreviation

$$k = 2\pi \sqrt{\frac{\gamma}{c}}\, w, \qquad (20.12)$$

which gives Eq. (20.11) the new form

$$\frac{1}{\omega^2} = \frac{1}{v^2} + \left(\frac{a/m}{vk}\right)^2. \qquad (20.13)$$

If we now plot $1/\omega^2$ against $(a/m)^2$, we should obtain a straight line that intersects the ordinate at the reciprocal of the square of the critical frequency, that is, at $1/v^2$. We see in Fig. 20.6 that the frequencies of the high-voltage winding obey these theoretical pre-

dictions exactly, while those for the low-voltage winding show some small deviations. From Fig. 20.6 the critical frequencies can be determined as 90 kc/sec for the high-voltage

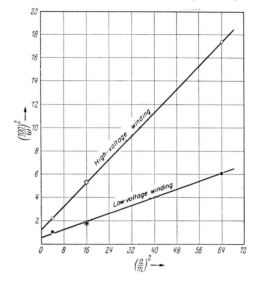

Fig. 20.6

winding and 130 kc/sec for the low-voltage winding.

Before considering frequencies above the critical, let us examine the behavior of traveling waves within the windings. Equation (20.6) gives the velocity of propagation as

$$v = \sqrt{\frac{1}{cl} - \frac{\gamma}{c}\, w^2 \omega^2} \qquad (20.14)$$

and we see that the upper limiting value

$$v_0 = \frac{1}{\sqrt{cl}} \qquad (20.15)$$

TABLE 20.1. Natural frequencies ω and velocities of propagation v observed in transformer windings of different lengths a.

Connection[a]	Number of waves in winding, m	High-voltage winding			Low-voltage winding		
		a (km)	ω (kc/sec)	v (m/μs)	a (km)	ω (kc/sec)	v (m/μs)
O══O	$\frac{1}{2}$	0.635	67	85	0.362	97	70
●══O	$\frac{1}{4}$	0.635	43.5	111	0.362	76	110
O—O—O	$\frac{1}{2}$	1.270	43.5	111	0.724	75	109
●—O—O	$\frac{1}{4}$	1.270	24	122	0.724	40.5	117

[a] O Open; ● grounded.

is valid only for waves of very low frequencies. However, as the exciting frequency ω increases, the velocity of propagation v decreases to zero, at which point no propagation at this critical or any higher frequency is possible within the winding.

To determine how closely the observations agree with these analytic conclusions, we rewrite Eq. (20.14) in the form

$$\left(\frac{v}{v_0}\right)^2 + \left(\frac{\omega}{v}\right)^2 = 1. \qquad (20.16)$$

Table 20.1 lists also the velocity of propagation in the transformer windings. It was determined

from the limiting values of v_0 and v obtained from the plots of Figs. 20.6 and 20.7 and from the observed values of v and ω in Table 20.1. We see again excellent and fair agreement for the high- and low-voltage windings respectively.

Figure 20.7 gives the limiting value of the velocity of propagation v_0 for low-frequency waves as 127 m/μs. An average combined magnetic and dielectric constant of 4 in the vicinity of the winding would give the velocity v_0 as 150 m/μs quite independently of the detailed geometry of the winding. The value 4 corresponds approximately to a dielectric constant of 3.3 for the oil-impregnated paper

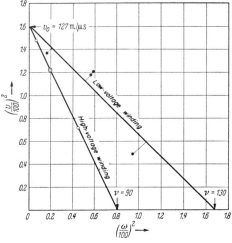

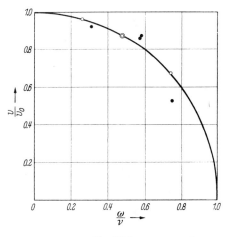

Fig. 20.7   Fig. 20.8

as the product of the observed natural frequency and the actual length of wire for each of the conditions and with due regard to m according to the boundary conditions as stated in Eq. (20.9) and shown in Fig. 20.5. We see that for higher natural frequencies the velocity of propagation of the traveling waves decreases significantly. In Fig. 20.7 v^2 is plotted against ω^2 and we see that the observations for each winding form a straight line, as demanded by Eq. (20.16). Again observations for the high-voltage winding are very much more consistent with the theory than those for the low-voltage winding.

The form of Eq. (20.16) also suggests a circle diagram, as in Fig. 20.8. The normalized values v/v_0 and ω/γ are plotted in Fig. 20.8

insulation and a magnetic permeability of 1.2 for the iron-core surface. We thus see that traveling waves of moderate frequency actually propagate at velocities approaching the appropriate free-space velocity of propagation. However, waves of higher frequency travel considerably more slowly on account of the interturn capacitance.

Figure 20.9 gives a number of observed values for the velocity of propagation as measured over a wide range of transformer windings of different constructions and operating voltages. With increasing operating voltage for the windings, the velocity actually approaches a limiting value of about 150 m/μs. The lower values observed can probably be ascribed to the ratio of internal (or interturn) to external (or ground) capacitance of the windings.

If the propagation velocity v of traveling waves is known empirically for the different types of transformers, the natural frequency of any winding can be simply determined as

$$\omega = 2\pi m \frac{v}{a},\qquad(20.17)$$

where a is the length of the winding and m is the order of the wave as defined in Eq. (20.9).

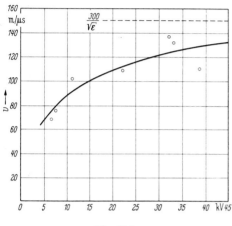

Fig. 20.9

and used in the second column of Table 20.1. If we now compare Eq. (20.17) with Eq. (20.11), we see that the denominator in Eq. (20.11) gives the velocity of propagation of the traveling waves for a given winding as

$$v = \frac{v_0}{\sqrt{1 + 4\pi^2 \dfrac{\gamma}{c}\left(\dfrac{w}{a}m\right)^2}}.\qquad(20.18)$$

Thus if we can deduce the capacitance ratio γ/c from the geometry of the winding it is possible to determine the velocity of propagation in and the natural frequencies of the winding on a purely theoretical basis.

If the frequency impressed upon the winding exceeds the critical frequency as defined by Eq. (20.8), the wave density α in Eq. (20.7) becomes imaginary and it is best to rewrite it in the form

$$\alpha = j\sqrt{\frac{c l \omega^2}{\gamma l w^2 \omega^2 - 1}} = j\beta.\qquad(20.19)$$

The solution of Eq. (20.3) for the voltage has now changed to

$$e = E\varepsilon^{j\omega t}\varepsilon^{-\beta x}.\qquad(20.20)$$

This indicates that above the critical frequency no standing or traveling waves can exist in the windings. Instead, there is merely an exponential decay of the voltage at the start of the winding as we progress into the winding. This was already indicated in Fig. 19.5.

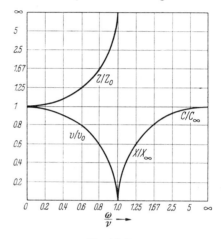

Fig. 20.10

The space constant X of this exponentially decaying voltage distribution is given by Eqs. (20.19) and (20.8),

$$X = \frac{1}{\beta} = w\sqrt{\frac{\gamma}{c}\left[1 - \left(\frac{v}{\omega}\right)^2\right]}.\qquad(20.21)$$

In the vincinity of the critical frequency v the space constant is fairly small, and it increases rapidly with higher frequencies to reach a maximum value

$$X_\infty = w\sqrt{\frac{\gamma}{c}}.\qquad(20.22)$$

Figure 20.10 shows the normalized increase of X with normalized frequency ω according to the equation

$$\left(\frac{X}{X_\infty}\right)^2 + \left(\frac{v}{\omega}\right)^2 = 1,\qquad(20.23)$$

which plots as a circle.

The current i in Eq. (20.1) is the conductor current within each subelement of the winding.

At the end of the winding, however, the total current drawn by the winding is the sum of this conductor current and the displacement current to the next adjacent turn. This is so because this displacement current cannot continue to flow toward the nonexistent next turn. Thus the total current is given by the first and last members of the first line of Eq. (20.1) as

$$i_0 = i + i_\gamma = c \int \frac{\partial e}{\partial t} \, dx. \qquad (20.24)$$

If we evaluate the integral with the aid of Eq. (20.3), the total current at the terminal for $x = 0$ becomes

$$i_0 = c \frac{\omega}{\alpha} E \varepsilon^{j\omega t} \varepsilon^{j\alpha x}. \qquad (20.25)$$

The surge impedance of the transformer at the terminal is the ratio of the voltage given by Eq. (20.3) to the current given by Eq. (20.25). Equations (20.7) and (20.8) thus give it here as

$$\frac{e_0}{i_0} = Z = \frac{\alpha}{c\omega} = \sqrt{\frac{l/c}{1 - \gamma l w^2 \omega^2}}$$

$$= \frac{Z_0}{\sqrt{1 - (\omega/v)^2}}, \qquad (20.26)$$

where

$$Z_0 = \sqrt{\frac{l}{c}} = \sqrt{\frac{L}{C}}, \qquad (20.27)$$

which is the surge impedance for low frequencies in the form used for long lines. This is the same value that we obtained for low frequencies in Chapter 18 using Maxwell's equations and that we used at the beginning of Chapter 17 to deduce numerical values from the dimensions of the windings. We see here that it is merely a lower limiting value for real transformers and that the actual surge impedance grows as the critical frequency is approached (Fig. 20.10), becoming infinite at the critical frequency.

For frequencies above the critical, the ratio of voltage to current becomes imaginary, as can be seen from Eq. (20.26). It is better to express the ratio of current to voltage as

$$\frac{i_0}{e_0} = \frac{\omega c}{\alpha} = j\omega w \sqrt{\gamma c \left[1 - \left(\frac{v}{\omega}\right)^2\right]} = j\omega C.$$

$$(20.28)$$

The winding acts in this region as a pure capacitance C that is small near the critical frequency but rises rapidly with increasing frequency to a maximum of

$$C_\infty = w \sqrt{\gamma c}. \qquad (20.29)$$

This is the equivalent capacitance at the end of an infinitely long ladder of series capacitances γ and ground capacitances c, each extending over the length w of each subelement. This was previously shown in Fig. 19.5.

In Fig. 20.10 all these relations are presented in normalized form as a function of frequency. The normalization used leads to very great compression at the infinity ends of both scales. For very low frequencies the winding behaves like a long line and for very high frequencies it behaves like a long capacitance ladder network. In the vicinity of the critical frequency the characteristic values of v, X, and C approach zero and that of Z approaches infinity. This indicates that the winding is in a state of resonance between adjacent subelements.

For exact numerical calculations at high frequencies it is important to consider the behavior of the characteristic values of the subelements as treated in Chapter 18, Sec. (e).

(b) *Transmitted and reflected voltage pulses.* The usual excitation of a transformer or machinery winding is the result of a traveling pulse that arrives on the power line, as shown in Fig. 20.11 for a step pulse. The pulse may

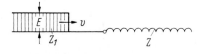

Fig. 20.11

have a variety of shapes; however, excitation by a single sinusoidal high-frequency oscillation occurs only rarely. Any pulse can be represented, according to Fourier's theorem, as the sum of a series of sinusoidal components with frequencies all the way from zero to infinity and of definite amplitudes. The step pulse shown in Fig. 20.11 is given in this form by the Fourier integral

$$e = \frac{E}{2} + \frac{E}{\pi} \int_0^\infty \frac{\sin \omega t}{\omega} \, d\omega, \quad (20.30)$$

as we saw earlier in Eq. (1.56).

The constant term $E/2$ elevates the pulse above the zero level and the integral term represents the summation of sinusoidal terms. We see that the amplitudes of these terms are inversely proportional to the frequency, as shown in the spectral density-distribution diagram of Fig. 20.12.

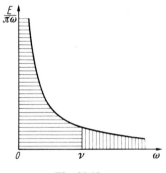

Fig. 20.12

If such a pulse were to impinge on the junction between two lines it would be reflected and transmitted without change of shape, as shown earlier. However, we saw that in a winding only components of frequency below the critical can be propagated. Thus only the portion below v in the amplitude distribution of Fig. 20.12 can be propagated into the winding; it is shown by horizontal shading. Vertical shading shows the frequencies above v, which cannot penetrate into the winding. They merely give rise to an exponentially decaying standing wave in the first elements of the winding and thus also to reflections into the power line.

This decomposition of an incident pulse into two distinct and differently behaving portions is the main difference between the behavior of windings and of homogeneous lines toward traveling pulses. Thus we see that in Eq. (20.30) only the constant term and the portion of the integral below v given by

$$e' = \frac{E}{2} + \frac{E}{\pi} \int_0^v \frac{\sin \omega t}{\omega} \, d\omega, \quad (20.31)$$

travels into the winding, while the portion of the integral above v, given by

$$e'' = \frac{E}{\pi} \int_v^\infty \frac{\sin \omega t}{\omega} \, d\omega, \quad (20.32)$$

will be totally reflected as at the end of an open line.

These definite integrals are well-known mathematical functions, for which tables and curves exist. The "sine integral"

$$\text{Si}(vt) = \int_0^v \frac{\sin \omega t}{\omega} \, d\omega \quad (20.33)$$

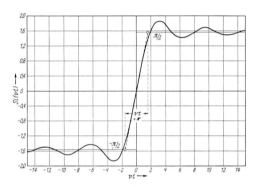

Fig. 20.13

is shown as a curve in Fig. 20.13. It is a pure function of vt, since the variable ω disappears by the integration. Since v can be regarded as a definite and known frequency for every winding, we realize that this function for every case is solely dependent on time. Thus we see that the abrupt step front of the pulse is given a less steep slope, as shown in Fig. 20.13, upon propagation into the winding. In addition to this flattening or stretching out of the steep front rise, we see that there are also some damped oscillations on both sides of the rise.

The sine integral has very simple asymptotic approximations. For small values, up to about $vt = 1$,

$$\text{Si}(vt) = vt, \quad (20.34)$$

and for large values, above about $vt = 4$,

$$\text{Si}(vt) = \frac{\pi}{2} - \frac{\cos vt}{vt}. \quad (20.35)$$

If we express the slope of the rise of this function by the tangent at the origin, we see from Fig. 20.13 that the rise τ of the voltage is given by the intersections of this tangent with the positive and negative values at infinity, and thus according to Eqs. (20.34) and (20.35),

$$v\tau = 2\frac{\pi}{2} = \pi \quad \text{or} \quad \tau = \frac{\pi}{v} = \pi w \sqrt{\gamma l}. \quad (20.36)$$

We see that the rise time is determined solely by the critical frequency of the winding. We

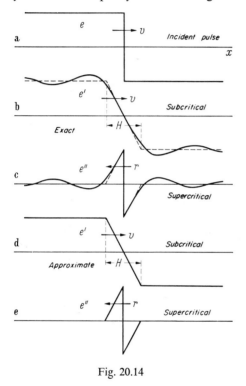

Fig. 20.14

also see from Eq. (20.35) that the oscillations on either side of the rise have a frequency equal to the critical frequency v and decay rapidly.

If we compare the incident step pulse with the pulse shape of Fig. 20.13 within the winding, we see that, owing to the existence of the critical frequency, the original step rise has been flattened or stretched over a definite rise time. Thus every element of the winding is more gradually subjected to a voltage rise.

However, because of the decaying oscillations, the peak value of the voltage is about 12 percent higher than that of the incident step pulse form. If we draw the shape of the pulse in space and not in time we see that the discontinuous incident step or jump pulse of Fig. 20.14a splits into the flattened pulse of Fig. 20.14b, which is transmitted into the winding, and the remainder pulse of Fig. 20.14c, which is reflected back into the line at the terminals. The former we may call the subcritical and the latter the supercritical component.

Although the shape of the subcritical transmitted pulse of Fig. 20.14b has a rise time determined by the value of the critical frequency, the subcritical pulse, by definition and as shown in Fig. 20.12, contains predominantly components of lower frequency. As these lower frequencies propagate with the common velocity v_0, given by Eq. (20.15), the spatial head- or front-rise length H of the subcritical transmitted pulse is, to a good approximation,

$$H_0 = v_0\tau = \pi \frac{v_0}{v} = \pi w \sqrt{\frac{\gamma}{c}}, \quad (20.37)$$

in which the expression for the critical frequency has been introduced from Eq. (20.8). We see that the flattened head of the pulse is a multiple of the turn length w if the internal capacitance γ is larger than the capacitance to ground c. We recognize that the total effect of the interaction of all turns or coils of a winding results in a flattening of pulses in which steep portions are eliminated. However, for travel within the winding and reflection and refraction at the ends the laws developed for homogeneous lines can be used.

To follow the travel of the pulses with the shape of the exact solution as shown in Figs. 20.14b, c will not always be convenient. Thus it may be useful to use the approximate shapes shown in Fig. 20.14d, e, which involve only straight-line segments. They show that the subcritical transmitted pulse has a head rise of length H and a constant-amplitude back, while the supercritical reflected pulse consists of two opposing triangular spikes joined to form the rapid step of the incident step pulse at the center.

With these preparations, it is easy to picture the performance at the beginning of a winding if a steep pulse impinges upon it. We have only to superpose at every instant the traveling pulse e' within the winding, which includes the steady voltage $E/2$, and the exponentially decaying distribution caused by the reflected pulse e'' within the end turns. For the sake of simplicity we can take for the space constant

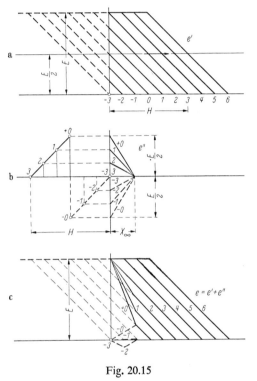

Fig. 20.15

of this component e'' the asymptotic value of Eq. (20.22), because Fig. 20.10 shows that most of the frequency components in this supercritical region are close to it. Numerically we can establish the relation between the head length of the transmitted pulse as given by Eq. (20.37) and the space constant of the reflected pulse as given by Eq. (20.22); it is

$$H_0 = \pi X_\infty. \qquad (20.38)$$

Figure 20.15a shows the propagation of the pulse e' through the entrance terminal in a sequence of successive steps. Figure 20.15b shows the decay of the reflected pulse e'' in the end turns for the same successive steps

in time. The two figures use the straight-line approximations of Fig. 20.14d, e and all events at times prior to 0 are indicated by broken lines. Figure 20.15c shows the superposition of the transmitted pulse and penetration of the reflected pulse. We see that the impinging step at time 0 raises the voltage in the end turns suddenly to its full value but that the steepness of the rise is gradually changing from the end turns toward the interior, where it eventually assumes the flattened value of the transmitted pulse e'.

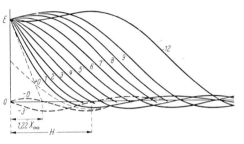

Fig. 20.16

After this approximate picture is clear, substitution of the exact shapes from Fig. 20.14b, c is easier and thus the exact solution as shown in Fig. 20.16 can be obtained. The small voltage excursions that precede the arrival of the step pulse are now more rounded; they are shown by broken lines here also. They are mainly present in our analysis because of our assumptions, particularly that of a constant velocity v_0 for the entire pulse, which we shall examine in detail later. We see in Fig. 20.16 that at time 0 the pulse in the winding has a shape that decays with a space constant about 22 percent greater than X_∞. This voltage appears at the instant of arrival of the impinging pulse at the entrance junction. This is due to the action of the interturn and ground capacitances of the winding. This initial shape changes, at first slowly and then with increasing speed, into a traveling pulse of rounded shape which remains constant at the entrance junction but oscillates by about 12 percent about its final value. As a result, not only the ends but also the interior of the winding experience voltage stresses far higher than those due to the steady-state operating voltages.

For the last cases the reflection and transmission coefficients at the terminal junction, for e'' and e' respectively, were assumed equal. In most practical cases, however, the surge impedance Z of the winding is higher than that of the supply line and increases even further in the vicinity of the critical frequency, as shown in Fig. 20.10. Thus the transmission factor approaches the value 2 in many cases. On the other hand, the entrance capacitance at the terminals, for frequencies above the critical, has a low value for frequencies only a little above the critical and approaches a moderately large value for most windings at higher frequencies. Thus the reflection factor is close to 1. More rigorous examination shows that these approximations for Z and C do not seriously disturb the results obtained. The final value of the voltage E within the winding due to the incident voltage pulse E_1 is

$$\frac{E}{E_1} = \frac{2Z}{Z + Z_1}, \qquad (20.39)$$

where Z_1 is the surge impedance of the supply line.

Figure 17.3 showed that the surge impedance at the end of the winding is only one-half of that inside the winding but rises rapidly to the full value. Thus very steep pulses would be transmitted with somewhat higher values than that given by Eq. (20.39), because with gradually increasing Z they would experience little or no reflection, as was demonstrated in Chapter 10 for spatially varying surge impedance. However, the halving of the surge impedance at the entrance is accompanied by a doubling of the effective capacitance, which will flatten out very steep pulse fronts and thus reduce somewhat the basic cause for excess voltage. For practical designs these two detailed features, supporting opposing trends, may cancel each other sufficiently that we may omit them from consideration.

When the flattened front of the transmitted pulse has traveled to the far end of the winding, such as the neutral star point of a three-phase winding, it will be reflected in accordance with rules established earlier. Figure 20.17 shows several sequences of such a reflection for a grounded end and Fig. 20.18 for an insulated end. For both figures we have chosen the pulse-front rise time to be one-third the time taken to traverse the winding. However, according to Eq. (20.37), the front rise may sometimes be longer than an entire winding,

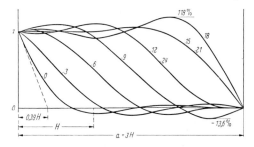

Fig. 20.17

depending on the square root of the capacitance ratio. In such cases the concentrated reflections shown in Figs. 20.17 and 20.18 would be distributed over several phase

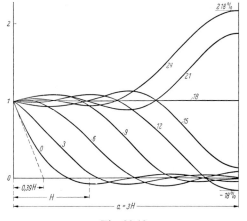

Fig. 20.18

lengths (that is, multiple internal reflections would superpose). This is often the case in transformers, while in the windings of rotating machinery the high capacitance to ground will usually make the pulse front shorter than the phase length of the winding.

In any case, we recognize from Figs. 20.17 and 20.18 that with a grounded end the voltage stress within the windings near the neutral point is doubled, while with an insulated end the voltage stress with respect to ground is doubled. Owing to the overshoot of the sine-integral curves shown in Fig. 20.13 a further

18-percent voltage excess over the final state is caused for both the grounded and the insulated neutral case. In both cases small precursor voltages appear and give rise to negative voltage stresses between 10 and 20 percent at or near the neutral point. The shape and speed of the traveling pulse front do not remain as uniform in the vicinity of the neutral point as they are in the central portion of the windings, since the reflected precursor voltages cause some distortions.

These calculated voltage distributions can be compared with the cathode-ray oscillograms obtained by J. F. Calvert and T. E. Allibone shown in Figs. 20.19 and 20.20. We see that

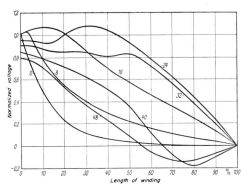

Fig. 20.19

there is excellent agreement between our analytic results and those obtained from these measurements. Even very detailed features, such as the overshoot of the pulse voltage and the negative precursor voltages, show up as analyzed in the measurements. Observations of these features had in the past been somewhat puzzling.

In the literature the behavior of windings after the incidence of step pulses is often treated by determining the initial voltage distribution in the capacitance ladder shown in Fig. 19.5. Next the amplitudes of all natural oscillations of the winding, as determined by Eq. (20.9) for either a grounded or an insulated end, are determined from the initial and final steady-state linear voltage distribution in the winding after the pulse has decayed. Finally, this infinite series, with its different frequencies and wavelengths, is summed numerically for a number of successive in-

stants of time. Because these series converge very slowly, a large number of terms is necessary and precise calculations are possible only for selected numerical examples; even then considerable computational effort is required. Even though this treatment by the summation of standing waves is in principle as correct and accurate as the use of traveling waves developed here, we shall in the following further examination use this latter method for the sake of simplicity and the insight into the phenomena that it makes possible.

The equations and curves developed so far give primarily the voltage differences between

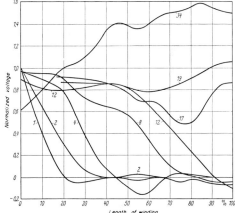

Fig. 20.20

points in the winding and ground. However, the equally important stress of the insulation between adjacent turns, layers, or coils composing the winding is given by the voltage difference between these elements of the winding.

If we wish to obtain this voltage difference over a distance w equal to the turn length of a winding, its peak value for a traveling pulse with a front-rise length H_0 is, from Eq. (20.37),

$$e' = \frac{w}{H_0} E = \frac{E}{\pi} \sqrt{\frac{c}{\gamma}}. \qquad (20.40)$$

Every internal element of the winding is successively exposed to this excess voltage. The duration of the exposure is determined by the rise time τ given by Eq. (20.36) and is always a very small fraction of a second.

The end turns or coils, however, are exposed, in addition, to the stress of the standing exponentially decaying voltage distribution. This, as shown in Fig. 20.15*b, c* has a space constant X and an amplitude $E/2$ and thus, from Eq. (20.22), is

$$e'' = \frac{w}{X_\infty} \frac{E}{2} = \frac{E}{2} \sqrt{\frac{c}{\gamma}}. \qquad (20.41)$$

Thus the total stress near the terminal is

$$e = e' + e'' = \left(\frac{1}{2} + \frac{1}{\pi}\right) E \sqrt{\frac{c}{\gamma}}$$

$$= 0.818 E \sqrt{\frac{c}{\gamma}}, \qquad (20.42)$$

and its ratio to the stress inside the winding is

$$\frac{e}{e'} = 1 + \frac{\pi}{2} = 2.57, \qquad (20.43)$$

a constant.

The duration of this excess stress at the terminal is only $\tau/2$, since, as we see in Fig. 20.15*b, c*, it lasts only one-half the time required for the front rise to travel past the entrance terminal.

In every case we must remember that E is the final value of the voltage transmitted into the winding. It may easily be twice the value of the impinging step pulse, depending on the surge impedances of the winding and power line, as related in Eq. (20.39). The numerical factor in Eq. (20.42) will in practice be a little larger than indicated, because to derive it we used the maximum values of X and H in the denominators of Eqs. (20.40) and (20.41), which apply at the extremes of frequency only while the average values applicable are, according to Fig. 20.10, somewhat smaller.

In Chapter 17, Sec. (*d*), we saw that in addition to the air-space and stray-field flux-linkage oscillations there are also very low-frequency oscillations that may arise owing to the core flux and that are governed by somewhat different relations, as indicated in Fig. 17.9. Such core-flux oscillations cause a linear voltage distribution over the length of the winding and thus do not contribute to concentration of voltage-distribution discontinuities. Hence we can regard them as additional low-frequency components of the

spectral density distribution shown in Fig. 20.12 that will not substantially change the effect and strength of the dangerous high-frequency components that determine the shape of the pulse front.

Our concept of pulses traveling in the winding has given us in Eqs. (20.36), (20.40), and (20.42) very simple expressions for the magnitude and duration of the excess voltage in the body as well as the end turns of windings. Large internal coil capacitance reduces the magnitude but increases the duration, while low ground capacitance reduces the amplitude without changing the duration. Thus for the

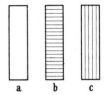

Fig. 20.21

best possible self-protection of the winding, every turn, every layer, and every coil should present the widest possible sides toward each other and the narrowest possible sides toward the ground. If the winding area given is oblong, as in Fig. 20.21*a*, it is thus evidently better to arrange the elements not short-sided, as in Fig. 20.21*b*, but long-sided, as in Fig. 20.21*c*. If, on account of large internal capacitance, the head length H and the space constant X become much larger than the winding length a, the excess voltages decrease toward zero and the problem of the distribution of voltage thus becomes a static one. This is the aim of every design of surge-proof windings.

If the incident pulse does not have a discontinuity, as the step pulse shown in Fig. 20.11 does, but instead has a front already somewhat flattened, then the spectral distribution is somewhat different from that shown in Fig. 20.12. However, as long as the front is steeper than the head H that enters the winding, only the higher frequencies above the critical frequency are influenced. The subcritical frequencies retain the same amplitudes as in Fig. 20.12 and travel undisturbed through the winding. Thus we recognize that for an incident pulse of any shape those parts that are steeper than the corresponding critical head length H in the winding are cut off and reflected at the terminals.

A capacitance C_s connected from the entrance terminal to ground for the protection of the winding, as in Fig. 20.22, will flatten the rise of any step pulse into a pulse with a gradual rise front, with the well-known time constant

$$T_C = \frac{C_s}{\dfrac{1}{Z_1} + \dfrac{1}{Z}}. \qquad (20.44)$$

Thus we can obtain protection against the excess voltage e'', if we choose this condenser

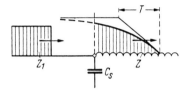

Fig. 20.22

so that its time constant T_C will be equal to or larger than the rise time τ. Then only the transmitted part of the pulse remains, and this gives rise to the uniform voltage stress e' over the entire winding. Thus setting Eqs. (20.44) and (20.36) equal we obtain the minimum value of such a safety condenser,

$$C_s = \pi \left(1 + \frac{Z}{Z_1} \right) w \sqrt{c\gamma}. \qquad (20.45)$$

We see that this is many times the capacitance C_∞ of the winding as given by Eq. (20.29).

Because the internal capacitance γ is so very important for the behavior of traveling pulses in windings of all kinds and thus for the design of surge-proof transformers and rotating machines, it will be very useful to obtain an idea of the range of values encountered. The best grasp on this range can be obtained by evaluating the capacitance ratio γ/c of the internal capacitance to the ground capacitance of turns, layers, or coils.

In both transformers and rotating machines there are two distinct types of winding. The actual coils of rectangular copper strips can be wound flat as in Fig. 20.23 or long as in Fig. 20.24. The layers of the coils usually are

disposed in flat parallel form as shown in Fig. 20.23. The coils finally can be arranged as flat pancakes, as in Fig. 20.23, or long cylinders, as in Fig. 20.24.

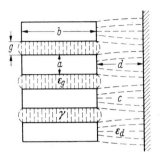

Fig. 20.23

The capacitance between adjacent conductor surfaces is always proportional to a dimensional constant, the dielectric constant ε of the insulation in the gap, the surface area, and the reciprocal of the separation between the

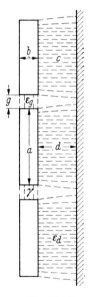

Fig. 20.24

surfaces. Thus for the dimensional proportions shown in Figs. 20.23 and 20.24 we obtain, leaving out the constant dimensional factor, for the ground capacitance of an element,

$$c = \varepsilon_d \frac{a}{d}, \qquad (20.46)$$

and for the internal capacitance between elements,

$$\gamma = \varepsilon_g \frac{b}{g}. \qquad (20.47)$$

Thus the ratio of internal to ground capacitance is

$$\frac{\gamma}{c} = \frac{b}{a} \cdot \frac{d}{g} \cdot \frac{\varepsilon_g}{\varepsilon_d}. \qquad (20.48)$$

This important ratio can vary over a wide range for the different types of construction.

For the flat-pancake type of Fig. 20.23 the ranges of values are about as follows:

$$\frac{b}{a} \cong 2 \text{ to } 20; \quad \frac{d}{g} \cong 5 \text{ to } 50; \quad \frac{\varepsilon_g}{\varepsilon_d} \cong 1 \text{ to } 30.$$

Here the ratio of ε values is based on the use of the older insulating materials such as air, oil, paper, or mica compounds in the spaces between the winding as a whole and ground, while newer insulating materials of much higher dielectric constants are available for insulation within the winding. This advantage has as yet been only little exploited in the construction of transformers and rotating machines. These ranges then lead to a range of the capacitance ratio

$$\frac{\gamma}{c} = 10 \text{ to } 30{,}000.$$

For the long-cylindrical type of Fig. 20.24 the ranges of values are about as follows:

$$\frac{b}{a} \cong \frac{1}{20} \text{ to } \frac{1}{2}; \quad \frac{d}{g} \cong 5 \text{ to } 50; \quad \frac{\varepsilon_g}{\varepsilon_d} \cong 1.$$

Here it will not be possible to make ε very different in the insulating spaces. Thus the range of capacitance ratios is

$$\frac{\gamma}{c} = \frac{1}{4} \text{ to } 25.$$

We thus see that this type of winding is much less safe with respect to surges. However, it unfortunately cannot always be avoided. In general one will encounter much lower capacitance ratios in rotating machines than in well-designed transformers. Thus rotating machines are much more sensitive to surge voltages than transformers and cannot be

designed for arbitrarily high voltages, quite apart from the limited space available for insulation in the rotor and stator slots.

(c) Flattening of the propagating pulse front. The velocity of propagation v of every partial oscillation in the winding depends on its frequency ω as shown in Eq. (20.14). Thus pulses composed of a large number of oscillations superposed show the effect of dispersion due to the change in velocity of propagation with frequency. While long waves of low

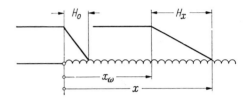

Fig. 20.25

frequency travel with a definite high velocity, given by Eq. (20.14), the higher-frequency partial waves travel with lower speed and therefore fall behind the longer waves. To determine the effect upon the shape of the pulse front we shall use here a geometric presentation, which is not rigorous but only approximate. However, it is simple enough to give us an overview of the phenomena.

If a pulse with uniformly rising front, as shown in Fig. 20.25, travels along the winding a distance x, it is distorted in shape because only the fast-traveling components with velocity v_0 propagate the start of the pulse. In a given time t the start of the pulse has propagated over a distance

$$x = v_0 t. \qquad (20.49)$$

During the same time, however, the slower partial waves, for example those of higher frequency ω and velocity v, have been propagated only a distance x_ω which, according to Eqs. (20.14) and (20.16), is

$$x_\omega = vt = v_0 t \sqrt{1 - \left(\frac{\omega}{v}\right)^2}. \qquad (20.50)$$

The front of the pulse ahead of this distance x_ω cannot contain components of frequency higher than ω. Thus the length of the pulse front H_x shown in Fig. 20.25 is

$$H_x = x - x_\omega = x\left[1 - \sqrt{1 - \left(\frac{\omega}{\nu}\right)^2}\right].$$

(20.51)

Now the pulse-front length of the flattened pulse is in principle determined by the highest frequency that is contained within the pulse spectrum. For the critical frequency ν the relation for the front length H_0, based on all frequency components lower than ν, as shown in Figs. 20.12 and 20.13, is given by Eq. (20.37). If, however, the pulse front is composed only of frequency components below

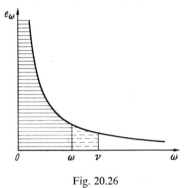

Fig. 20.26

ω, as shown in Fig. 20.26, the corresponding front length at x becomes

$$H_x = \frac{\pi v_0}{\omega} = \frac{\nu}{\omega} H_0.$$

(20.52)

Here the last term normalizes the length to that at the entrance of the winding H_0. If we now enter the frequency ratio ν/ω from Eq. (20.52) in Eq. (20.51) we obtain

$$H_x = x\left[1 - \sqrt{1 - \left(\frac{H_0}{H_x}\right)^2}\right].$$

(20.53)

This expression contains only the original front length H_0, the flattened front length H_x, and the distance x over which the start of the pulse has been propagated into the winding. This is an equation of the fourth degree in H_x and thus difficult to solve. However, if we normalize x in terms of the original front length H_0, we obtain the far simpler expression

$$\frac{x}{H_0} = \frac{H_x/H_0}{1 - \sqrt{1 - (H_0/H_x)^2}}$$

$$= \left(\frac{H_x}{H_0}\right)^3\left[1 + \sqrt{1 - \left(\frac{H_0}{H_x}\right)^2}\right].$$

(20.54)

This normalized expression is shown graphically in Fig. 20.27 and we see that the front length increases only relatively slowly with increasing distance traveled by the start of the pulse. To double the front length, the start of the pulse must travel a distance 15 times its original length. To flatten the front very significantly, distances of the order of hundreds to thousands of the original front length must be covered.

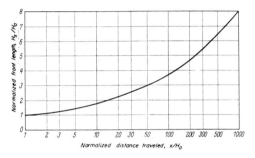

Fig. 20.27

The initial front length of pulses in transformer or machinery windings extends over a range of fractions to small multiples of the winding length. Thus the increased flattening is very moderate in a single traversal of a winding in both directions. However, with multiple reflections leading to repeated traversals of the winding, flattening of the pulse will become increasingly important. Thus oscillograms of voltage or current pulses in such windings show sharp features under the influence of the original short H_0 only for the first few oscillations. The later oscillations become increasingly flattened in shape and ultimately approach sinusoidal shape owing to the increasing front length H_x.

For large distance of travel x the ratio H_0/H_x in Eq. (20.54) will become very small and this allows us to develop an approximation for the front length, namely

$$\frac{H_x}{H_0} = \sqrt[3]{\frac{1}{2}\frac{x}{H_0}}.$$

(20.55)

Thus the flattening ultimately only grows with one-third power of the distance traveled.

Quite similarly, the voltage stress to which

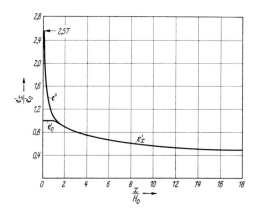

Fig. 20.28

the internal insulation of the windings is exposed also varies for large values of x as

$$\frac{e_x'}{e_0'} = \frac{H_0}{H_x} = \sqrt[3]{2\frac{H_0}{x}}, \qquad (20.56)$$

because it is inversely proportional to the front length.

Figure 20.28 shows the exact distribution of the stress for all values of x down to zero, including also the exponentially decaying stress e'' caused by the reflected pulse given by Eqs. (20.41) and (20.43). This total curve agrees very well with the innumerable experimental results published for the voltage stress distribution in the windings of transformers and rotating machines.

If one attempts to determine the velocity of propagation of pulses, some care is required to obtain the exact start of the pulse because it alone travels with a constant velocity, which also represents the upper limit. The central portion, the end of the front rise, or other significant features of more complicated pulses all have a lower velocity of propagation along the winding, depending on the position of the observed feature within the pulse. Only sinusoidal pulses have a well-defined velocity of propagation, as given by Eq. (20.14). Thus such pulses, or, better, waves, are eminently suited to the measurement of the velocity of propagation in windings.

CHAPTER 21

OSCILLATIONS AND PULSES IN SUBDIVIDED WINDINGS

In the last two chapters it was seen that uniformly distributed windings in chokes, transformers, or rotating machines differ from homogeneous lines in their behavior under traveling pulses. The front of an impinging steep pulse is flattened by the effect of the internal capacitance of the winding and only this flattened part can propagate deep into the winding. The remainder of the

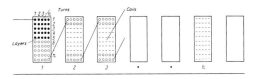

Fig. 21.1

incident pulse is reflected within the terminal elements of the winding back into the feeding line.

In actual practice windings are always built up of many distinct elements. This is shown in Fig. 21.1 for a transformer. Here several turns form a layer, several layers form a coil and a number of coils form the entire winding. Figure 21.2 shows a corresponding structure for the stator or rotor winding of a machine.

Thus for actual windings there appear two new problems that must be analyzed: first, the effect of a finite number of elements in a group, such as 6 coils forming the whole winding, or 20 layers forming a complete coil, or 12 turns forming one layer; second, the effect of pulses or waves passing from one group to another one with different elements, as from turns to layers, or from layers to coils.

(a) *A finite number of elements.* We denote by e_n and i_n the mean voltage and the mean current, respectively, in the nth element of a

group, by l and c the mean self-inductance and ground capacitance per unit length of the wire, and by γ the internal capacitance per unit length between any two adjacent elements. Note that l and c have to be regarded again as the contribution of each element of length

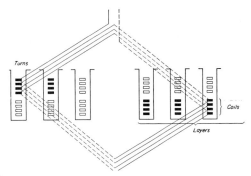

Fig. 21.2

w to the total self-inductance and capacitance of the winding. In the differential equations of this problem we must now retain the finite differences Δ of current and voltage over an element of length $w = \Delta x$ in the winding, as developed in Eqs. (19.7) and (19.14), namely

$$-\Delta i_n = cw \frac{\partial e_n}{\partial t} - \gamma w \frac{\partial \Delta^2 e_n}{\partial t},$$
$$-\Delta e_n = lw \frac{\partial i_n}{\partial t}, \tag{21.1}$$

in which we have again neglected as insignificant the term involving mutual inductance between adjacent subelements. Here t is the time, Δ denotes the difference of current or voltage between two adjacent elements, and

$$\Delta^2 e_n = e_{n+1} - 2e_n + e_{n-1} \tag{21.2}$$

is the second difference between three consecutive elements.

203

By differentiating the first of Eqs. (21.1) with respect to t and forming the second space difference, using the second of Eqs. (21.1), we can eliminate the current i and thus obtain for the voltage e_n of the nth element

$$\frac{\Delta^2 e_n}{w^2} - lc\frac{\partial^2 e_n}{\partial t^2} + l\gamma\frac{\partial^2}{\partial t^2}(\Delta^2 e_n) = 0. \quad (21.3)$$

This is valid for any finite number of winding elements of the same kind.

We solve this equation first again for standing harmonic oscillations of amplitude E by the trial form

$$e_n = E\varepsilon^{j\omega t}\varepsilon^{j\alpha n w}, \quad (21.4)$$

in which ω is the frequency and α the wave density in space. Instead of a continuous coordinate x along the wire, we have introduced in the last term the variable

$$nw = n\Delta x, \quad (21.5)$$

which increases in finite steps if n is an integer denoting the number of the element considered. For a winding consisting of 6 coils, for example, n increases from 1 to 6.

According to the definition of Eq. (21.2), the second difference of the coil voltage e_n is

$$\Delta^2 e_n = (\varepsilon^{j\alpha w(n+1)} - 2\varepsilon^{j\alpha w n} + \varepsilon^{j\alpha w(n-1)})E\varepsilon^{j\omega t}$$
$$= (\varepsilon^{j\alpha w} - 2 + \varepsilon^{-j\alpha w})e_n$$
$$= (\varepsilon^{j\alpha w/2} - \varepsilon^{-j\alpha w/2})^2 e_n \quad (21.6)$$
$$= -4e_n \sin^2\left(\alpha\frac{w}{2}\right).$$

By introducing this expression into Eq. (21.3) and differentiating, we obtain the characteristic equation

$$-\frac{4}{w^2}\sin^2\left(\alpha\frac{w}{2}\right) + lc\omega^2 + l\gamma\omega^2$$
$$\cdot 4\sin^2\left(\alpha\frac{w}{2}\right) = 0. \quad (21.7)$$

If we compare the formation of a difference with the differentiation as in Eq. (20.4) of a harmonic function, we obtain as multiplier, instead of the wave density α, the somewhat more complicated function

$$\frac{2}{w}\sin\left(\alpha\frac{w}{2}\right).$$

If we change from the standing-wave solution of Eq. (21.4) to one for traveling waves, and let

$$e_n = E\varepsilon^{j\omega(t-nw/v)}, \quad (21.8)$$

we see that the velocity of propagation v along the wire is given simply by

$$v = -\frac{\omega}{\alpha}, \quad (21.9)$$

which can be computed with the aid of the solution of Eq. (21.7).

For the wave density α and the frequency ω we obtain the solution of Eq. (21.7),

$$\frac{2}{w}\sin\left(\alpha\frac{w}{2}\right) = \sqrt{\frac{c l\omega^2}{1 - \gamma l\omega^2 w^2}}$$

or $\qquad\qquad (21.10)$

$$\omega = \frac{\frac{2}{w}\sin(\alpha w/2)}{\sqrt{cl + \gamma l\,4\sin^2(\alpha w/2)}},$$

which is transcendental in α but nevertheless easy to compute numerically. For increasing ω, α at first increases proportionally. However, since the sine cannot exceed the value 1, the square root of the first of Eqs. (21.10) has, with further increase in ω, a definite limit long before its denominator becomes zero. Our trial solution Eq. (21.4) is valid only up to this definite limit. The frequency at this limit constitutes a critical value ν which is, according to the second of Eqs. (21.10),

$$\nu = \frac{2/w}{\sqrt{cl + 4\gamma l}} = \frac{1}{w\sqrt{\gamma l + cl/4}}, \quad (21.11)$$

since the sine is now 1.

The critical frequency of a finite number of elements is thus determined by the sum of the internal capacitance γ and one-fourth of the ground capacitance c, and is therefore smaller than for an infinite number of elements, where the fraction $c/4$ does not exist.

Even if there is no internal capacitance, there remains a definite critical frequency, namely,

for $\qquad \gamma \cong 0, \quad \nu_0 = \frac{2}{w\sqrt{cl}}, \quad (21.12)$

which is due solely to the lumped arrangement

of the winding elements of length w. If, on the other hand, the internal capacitance is overwhelming, we obtain the extreme value

for
$$\gamma \cong \infty, \quad \nu_\infty = \frac{1}{w\sqrt{\gamma l}}, \qquad (21.13)$$

which coincides with the value for uniform coils. For machine windings with small γ, the critical frequency is in the neighborhood of ν_0; for transformer windings with large γ, it approaches ν_∞. The actual critical frequency is given by

$$\nu^2 = \frac{1}{1/\nu_0^2 + 1/\nu_\infty^2} \qquad (21.14)$$

and is always smaller than either ν_0 or ν_∞.

The wave density α_ν in the critical state of the windings follows from

$$\sin\left(\alpha\frac{w}{2}\right) = 1 \quad \text{as} \quad \alpha_\nu = \frac{\pi}{w}, \qquad (21.15)$$

and therefore the wavelength becomes, as the reciprocal of the density,

$$\lambda_\nu = \frac{2\pi}{\alpha_\nu} = 2w. \qquad (21.16)$$

This means that every element having the length w oscillates against its two adjacent

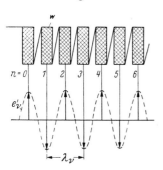

Fig. 21.3

elements, forming a half-wavelength of this resonance state, as represented in Fig. 21.3. A finite number of coils, layers, or turns thus prevents the wavelength of harmonic oscillations from decreasing toward zero, as it would do with infinite subdivision.

The natural frequencies of a winding with open or closed ends are given by the number m of full waves over the total wire length a, so that the wave density becomes

$$\alpha = \frac{2\pi}{\lambda} = 2\pi\frac{m}{a}. \qquad (21.17)$$

Inserting this in the second of Eqs. (21.10), we get

$$\omega = \frac{2}{w\sqrt{cl}} \cdot \frac{1}{\sqrt{4\gamma/c + 1/\sin^2(\pi m w/a)}}. \qquad (21.18)$$

If in this equation we increase the number m from its fundamental value of $\frac{1}{4}$ to integral multiples of this value, we obtain the entire series of natural frequencies. However, we cannot increase m beyond the value

$$m_\nu = \frac{a}{2w}, \qquad (21.19)$$

because otherwise the argument of the sine in Eq. (21.18) would exceed $\pi/2$. The number of natural frequencies of every group of elements is therefore limited and, since a/w is identical with the number of elements, we see that there develop only a finite number of natural frequencies, equal to twice the number of elements in every group.

For example, a winding with both ends open, consisting of $a/w = 6$ coils, has only six different natural frequencies, the lowest one given by $m = 1/2$, the highest by $m = 6/2$,

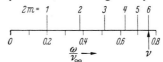

Fig. 21.4

according to Eq. (21.19). If one end of the winding is grounded, the lowest natural frequency corresponds to $m = 1/4$, the highest to $m = 11/4$, the number of frequencies remaining 6. The two groups together thus form 12 possible natural frequencies. All higher frequencies of the two groups are cut off by the effect of the lumped elements and the spectrum of frequencies is limited as shown in Fig. 21.4. The highest natural frequency of each group either coincides with or is adjacent to the critical frequency, as we see by comparison of Eqs. (21.18) and (21.11).

The fundamental frequency is given by $m = 1/4$ or $1/2$, and, if the number of elements a/w is not too small, the argument of the sine in Eq. (21.18) for this lowest frequency is a small fraction of $\pi/2$, and can thus be taken instead of the sine itself. In this case there is no substantial difference between the rigorous calculation and the results for uniform windings.

The wave velocity can be computed by inserting α from Eq. (21.9) in the first of Eqs. (21.10). We obtain

$$\sin\left(\frac{\omega w}{2v}\right) = \frac{\omega w}{2}\frac{1}{\sqrt{(1/cl) - (\gamma/c)\omega^2 w^2}} = \frac{\omega w}{2v_\infty},$$

$$(21.20)$$

where the right-hand member introduces the velocity v_∞ for infinite subdivision. For low

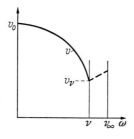

Fig. 21.5

frequencies, the velocity simplifies in any case to

$$v_0 = \frac{1}{\sqrt{lc}}. \qquad (21.21)$$

For higher frequencies, we realize that, in addition to the decrease due to internal capacitance, the velocity of propagation is always diminished by the effect of lumped elements. Its variation is represented by Fig. 21.5. At the critical frequency the argument of the sine in Eq. (21.20) becomes $\pi/2$ and thus the limiting velocity is

$$v_v = \frac{vw}{\pi} = \frac{2/\pi}{\sqrt{cl + 4\gamma l}}. \qquad (21.22)$$

The critical velocity of waves therefore remains at a finite value, which, with the limiting value v_0 for low frequencies, as given by Eq. (21.21), is

for $\qquad \gamma \lessgtr c, \quad v_v = \frac{2}{\pi}v_0;$

$$(21.23)$$

for $\qquad \gamma \gtrless c, \quad v_v = \frac{v_0}{\pi}\sqrt{\frac{c}{\gamma}}.$

This remaining velocity corresponds to the fact that in the critical state there are well-formed standing waves on the winding.

The current at the entrance terminals of the winding, which is composed of the conductive current and the internal-capacitance displacement current, is, by Eq. (20.24),

$$i_0 = cw \Sigma \frac{\partial e_n}{\partial t}. \qquad (21.24)$$

For a finite number of elements the summation is the inverse operation to the formation of the differences. This puts a sine expression

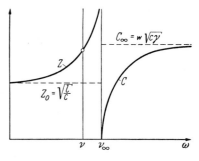

Fig. 21.6

into the denominator and so we obtain from the voltage in Eq. (21.4)

$$i_0 = \frac{c}{(2/w)\sin(\alpha w/2)}E\varepsilon^{j\omega t}\varepsilon^{j\alpha nw}, \qquad (21.25)$$

which is valid only for $n = 0$. Thus the surge impedance becomes

$$Z = \frac{e_0}{i_0} = \frac{2}{w}\frac{\sin(\alpha w/2)}{c\omega}$$

$$= \sqrt{\frac{l/c}{1 - \gamma l\omega^2 w^2}} = \frac{Z_0}{\sqrt{1 - (\omega/v_\infty)^2}}. \qquad (21.26)$$

We have inserted here the first of Eqs. (21.10) and see that the surge impedance is independent of the subdivision and increases hyperbolically with ω from an initial value Z_0 that is identical with the surge impedance commonly used. The variation is shown in Fig. 21.6.

(b) *Phenomena at supercritical frequencies.* If the frequency ω exceeds the critical frequency ν, the determining sine in Eqs. (21.10) becomes larger than 1. This is possible only for a complex argument of the sine and we extend this therefore to $\alpha + j\delta$. We can then develop the sine into

$$\sin(\alpha + j\delta)\frac{w}{2} = \sin\left(\alpha\frac{w}{2}\right)\cosh\left(\delta\frac{w}{2}\right)$$

$$+ j\cos\left(\alpha\frac{w}{2}\right)\sinh\left(\delta\frac{w}{2}\right). \quad (21.27)$$

The real part of this is determined by the square root in the first of Eqs. (21.10), while the imaginary part must vanish. As δ certainly remains finite, this second condition gives

$$\cos\left(\alpha\frac{w}{2}\right) = 0 \quad \text{or} \quad \alpha w = \pi. \quad (21.28)$$

The sine in the first term in the right-hand member of Eq. (21.27) is therefore equal to 1 and thus there remains the condition

$$\cosh\left(\delta\frac{w}{2}\right) = \frac{w}{2}\sqrt{\frac{c l\omega^2}{1 - \gamma l\omega^2 w^2}}, \quad (21.29)$$

from which δ can be computed. Every value between 1 and ∞ for the expression in the right-hand member of Eq. (21.29) gives a definite value of δ. The corresponding variation of ω is from $\omega = \nu$, according to Eq. (21.11), to $\omega = \nu_\infty$, according to Eq. (21.13).

The complex argument in Eq. (21.27) alters the form of the harmonic wave within the winding, as given by Eq. (21.4). We now obtain for the spatial factor, if we use for α Eq. (21.28),

$$\varepsilon^{j(\alpha + j\delta)nw} = \varepsilon^{jn\pi}\varepsilon^{-\delta nw}, \quad (21.30)$$

and, since the first term in the right-hand member for various integers n is alternately $+1$ and -1, the voltage distribution becomes

$$e_n''' = E\varepsilon^{j\omega t}(-1)^n\varepsilon^{-\delta nw}. \quad (21.31)$$

This distribution is shown in Fig. 21.7 and consists of a standing attenuated alternating voltage along the elements of the winding.

For the critical frequency ν, the damping factor δ is zero, according to Eq. (21.29), and the distribution then coincides with that of Fig. 21.3. For higher frequencies δ increases,

and for the frequency ν_∞ it becomes infinite, so that only the first coil receives any voltage. As the voltages of adjacent coils are of opposite sign to each other, the winding is in a kind of resonance at frequencies between ν and ν_∞, and an excitation by any frequency ω within this range appears particularly dangerous.

The ratio of terminal voltage to current within this range, ν to ν_∞, always remains real and still follows Eq. (21.26). The surge

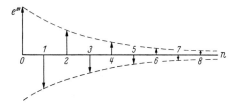

Fig. 21.7

impedance within this range, however, is large and rapidly approaches infinity, according to Fig. 21.6, indicating an actual state of resonance.

We may conceive of the standing alternating distribution in Fig. 21.7 as produced by traveling waves, although this concept would not be useful here. The velocity of the necessary pair of exponentially decreasing or increasing traveling waves would then be given by Eqs. (21.9) and (21.28) as

$$v = \pm\frac{\omega w}{\pi}, \quad (21.32)$$

which is represented by a broken line in Fig. 21.5 and closely joins the end of the previous curve for v, as seen from Eq. (21.22).

For exciting frequencies exceeding ν_∞, the square root in the first of Eqs. (21.10) becomes imaginary. Thus the argument of the sine is also imaginary, and we now write $j\vartheta$ instead of α. If we now introduce

$$\sin\left(j\vartheta\frac{w}{2}\right) = j\sinh\left(\vartheta\frac{w}{2}\right), \quad (21.33)$$

our characteristic Eq. (21.10) appears in the real form

$$\sinh\left(\vartheta\frac{w}{2}\right) = \frac{w}{2}\sqrt{\frac{c l\omega^2}{\gamma l\omega^2 w^2 - 1}}. \quad (21.34)$$

The voltage distribution of Eq. (21.4) now changes to

$$e_n'' = E\varepsilon^{j\omega t}\varepsilon^{-\vartheta nw}, \qquad (21.35)$$

which represents a uniformly attenuated standing-voltage-wave distribution. The space constant of this exponential curve is given by the inverse value of ϑ. For every value of ω between ν_∞ and ∞ the square root in Eq. (21.34) has a real value, decreasing from ∞, and therefore ϑ can always be computed numerically.

For the frequency ν_∞ at the lower limit of the range, the value of ϑ becomes infinite and therefore the space constant X is zero. For higher frequencies, the right-hand member of Eq. (21.34) rapidly approaches an asymptotic value, the space constant of which is given by

$$\sinh\left(\frac{w}{2X_\infty}\right) = \frac{1}{2}\sqrt{\frac{c}{\gamma}}. \qquad (21.36)$$

For overwhelming internal capacitance γ the value of Eq. (21.36) is very small and therefore, to a good approximation, the space constant becomes

$$X_\infty = w\sqrt{\frac{\gamma}{c}}, \qquad (21.37)$$

which coincides with the expression for uniform windings. For small values of γ/c, we may develop the hyperbolic sine of Eq. (21.36) into an exponential function and derive the approximation

$$X_\infty = \frac{w}{2\ln\sqrt{c/\gamma}}. \qquad (21.38)$$

Thus the space constant still decreases with γ, but more slowly than in a continuous winding.

Equation (21.26) for the ratio of terminal voltage to current remains valid even for frequencies above ν_∞, because the transcendental function has disappeared. However, the square root becomes imaginary for this upper range of frequencies and therefore the winding acts under this condition as a pure capacitance. We therefore invert Eq. (21.26), obtaining

$$\frac{i_0}{e_0} = j\omega w\sqrt{c\gamma\left[1 - \left(\frac{\nu_\infty}{\omega}\right)^2\right]} = j\omega C,$$

$$(21.39)$$

which agrees completely with the expression for uniform windings. Neither the surge impedance nor the apparent capacitance for supercritical frequencies is altered by the subdivision of the winding into elements. With increasing ω the capacitance rapidly approaches a maximum value, shown in Fig. 21.6, which expresses quite generally the terminal capacitance at the higher frequencies for every kind of winding.

If we survey the whole range of frequencies 0 and ∞, we see that there are now two different critical frequencies, ν and ν_∞, which

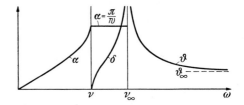

Fig. 21.8

determine the behavior of the winding. We may transform the characteristic Eqs. (21.10), (21.29), and (21.34) into the unified relation

$$\sin\left(\alpha\frac{w}{2}\right) = \cosh\left(\delta\frac{w}{2}\right)$$

$$= j\sinh\left(\vartheta\frac{w}{2}\right) = \frac{\omega/\nu_0}{\sqrt{1 - (\omega/\nu_\infty)^2}},$$

$$(21.40)$$

in which the right-hand member is condensed by introducing the two critical frequencies ν_0 and ν_∞ from Eqs. (21.12) and (21.13), the former being critical only for subdivided windings without internal capacitance, and the latter being critical in the general case. Figure 21.8 then shows the variation with frequency of the wave density α and of the attenuation constants δ and ϑ.

The first member of Eq. (21.40) is valid so long as the value of the right-hand member is less than 1. This is the subcritical range, the waves of which penetrate into the winding and travel undistorted in it. The second term is valid for values of the right-hand member greater than 1 but still finite. This occurs for frequencies between ν and ν_∞. These two critical values separate the entire frequency spectrum into three regions. The waves in this middle range do not travel through the winding. Instead they form an alternating standing distribution over the end turns or end coils that is spatially increasingly attenuated with increasing frequency. The third term in Eq. (21.40) refers to supercritical frequencies higher than v_∞, which render the right-hand member imaginary. In this range, too, no oscillations can penetrate entirely through the winding. Rather they form a spatially attenuated standing voltage distribution, which decreases exponentially along the terminal elements. This nonalternating distribution results from the internal capacitance, while the former alternating distribution is due to the finite subdivision of the winding into distinctive elements.

(c) *Stress on the insulation.* Any incident step pulse of amplitude E can be represented by the Fourier integral

$$e = \frac{E}{2} + \frac{E}{\pi} \int_0^\infty \frac{\sin \omega t}{\omega} \, d\omega. \quad (21.41)$$

This spectral distribution is shown in Fig. 21.9. Since the behavior of the winding is different in the three frequency regions, distinguished by different shadings in Fig. 21.9, let us

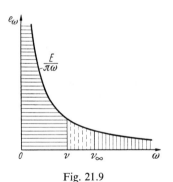

Fig. 21.9

separate the sine integral of Eq. (21.41) into three portions:

$$\int_0^\infty \frac{\sin \omega t}{\omega} \, d\omega = \int_0^v \frac{\sin \omega t}{\omega} \, d\omega$$
$$+ \int_v^{v_\infty} \frac{\sin \omega t}{\omega} \, d\omega + \int_{v_\infty}^\infty \frac{\sin \omega t}{\omega} \, d\omega.$$
$$(21.42)$$

Only the voltage produced by subcritical frequency components, corresponding to the first integral, can penetrate the winding. This sine-integral function

$$\text{Si} \, (vt) = \int_0^v \frac{\sin \omega t}{\omega} \, d\omega \quad (21.43)$$

is shown graphically as a function of time in Fig. 21.10. Physically it represents a step pulse

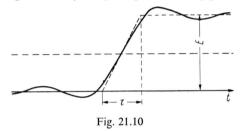

Fig. 21.10

with a flattened front of rise time τ. Numerical values of the function were given in Fig. 20.13.

The sine integral, which actually produces a linear gradual rise with some oscillations on either side, can be represented more simply by the broken straight line shown in Fig. 21.10. The rise time τ is given as in Eq. (20.36) by the upper limit of the integral, with Eq. (21.11):

$$\tau = \frac{\pi}{v} = \pi w \sqrt{\gamma l + cl/4}. \quad (21.44)$$

Thus it is determined completely by the critical frequency v.

The voltages corresponding to the second and third integrals in the right-hand member of Eq. (21.42) can be similarly represented graphically, either by the rigorous curve or by approximate straight lines. So we see that the three frequency regions of the spectral distribution of Fig. 21.9 produce three voltage distributions that perform differently within the winding. Figure 21.11a shows the incident step pulse in spatial distribution and, by the slanting lines, its splitting into three components. First (Fig. 21.11b) is the subcritical part e' with frequencies below v, forming a flattened traveling pulse throughout the winding. Second (Fig. 21.11c) is the supercritical part e'' with frequencies above v_∞, containing the incident discontinuous step, and some adjacent steep portions of the front; this part, consisting of rapid oscillations only, will be reflected into the incident line, exciting exponentially attenuated voltages in the terminal elements of the winding. Finally (Fig. 21.11*d*), the third or intercritical part consists of the frequencies between ν and ν_∞, corresponding to the middle integral in the right-hand member of Eq. (21.42); this part, which is the remainder formed between the three previous components, is also reflected

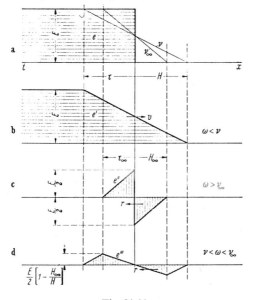

Fig. 21.11

into the line, exciting alternate voltages within the terminal elements of the winding.

If we use the rigorous sine-integral and exponential curves in time and space, rather than the approximate straight lines of Fig. 21.11, and superpose all voltage components, we obtain the complete pulse shapes of Fig. 21.12 along the winding for several consecutive times following the arrival of the step pulse. There appears instantaneously an attenuated voltage distribution over the terminal elements and from this there emerges, first slowly and then progressively faster, the smooth traveling pulse with its oscillating forerunners.

Instead of expressing the rise of the subcritical traveling pulse in time by Eq. (21.44), we may determine the length in space associated with the rise time as shown in Fig. 21.11*b*. Since the wave components of the pulse

propagate substantially with the velocity v_0 of Eq. (21.21), the length, by Eq. (21.11), is

$$H = v_0\tau = \pi\frac{v_0}{\nu} = \pi w \sqrt{\frac{\gamma}{c} + \frac{1}{4}}. \quad (21.45)$$

Because of the subdivision of the winding, the front of this penetrating pulse is lengthened, since the term $\frac{1}{4}$ under the square root disappears for uniform windings.

The length H_∞, giving the slope of the supercritical pulse e'' in Fig. 21.11*c*, is correspondingly

$$H_\infty = v_0\tau_\infty = \pi\frac{v_0}{\nu_\infty} = \pi w \sqrt{\frac{\gamma}{c}}. \quad (21.46)$$

For windings with high internal capacitance γ, as in transformers, the difference between

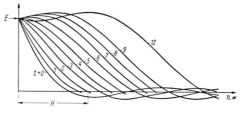

Fig. 21.12

the two head lengths is insignificant. Then the intercritical remainder voltage e''' represented in Fig. 21.11*d* is very small and only the voltages e' and e'' play a part. However, for small internal capacitance, as in most rotating-machine windings or widely spaced single-layer choke coils, the head length H has a lower limiting value

$$H_{min} = \frac{\pi}{2} w = 1.57 w, \quad (21.47)$$

and so the head can never be shorter than this. Thus we see that the subdivision of a winding into distinct elements, whether coils or layers or turns, always causes a minimum flattening of rectangular pulses that impinge upon its terminals.

The subcritical voltage e' of Fig. 21.11*b* proceeds through the entire winding and stresses the internal insulation between adjacent turns, layers, or coils by the steepness of its front. The maximum internal stress e' is given by the voltage difference over a length w of an element near the center of the rise, as shown in Fig. 21.13. This is expressed by the sine integral of Eqs. (21.41) and (21.43) as

$$e' = \frac{E}{\pi}\left[\operatorname{Si}\left(+\frac{vt}{2}\right) - \operatorname{Si}\left(-\frac{vt}{2}\right)\right]$$

$$= \frac{2}{\pi} E \operatorname{Si}\left(\frac{vt}{2}\right). \qquad (21.48)$$

The time interval t corresponding to a wire length w is approximately w/v_0 and therefore

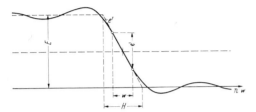

Fig. 21.13

the argument in Eq. (21.48) becomes, by Eqs. (21.11) and (21.21),

$$\frac{vt}{2} = \frac{vw}{2v_0} = \frac{1}{\sqrt{4(\gamma/c) + 1}}. \qquad (21.49)$$

Thus the maximum internal stress is

$$e' = \frac{2}{\pi} E \operatorname{Si}\left(\frac{1}{\sqrt{4(\gamma/c) + 1}}\right) \simeq \frac{E/\pi}{\sqrt{(\gamma/c) + 1/4}}. \qquad (21.50)$$

The right-hand member of this equation is an approximate value for small arguments, for which the sine integral is equal to the argument. Even at the maximum value of the argument, which is 1 for $\gamma = 0$, the error of this approximation is less than 5 percent.

Since w/H is for the most part a relatively small fraction, the internal strain may be computed alternatively by the derivative of the sine integral, which forms a curve with the shape $\sin(vt)/vt$ shown in Fig. 21.14a and gives the entire change of the voltage stress e' as a function of time. As this curve passes through zero for $vt' = \pm\pi$, we see that the entire time of exposure of the internal insulation to the excess voltage, from Eq. (21.44), is

$$t = 2t' = \frac{2\pi}{v} = 2\tau. \qquad (21.51)$$

Thus for low critical frequencies v this exposure will be longer. For the most dangerous part of this stress, with a maximum as given

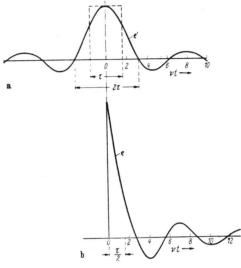

Fig. 21.14

by Eq. (21.50), we may take, however, an effective duration of about τ, as indicated in Fig. 21.14a, which is always a very small fraction of a second.

The terminal elements of the winding are exposed, in addition to this, to the strain of

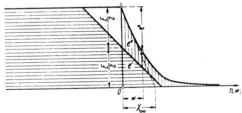

Fig. 21.15

the standing exponential distributions caused by the voltages e'' and e'''. The first one, e'', has, according to Fig. 21.11c, a positive maximum $E/2$ at the moment of incidence of the step pulse and decreases exponentially into the winding, as shown in Fig. 21.15, substantially as determined by the asymptotic space constant X_∞ derived in Eq. (21.36). This gives over the wire length w a voltage difference of approximately

$$e'' = \frac{E}{2}(1 - \varepsilon^{-w/X\infty})$$

$$= \frac{E}{2}(1 - \varepsilon^{-2\sinh^{-1}\sqrt{c/4\gamma}}). \qquad (21.52)$$

If we replace the hyperbolic sine function by a logarithm, this value can be expressed algebraically as

$$e'' = E\sqrt{\frac{c}{4\gamma}}\left(\sqrt{1 + \frac{c}{4\gamma}} - \sqrt{\frac{c}{4\gamma}}\right) \simeq \frac{E}{2}\sqrt{\frac{c}{\gamma}}. \qquad (21.53)$$

The right-hand member is an approximate value that is valid only for very large γ/c, of the order of 100 or more.

We see from Fig. 21.11c, d that the duration of the combined positive excess voltages e''

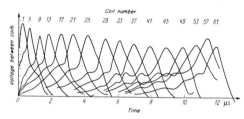

Fig. 21.16

and e''' with simplified triangular shapes is $\tau/2$. The internal insulation is thus exposed to the danger due to the additional terminal stresses $e = e' + e'' + e'''$ only over a period less than $\tau/2$, which is favorable with respect to the time lag of breakdown. Figure 21.14b shows the change with time, for small w/H, of the internal terminal stress e after arrival of a step pulse, as composed of the stress e' of Fig. 21.14a and the rigorous time variation of the stress e'' corresponding to Fig. 21.11c; e''' is negligibly small.

Many measurements, published in the extensive literature on this subject, show a change of voltage in time that follows closely the shapes of Fig. 21.14a, b, for interior as well as terminal turns or coils in machines and transformers. Of particular interest are the measurements by E. T. Norris (Fig. 21.16),

which give the time dependence of the voltage between a number of successive coils in the winding of a transformer, as observed by means of a cathode-ray oscillograph. The wavelike propagation and shape of the voltage shows up very well, initially as in Fig. 21.14b and later as in Fig. 21.14a.

The intercritical voltage e''' produces only very little additional stress at the time of arrival of the incident step pulse, since e''' passes through zero just then, as seen in Fig. 21.11d. The amplitude of this voltage, as given by a simple geometric consideration of the straight-line diagrams in Fig. 21.11, has the value

$$e''' = \frac{E}{2}\left(1 - \frac{H_\infty}{H}\right) = \frac{E}{2}\left(1 - \frac{1}{\sqrt{1 + c/4\gamma}}\right). \qquad (21.54)$$

For a ratio $\gamma/c = 10$ this is only 1 percent and even for $\gamma/c = 1$ it is still only 10 percent. Since this positive amplitude, according to Fig. 21.11d, appears at the terminal of the winding somewhat later than the incident step, that is, when the voltage e'' is diminishing toward zero, it is justifiable to neglect the internal stress e''' for any finite value of γ/c.

However, in the extreme case of $\gamma = 0$ the two voltages e'' and e''' reverse their roles. Now H_∞ becomes zero according to Eq. (21.46) and therefore e'' of Fig. 21.11c vanishes completely. On the other hand, e''' of Fig. 21.11d expands from both sides and now contains the entire step in its center, as e'' formerly did. The positive amplitude of e''' in this case is $E/2$, according to Eq. (21.54). Figure 21.7 shows that the stress e''' between the terminal coils, due to the intercritical alternating voltage e''', is always given by the sum of adjacent voltages. This yields for $\gamma \simeq 0$, by Eq. (21.54),

$$e'''_{\max} = 2e''' = E. \qquad (21.55)$$

This is the highest possible supplementary stress between terminal elements, and is in addition to the stress e' from Eq. (21.50), which is uniformly experienced by all elements and is nearly $2E/\pi$ for $\gamma = 0$.

The total internal voltage difference between adjacent terminal elements for any γ thus is composed of the sum of e', as given in Eq.

(21.50), and e'', as given in Eq. (21.53) or, for $\gamma = 0$, e''' instead, as given in Eq. (21.55). A numerical computation shows that we can combine all these expressions in a form similar to Eq. (21.50), and so the maximum internal stress becomes, to a good approximation,

$$e = \left(\frac{1}{\pi} + \frac{1}{2}\right) \frac{E}{\sqrt{(\gamma/c) + 1/4}}$$

$$= \frac{0.818}{\sqrt{(\gamma/c) + 1/4}} E, \qquad (21.56)$$

which is valid for any value of γ. The maximum possible stress for $\gamma = 0$ has the finite value

$$e_{max} = 2\left(\frac{1}{\pi} + \frac{1}{2}\right) E = 1.64\, E, \qquad (21.57)$$

which can never be exceeded.

If we compare the results of our computations with those obtained in Chapter 20 for uniform windings, we see that we may take into consideration the effect of the finite subdivision merely by adding $c/4$ to the value of the internal capacitance γ, thus obtaining an apparent or effective capacitance

$$\gamma_c = \gamma + \frac{c}{4}. \qquad (21.58)$$

This gives the correct boundary values for $\gamma = 0$ and forms a transition to the simpler formulas for large values of γ. We see that the use of this effective capacitance γ_c is valid for the critical frequency v in Eq. (21.11), for the head length H in Eq. (21.45), and for the voltage stresses e according to Eqs. (21.50) and (21.56), but is not valid for surge impedance Z and terminal capacitance C, which, according to Eqs. (21.26) and (21.39), are independent of the finite subdivision of the winding into lumped elements.

(d) *Interaction of coils, layers, and turns.* We shall now survey the performance of a complete winding built up of a group of coils, each one consisting of a number of layers, each layer containing many turns. A step pulse of voltage E_0 may arrive at the terminal, as shown in Fig. 21.17, and impinge at first

upon the turns of the first layer. The layers, with the turn length w_1, have a definite, fairly high critical frequency

$$v_1 = \frac{1}{w_1\sqrt{l\gamma_1}}, \qquad (21.59)$$

where l is the average self-inductance per unit length of the entire winding and γ_1 is the capacitance, also per unit length, between every two turns of the layer, including one-fourth of the ground capacitance as in Eq. (21.58). As we have seen previously, the steep front of the pulse, according to Eq.

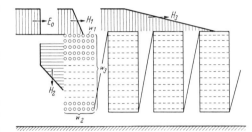

Fig. 21.17

(21.45), is converted to a flattened head of length

$$H_1 = \pi \frac{v}{v_1} = \pi v w_1 \sqrt{l\gamma_1} = \pi w_1 \sqrt{\frac{\gamma_1}{c}}, \qquad (21.60)$$

which travels over the first layer. Even if the interturn capacitance γ_1 is only of the order of the average winding capacitance c to ground, the head length comprises several turn lengths on account of the factor π in Eq. (21.60), and in general the head length is considerably larger.

If the pulse front travels from the first to the second and to further layers, as shown in Fig. 21.17, it cannot retain its relatively short head length, for the layers of a coil have a lower critical frequency

$$v_2 = \frac{1}{w_2\sqrt{l\gamma_2}}, \qquad (21.61)$$

where w_2 is now the wire length of the layers that form the elements of the coil, γ_2 being the capacitance per unit length between every two layers including $c/4$. The head length over the layers is therefore

$$H_2 = \pi \frac{v}{v_2} = \pi v w_2 \sqrt{l \gamma_2} = \pi w_2 \sqrt{\frac{\gamma_2}{c}}.$$

(21.62)

Because this is a multiple of the wire length w_2 of a layer, it is in all practical cases larger than the first head length H_1 determined by Eq. (21.60).

The transformation of H_1 into H_2 is shown in detail in Fig. 21.18, in which again the approximate straight-line front shape has been used.

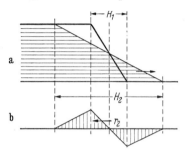

Fig. 21.18

Since only the flatter pulse H_2 can penetrate into the deeper layers of the first coil, the difference of the shapes H_1 and H_2, shown in Fig. 21.18b as r_2, is reflected within the first few layers and causes in these an additional standing distribution of voltage. Since the frequencies of the pulse r_2, however, are lower than the critical frequency v_1 of the turns, the attenuated standing distribution across the layers consists of traveling pulses over the turns forming those layers.

The pulse now travels with its new front length H_2 over the whole first coil, and then penetrates into the second and later into the third and further coils. But these coils have altogether a critical frequency

$$v_3 = \frac{1}{w_3 \sqrt{l \gamma_3}},$$

(21.63)

which is still lower, because w_3 is the wire length of every complete coil, while γ_3 is the internal capacitance between adjacent coils, including again $c/4$. The front is thus distributed over the head length

$$H_3 = \pi \frac{v}{v_3} = \pi v w_3 \sqrt{l \gamma_3} = \pi w_3 \sqrt{\frac{\gamma_3}{c}},$$

(21.64)

as shown in Fig. 21.17. The transformation of the wave front into this new head H_3 follows the same scheme as in Fig. 21.18. The reflected pulse r_3 causes a standing voltage distribution over the first few coils, in addition to the pulse front H_3 traveling over these coils. Here, correspondingly, the pulse r_3 may be formed by traveling-wave components within each one of the first coils, as the frequency components of r_3 are smaller than v_2. The average of these pulses traveling to and fro across the layers of each coil forms the attenuated standing distribution over the coils.

So we see that in actual transformer or machine windings that consist of the three elements turns, layers, and coils, each having different wire length and different internal capacitance, an original incident steep-front pulse is flattened out in three steps, the head lengths being in the proportion

$$H_1 : H_2 : H_3 = w_1 \sqrt{\gamma_1} : w_2 \sqrt{\gamma_2} : w_3 \sqrt{\gamma_3}. \quad (21.65)$$

Since the maximum internal stress between adjacent elements is determined by the quotient of their respective wire lengths and head lengths, we can see that the internal stress between turns, layers, and coils will be in the proportion

$$e_1 : e_2 : e_3 = \frac{1}{\sqrt{\gamma_1}} : \frac{1}{\sqrt{\gamma_2}} : \frac{1}{\sqrt{\gamma_3}}. \quad (21.66)$$

Thus it is evident that the voltage between adjacent elements is lower the larger the square root of the effective internal capacitances per unit length between the elements, while all other characteristics of the design do not matter at all. Usually the internal specific turn capacitance γ_1 is much larger than the capacitance γ_2 between layers and this is again larger than the specific coil capacitance γ_3. By favorable design of the geometric shape within the cross section of the whole winding, it appears possible to equalize substantially the effect of these stresses on the internal insulation.

In addition to the increased flattening of

the pulse front at the transition between turns and layers or layers and coils, the surge impedance Z given by Eq. (21.26) may be different for the turns, the layers, and the coils, because of different critical frequencies of these respective groups, so that refraction may occur. Figure 21.19 shows the conversion

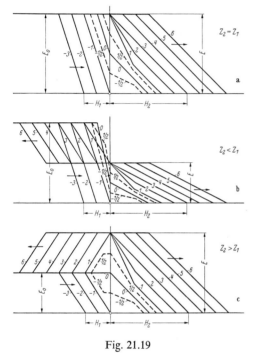

Fig. 21.19

of a flat pulse front with a head length H_1 into a front with a head length H_2 three times as large as H_1, under three different assumptions as to the relative surge impedances on the two sides of the transition point. The transmitted and reflected pulses, together with the standing exponential distributions, are superposed step by step; for simplicity, however, the straight-line approximate diagrams have been used. Figure 21.19a shows that for uniform surge impedances the steep incident front is transformed to the flat penetrating front without the occurrence of any reflected pulse on the first unit. In Fig. 21.19b, for $Z_2 < Z_1$, the voltage difference r_2 of Fig. 21.18b causes only an additional steep terminal distribution in the second unit, but even this is now gradually built up, as we

see by comparing Fig. 21.19 with the rigorous instantaneous development in Fig. 21.12.

For uniform surge impedance Z the maximum value E of the penetrating pulse is equal to E_0 of the incident pulse. If we assume in Fig. 21.19b a decrease in the surge impedance from Z_1 to $Z_2 = Z_1/3$, the graphical construction leads to the result that the converted pulse in the second unit is refracted into a smaller value, while its general shape remains exactly the same, as just explained in Fig. 21.19a. Simultaneously, there now appears a reflected wave reducing the voltage on the first unit, the front of the reflected wave having the same head length as the original incident pulse. The surge impedances in Fig. 21.19b were chosen as 3:1 so that the penetrating pulse E becomes just one-half of the incident pulse E_0.

In Fig. 21.19c the surge impedance Z_2 is chosen large in comparison to Z_1, so that the amplitude of the penetrating pulse becomes twice that of the incident pulse. The converted pulse, in its traveling- and standing-wave components, shows again the same distribution as in the previous case, and we see that the general shape of the penetrating voltage therefore is independent of the ratio of the surge impedances. The reflected pulse on the first unit is now positive and increases the voltage to twice the value E_0.

As a whole, we see from Fig. 21.19 that the transition point between turns, layers, or coils acts on the incident pulse like an ordinary junction, with either uniform or changing surge impedance. The reflected pulse always retains the shape of the incident pulse. On the outgoing line, however, simultaneously with the refraction to lower or higher values, not only is the pulse front converted to a greater head length but the difference between the incident and the penetrating pulse fronts causes a transient voltage distribution near the transition points. The steepness or space constant of this is somewhere between the head lengths of the incident and the penetrating pulses.

On the basis of these considerations, Fig. 21.20 shows the three conversions of the pulse front as it travels over the whole winding. The interturn capacitance causes the first flattening within the first layer; the internal capacitance between layers causes the second flattening during the travel over the first coil; and the coil-to-coil capacitance causes the third flattening during the propagation over the whole winding.

If instead of an individual single-phase winding we have a three-phase winding with connections in star, zig-zag, or delta and with phase-to-phase capacitance γ_4 between the

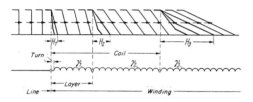

Fig. 21.20

phases, as shown in Fig. 21.21 for a transformer, there is one more transition point within the complete traveling course of the pulses. Finally, if protective coils with increased insulation, and thereby decreased internal capacitance γ_0, are used near the terminals, as in Fig. 21.22, the winding is subdivided still further into one more group

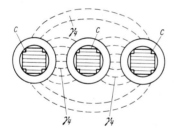

Fig. 21.21

of elements near its ends, and we have one additional transition point, making five in all.

It is useful to consider the quantitative advantage of such extra insulation for elements, whether turns, layers, or coils. We can extend our law of voltage stresses from Eq. (21.66) to any element of different insulation, for example, to the highly insulated terminal turns of Fig. 21.22. Instead of ordinary insulation with a thickness d_1, these turns may be covered with material of the same kind but greater thickness d_0. Thereby the internal capacitances between adjacent turns are

diminished nearly inversely as the thicknesses. Thus

$$\frac{\gamma_0}{\gamma_1} = \frac{d_1}{d_0}. \qquad (21.67)$$

The voltage gradient in this insulation, as the quotient of voltage and distance, is, from Eq. (21.66),

$$\frac{e_0}{d_0} : \frac{e_1}{d_1} = \frac{1/\sqrt{\gamma_0}}{d_0} : \frac{1/\sqrt{\gamma_1}}{d_1} = \frac{1}{\sqrt{d_0}} : \frac{1}{\sqrt{d_1}}, \qquad (21.68)$$

and thus we see that the gradient decreases only as the square root of the insulation thickness. Thus, if we wish to provide for protective extra insulation of the terminal layers or terminal coils, we see that in general we have to use four times as much insulation

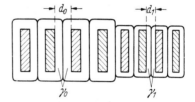

Fig. 21.22

if we are to reduce the internal surge stress by one-half. This square-root law of Eqs. (21.66) and (21.68), which governs the internal surge insulation of windings, causes many more difficulties in design than the linear law of the steady-state insulation. We can achieve some compensation for it in design if we introduce insulating materials of higher dielectric constant.

(e) Natural oscillations of composite windings. If for any pulse that penetrates the winding the surge impedance remains constant during its propagation along the entire wire length, there are reflection points only at the ends of the total winding. The spectrum of natural frequencies is then given directly by our previous considerations, which led to Fig. 21.4. In windings consisting of coils, layers, and turns, however, each part has its own critical frequency according to Eqs. (21.59), (21.61), and (21.63), which is represented on the axis

of abscissas in Fig. 21.23. Then the surge impedance Z of every group, according to Fig. 21.6, increases as we approach its critical frequency and therefore the surge impedance Z_3 of the whole winding increases, as shown in Fig. 21.23, faster than Z_2 of the first coil, and this increases faster than Z_1 of the first layer of the first coil.

We see from Fig. 21.23, in which, for simplicity, the small difference between v and v_∞ is neglected, that only the initial value Z_0 for very low frequencies is uniformly distributed over the winding, corresponding to the fact that l and c in Eq. (21.26) are average values throughout the entire wire length. But for any higher frequency, for example ω

ω_2 occurs within the first coils and ω_1 within the first layers adjacent to the terminal. If we consider three-phase windings with protective coils, there may be developed five such series, forming a fairly complex total spectrum of frequencies. Such local oscillations have often been observed experimentally.

While the maximum or critical frequencies v of the spectral series in Fig. 21.24 are given by Eqs. (21.59), (21.61), and (21.63), the fundamentals are

$$\omega_{10} = 2\pi m_1 \frac{v_1}{w_2}; \quad \omega_{20} = 2\pi m_2 \frac{v_2}{w_3};$$

$$\omega_{30} = 2\pi m_3 \frac{v_3}{a}, \quad (21.69)$$

where w_2, w_3, and a are the related wire lengths of the units and v_1, v_2, and v_3 the velocities of travel of their pulses, as shown by the

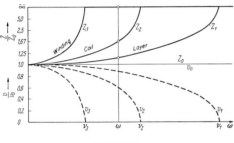

Fig. 21.23

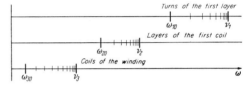

Fig. 21.24

in Fig. 21.23, the surge impedance of the first layer is relatively small, that of the first coil is much larger, and that of the whole winding appears already to be infinite.

Thus for waves of higher frequency we have to reckon with a considerable increase of the surge impedance at each of the transition points and, therefore, not only does an increased voltage of the penetrating pulse fronts develop by transmission, as in Fig. 21.19c, but in addition there is considerable reflection and a partial pulse travels back to the terminal. Each group of elements, layers, and coils, as well as the whole winding, is therefore excited in its own natural oscillations, and hence such composite transformer or machine windings will have three series of natural frequencies, as represented in Fig. 21.24.

Thus we see that in addition to the natural frequencies ω_3 of the entire winding, there develop at the transition points two series of more rapid local oscillations, of which

broken lines in the lower half of Fig. 21.23, these velocities being smaller than the limiting velocity v_0 of Eq. (21.21). The smallest number m of wavelengths along the wire lengths, corresponding to ω_{10} and ω_{20}, is in general equal to $\frac{1}{4}$, if, as usual, the first layer and first coil have an equivalent closed end at the terminal and an equivalent open end toward the remainder of the winding.

Under special conditions ω_{10} or ω_{20}, or even the lower harmonics of the corresponding series, are so small that there is no marked difference in their surge impedances Z at the two sides of the transition points. Such oscillations develop almost as in Fig. 21.19a without appreciable reflection and will be only weak if they are excited by switching processes or by resonance.

Let us compare the highest frequency v of one group of elements with the lowest possible frequency ω_0 of the following group. For example, the ratio of the lowest natural frequency ω_{20} of the layers to the highest frequency ν_3 of the coils is, according to Eqs. (21.63) and (21.69),

$$\frac{\omega_{20}}{\nu_3} = 2\pi m_2 v_2 \sqrt{l\gamma_3} \cong \frac{\pi}{2}\sqrt{\frac{\gamma_3}{c}}. \quad (21.70)$$

The approximate expression in the right-hand member is obtained by taking $m_2 = \frac{1}{4}$ and the velocity v_2 as the limiting value of Eq. (21.21). Since in most transformer windings the capacitance ratio γ_3/c for the coils, and for the layers as well, is much larger than 1, we must expect substantial gaps between the spectra of the various groups, as is indicated in Fig. 21.24. For machine windings, however, with γ_3 approaching $c/4$, these gaps may vanish, so that a more continuous spectrum is formed.

J. P. Newton conducted a series of measurements on a small model transformer so as to observe its frequency spectrum. It had 5 pancake coils of average diameter 11 cm, each containing 15 layers of 130 turns per layer and a total wire length $a = 3370$ m. By measuring the impedance across the entrance terminals it was possible, both for two open ends and for one grounded end, to observe three significant resonance zones: the lowest of about 10 to 30 kc/sec, a second of about 150 to 300 kc/sec, and the third of about 2000 to 5000 kc/sec. These last two zones gave a somewhat blurred change of impedance. The ratio of the resonance zones corresponds well to the ratio of wire lengths as given by Eq. (21.69). Thus the lowest-frequency zone represents the oscillations of coils, the middle one that of layers, and the highest one that of turns.

We must remember that there are many secondary effects, overlooked by our simplifying assumptions, that can blur the appearance of strong single resonant lines: circular coils have a slightly different length for successive turns or layers; honeycomb winding of the coils causes further complications; ohmic resistance in the windings produces a low Q for all oscillatory frequencies. All these effects contribute to the blurring and broadening of spectral lines, particularly at the higher frequencies.

While 5 coils could give rise to 10 possible natural frequencies, only 8 sharp resonance lines were observed; they are shown in Fig. 21.25. From the simple Eqs. (21.9) and (21.17) it is possible for each natural frequency ω to determine the velocity of propagation from a, the wire length of the winding, as

$$v = \frac{\omega a}{2\pi m}, \quad (21.71)$$

where m increases in integral multiples of $\frac{1}{4}$. In Fig. 21.25 the velocity of propagation is plotted against frequency for each of the observed resonances. We see that, except for

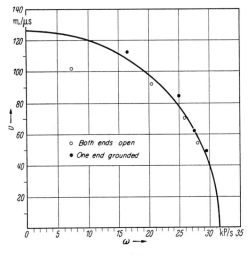

Fig. 21.25

one observation (the lowest frequency), the graph forms a circle diagram. The lowest frequency may be reduced by the effects of the iron surface, as developed in Chapter 17, Sec. (b). The velocity of propagation starts for low frequencies with an extrapolated value of about $v_0 = 126$ m/μs and decreases rapidly for increasing frequency. The velocity curve intersects the frequency axis at a critical frequency of about $\nu = 31.5$ kc/sec. Thus experimentally the same form of graph of velocity against frequency was observed as was deduced analytically in Fig. 21.5. From direct measurements of the self-inductance l, the capacitance to ground c, and the inter-coil capacitance γ, all per unit length, it was possible to calculate the critical frequency, from Eq. (21.13), as $\nu_\infty = 35.3$ kc/sec. This is indeed slightly above the experimentally observed value, as would be expected from Eq. (21.11).

Similar measurements were made by J. S. Cliff using a 50-MVA three-phase transformer group constructed with pancake coils. The high-voltage windings of the three single-phase transformers were connected in open delta so as to keep the spectrum to as low frequencies as possible. In one case both ends

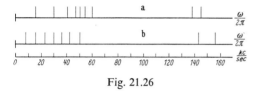

Fig. 21.26

of the open delta were insulated; in the other case one end was grounded. The low-voltage winding was always short-circuited and grounded. The 19-kV high-voltage winding was excited by a variable high-frequency generator. Resonance of the winding was noted by zero phase shift between voltage and current as observed on a cathode-ray oscilloscope. The length of the entire winding was $a = 3810$ m and the wave numbers m were assigned in sequence to the observed frequencies. However, this could not always be

done very reliably for the enclosed oil-filled transformer used.

The spectrum in Fig. 21.26 shows the observed resonant frequencies separately, in Fig. 21.26*a* for both ends open and in Fig. 21.26*b* for one end grounded. In both series we can observe the large gap between the first resonant region with $2 \cdot 7 = 14$ lines and the beginning of the second resonant region with higher frequencies. This is here also due to the subdivision of the winding into coils. The observations at the higher frequency of the first series were somewhat uncertain and for the second series higher frequencies could not be observed, owing to the frequency limits of the oscillator. If Eq. (21.71) is used to obtain the velocities of propagation and these are plotted against frequency, a graph almost the same as that in Fig. 21.25 is obtained.

Thus we see from these experimental observations that our analytical developments apply very well over a range of transformers from small to large with respect to the reduction of the velocity of propagation with increasing frequency, the finite upper limit of the spectrum of natural frequencies, and the separation of the resonant frequencies into distinct zones.

CHAPTER 22

PULSE TRANSFER BETWEEN WINDINGS

So far we have obtained an insight into the essential features of the behavior of coils and windings of all kinds under the impact of traveling pulses. In particular we have noted the differences and similarities of this behavior as compared with that of long homogeneous lines. We are now in a position to look into problems connected with the two windings of transformers when a portion of the incident voltage step pulse is transferred from the primary to the secondary winding. At first we shall limit our attention to traveling pulses propagating in both windings, which are coupled by their magnetic and electric fields. This leaves for later the treatment of the standing distributions of voltage and current, which act only statically; this treatment will be found to be quite simple. Thus we may initially neglect the effects of the internal capacitances of both windings and focus our attention mainly on the electromagnetic interactions between the fields associated with the pulses traveling in the coils.

(a) *Interaction of primary and secondary coils.* The two windings in a real transformer usually have a substantially different number of turns, which are coupled magnetically by mutual inductance and electrostatically by mutual capacitance. To examine the behavior of traveling pulses we shall extend our previous treatment and consider the equivalent mutual effects per unit length of wire in the coils. For very long coils they may be uniformly distributed over the length.

Then the spatial variation of voltages in the two windings, distinguished by the subscripts 1 and 2, is

$$-\frac{\partial e_1}{\partial x_1} = l_1 \frac{\partial i_1}{\partial t} + m \frac{\partial i_2}{\partial t},$$
$$-\frac{\partial e_2}{\partial x_2} = l_2 \frac{\partial i_2}{\partial t} + m \frac{\partial i_1}{\partial t}, \quad (22.1)$$

by analogy with Eq. (11.1). The effective self- and mutual inductances l and m per unit length of the conductors are related to the constants L and M of the entire windings under steady-state conditions by the equations

$$l_1 = \frac{L_1}{a_1}, \quad l_2 = \frac{L_2}{a_2}, \quad m = \frac{M}{\sqrt{a_1 a_2}}, \quad (22.2)$$

where a_1 and a_2 are the lengths of wire in the two windings.

Since the mutual inductance m is the same in both Eqs. (22.1), it must be formed symmetrically with respect to the two windings. Furthermore, it must be derived from the mutual inductance M between the entire windings by dividing by a length. These dimensional conditions are rather simply satisfied by the last of Eqs. (22.2).

The variation of voltage as a function of time for each winding is determined by the variation as a function of space of the currents in the two windings:

$$-\frac{\partial e_1}{\partial t} = p_1 \frac{\partial i_1}{\partial x_1} + g \frac{\partial i_2}{\partial x_1},$$
$$-\frac{\partial e_2}{\partial t} = p_2 \frac{\partial i_2}{\partial x_2} + g \frac{\partial i_1}{\partial x_2}, \quad (22.3)$$

where p and g are Maxwell's self- and mutual potential coefficients. These are reciprocal values of capacitances and thus are determined by the potential coefficients P and G of the whole winding:

$$p_1 = P_1 a_1, \quad p_2 = P_2 a_2, \quad g = G \sqrt{a_1 a_2}, \quad (22.4)$$

where the last expression again satisfies the dimensional considerations.

Figure 22.1 shows the self-constants l and p and the mutual constants m and g for the high-voltage and low-voltage windings of a transformer, both situated between the internal and external ground electrodes. In addition to these parameters, actual transformers have internal capacitances between adjacent turns of each winding, though these produce no interaction between the two windings. Therefore we shall not consider them and shall thus obtain the interaction of the transformer coils in the simplest form.

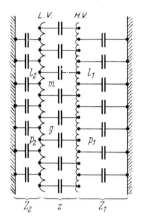

Fig. 22.1

The differential Eqs. (22.1) and (22.3) can be solved by linear dependences between voltages and currents in both windings, namely,

$$e_1 = Z_1 i_1 + z i_2,$$
$$e_2 = Z_2 i_2 + z i_1,$$

$$(22.5)$$

where Z_1 and Z_2 denote the surge self-impedances of the two coils while z is a surge mutual impedance between the coils as indicated at the bottom of Fig. 22.1.

To determine these constants we insert Eqs. (22.5) into the left-hand members of Eqs. (22.1) and (22.3), and obtain from the first equation of each pair

$$-\frac{\partial}{\partial x_1}(Z_1 i_1 + z i_2) = \frac{\partial}{\partial t}(l_1 i_1 + m i_2),$$

$$-\frac{\partial}{\partial x_1}(p_1 i_1 + g i_2) = \frac{\partial}{\partial t}(Z_1 i_1 + z i_2).$$

$$(22.6)$$

For pulses traveling with constant shape, time and space derivatives are proportional to each other. We can then divide the first of Eqs. (22.6) by the second to obtain

$$(Z_1 i_1 + z i_2)^2 = (l_1 i_1 + m i_2)(p_1 i_1 + g i_2). \quad (22.7)$$

Since this is valid for any currents i_1 and i_2,

the coefficients of the respective currents in the two members must be equal. We thus have the three conditions

$$Z_1^2 = l_1 p_1, \quad z^2 = mg, \quad 2Z_1 z = l_1 g + p_1 m. \quad (22.8)$$

The first two conditions give the surge impedances

$$Z_1 = \sqrt{l_1 p_1}, \quad z = \sqrt{mg}; \quad (22.9)$$

substituting these surge impedances in the third condition, we find

$$\frac{p_1}{l_1} = \frac{g}{m}. \quad (22.10)$$

If we insert the last of Eqs. (22.5) into the last lines of Eqs. (22.1) and (22.3) we obtain from the corresponding conditions the surge impedances

$$Z_2 = \sqrt{l_2 p_2}, \quad z = \sqrt{mg}, \quad (22.11)$$

and the additional condition

$$\frac{p_2}{l_2} = \frac{g}{m}. \quad (22.12)$$

The second of Eqs. (22.9) and (22.11) are identical and the equivalent Eqs. (22.10) and (22.12) indicate certain symmetry properties in the windings. Thus we have defined the three surge impedances that determine the interaction of currents and voltages in the coils according to our conditions established in Eq. (22.5).

The conditions given in Eqs. (22.10) and (22.12) are always fulfilled in a space having homogeneous dielectric and magnetic properties. Even in nonhomogeneous spaces such as occur in most actual devices, they are found to be applicable with good approximation.

To determine the velocity of propagation in the two windings, we differentiate the first of Eqs. (22.6) with respect to t and the second with respect to x_1. Subtraction eliminates the surge impedances, and arranging the terms with regard to the currents gives

$$l_1 \frac{\partial^2 i_1}{\partial t^2} - p_1 \frac{\partial^2 i_1}{\partial x_1^2} = -m \frac{\partial^2 i_2}{\partial t^2} + g \frac{\partial^2 i_2}{\partial x_1^2}. \quad (22.13)$$

Since this relation is valid for any value of

i_1 and i_2, both members must disappear independently of each other. Therefore each member forms the well-known wave equation, and thus the wave velocities along the wire length x_1 of the winding 1 are

$$v_1 = \sqrt{\frac{p_1}{l_1}}, \quad v_1'' = \sqrt{\frac{g}{m}}. \quad (22.14)$$

The first expression in Eqs. (22.14) gives the velocity of the self-induced waves, the second gives that of the mutually induced waves, and by inserting condition (22.10) in the second expression we see that the two velocities coincide.

By subjecting the second of Eqs. (22.1), (22.3), and (22.5) to the same operations as performed in the first, we obtain the velocities of wave propagation in the wires of winding 2,

$$v_2 = \sqrt{\frac{p_2}{l_2}}, \quad v_2' = \sqrt{\frac{g}{m}}. \quad (22.15)$$

Comparison with Eq. (22.12) shows that in the second winding there also exists only one well-defined velocity of wave propagation along the wires. Furthermore, we see by comparison of v_1'' and v_2' in Eqs. (22.14) and (22.15) that all the velocities along the wires of both windings are exactly the same.

These results show that the primary as well as the secondary winding of any transformer possesses one and only one wave velocity, and, furthermore, that it is the same in both windings, measured along the conductors of each winding. The presence of the other winding and of currents therein does not alter this electromagnetic property of every coil. This does not mean, as we shall see later, that the windings do not interact with each other. However, it has the result that along the axis of the winding an electromagnetic field will travel with a velocity corresponding only to the pitch of the winding under consideration, independent of the presence of other winding. Therefore we have to distinguish two different axial velocities, namely, those of the high-voltage and of the low-voltage windings. With the pitch s_1 for the high-voltage winding and s_2 for the low-voltage winding and a common axial coil length h, the two velocities of axial propagation are

$$v_H = \frac{s_1}{w} v = \frac{h}{a_1} v,$$

$$v_L = \frac{s_2}{w} v = \frac{h}{a_2} v, \quad (22.16)$$

where v now is the fundamental velocity of propagation along the wire in both coils according to Eqs. (22.14) and (22.15).

(b) Characteristics of two-winding transformers. With the same assumptions for coils and insulating spaces as previously made in Chapters 17 and 18, Fig. 22.2 represents the

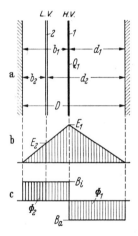

Fig. 22.2

cross section of a two-winding transformer with two long coils. Here b_1 and d_1 are the internal and external distances to ground of the high-voltage winding 1, while b_2 and d_2 are those of the low-voltage winding 2. The self-potential coefficients P are then reciprocal to Eq. (17.2),

$$\left.\begin{aligned} P_1 &= \frac{E_1}{Q_1} = \frac{4\pi v_0^2}{\varepsilon w h} \frac{b_1 d_1}{b_1 + d_1} \\ P_2 &= \frac{E_2}{Q_2} = \frac{4\pi v_0^2}{\varepsilon w h} \frac{b_2 d_2}{b_2 + d_2} \end{aligned}\right\} (\cdot\, 10^{-9} \text{ in farad}^{-1})$$

$$(22.17)$$

where Q is the charge on the windings.

As shown in Fig. 22.2b, the potential of winding 1 drops linearly from its own voltage E_1 to zero at the ground electrodes. If we introduce into this field a secondary winding at position 2 in Fig. 22.2a, it takes up by static induction a voltage

$$E_2 = \frac{b_2}{b_1} E_1 = \frac{4\pi v_0^2}{\varepsilon w h} \frac{b_2/b_1}{1/b_1 + 1/d_1} Q_1, \quad (22.18)$$

where the value of the voltage E_1 is inserted from Eq. (22.17). The mutual potential coefficient between the windings is, therefore,

$$G = \frac{E_2}{Q_1} = \frac{4\pi v_0^2}{\varepsilon w h} \frac{b_2 d_1}{b_1 + d_1} (\cdot 10^{-9} \text{ in farad}^{-1}).$$
$$(22.19)$$

The form of this expression is symmetrical for the two coils, as we see from Fig. 22.2, since all the denominators of Eqs. (22.17) and (22.19) have the same value

$$b_1 + d_1 = b_2 + d_2 = D, \quad (22.20)$$

which gives the distance between the two ground electrodes, independent of the positions of the coils.

The self-inductance of each of the windings follows from the Eq. (17.4); it is

$$\left. \begin{array}{l} L_1 = 4\pi\mu \dfrac{w}{h} N_1^2 \dfrac{b_1 d_1}{b_1 + d_1} \\[3mm] L_2 = 4\pi\mu \dfrac{w}{h} N_2^2 \dfrac{b_2 d_2}{b_2 + d_2} \end{array} \right\} (\cdot 10^{-9} \text{ in henrys}),$$
$$(22.21)$$

where N_1 and N_2 denote the number of turns in the two windings.

The distribution of the primary magnetic flux over the insulating spaces is shown in Fig. 22.2c. The secondary coil is linked with only a part of this flux within the inner space, namely,

$$\Phi_2 = \frac{b_2}{b_1} \Phi_1 = 4\pi\mu w \frac{N_1 I_1}{h} \frac{b_2/b_1}{1/b_1 + 1/d_1},$$
$$(22.22)$$

where the value of the flux Φ_1 is inserted from Eq. (17.3). Thus the mutual inductance of the coils becomes

$$M = \frac{N_2 \Phi_2}{I_1} = 4\pi\mu \frac{w}{h} N_1 N_2 \frac{b_2 d_1}{b_1 + d_1} 10^{-9} \text{ henry}.$$
$$(22.23)$$

This expression is also geometrically symmetrical for the two coils, as we see from Fig. 22.2.

Now we can determine the specific constants per unit length according to Eqs. (22.2) and (22.4), by dividing or multiplying, as appropriate, the total values by the wire lengths

$$a_1 = N_1 w, \quad a_2 = N_2 w. \quad (22.24)$$

We see immediately that all these constants, inserted in Eqs. (22.2) and (22.4), give only a single value for the velocity of wave propagation along the wires, namely,

$$v = \frac{v_0}{\sqrt{\varepsilon\mu}}, \quad (22.25)$$

which is again identical with the universal velocity of propagation of light, adapted to the electromagnetic constants of the transformer insulating space.

On the other hand, insertion of the constants in Eqs. (22.9) and (22.11) gives the three surge impedances

$$\left. \begin{array}{l} Z_1 = 4\pi v_0 \sqrt{\dfrac{\mu}{\varepsilon}} \dfrac{N_1}{h} \dfrac{b_1 d_1}{D} \\[4mm] Z_2 = 4\pi v_0 \sqrt{\dfrac{\mu}{\varepsilon}} \dfrac{N_2}{h} \dfrac{b_2 d_2}{D} \\[4mm] z = 4\pi v_0 \sqrt{\dfrac{\mu}{\varepsilon}} \dfrac{\sqrt{N_1 N_2}}{h} \dfrac{b_2 d_1}{D} \end{array} \right\} (\cdot 10^{-9} \text{ in ohms}),$$
$$(22.26)$$

which are dependent on winding density and insulating distances. From the last expressions we derive the ratio of the surge mutual impedance to both of the surge self-impedances, which we shall use frequently, as

$$\frac{z}{Z_1} = \sqrt{\frac{N_2}{N_1} \frac{b_2}{b_1}},$$
$$\frac{z}{Z_2} = \sqrt{\frac{N_1}{N_2} \frac{d_1}{d_2}}.$$
$$(22.27)$$

If ε and μ are not uniformly distributed with respect to both windings, equivalent distances b and d can be determined easily, as shown in Chapter 17, Sec. (b).

From Eq. (22.5) we can show that the ratio of voltages is determined directly by the ratio of surge impedances. The coupling ratio for traveling pulses in a transformer is given, therefore, according to Eqs. (22.27), by the square root of the turns ratio, in contrast to the steady-state coupling, which is given directly by the ratio of turns. This is advantageous when a pulse is transmitted from the low-voltage to the high-voltage side, but very unfavorable for the transmission of pulses from the high-voltage to the low-voltage side of a transformer.

The coupling ratios in Eqs. (22.27) are determined further by a geometric factor

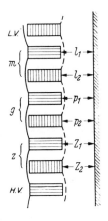

Fig. 22.3

given simply by the distances. The ratio of the inner distances b is always materially smaller than 1, while the ratio of the outer distances d for the most part is slightly smaller than 1. In order to avoid high transmitted surges from the high-voltage to the low-voltage winding, it is thus preferable to arrange the low-voltage winding as the inner coil and the high-voltage as the outer coil, which actually is the usual practice with transformers having only two coils.

Interleaved pancake windings, as shown in Fig. 22.3, also satisfy the general coupling diagram of Fig. 22.1 and can therefore be treated in the same way as concentric windings. However, here the subdivision into partial coil sections, as examined in Chapter 21, is much more pronounced and the mutual interactions between primary and secondary windings are much stronger, owing to the closer physical interleaving. This leads to the result that traveling pulses of much higher voltage can be transferred to the low-voltage side and thus this method of construction is

finding less and less use in high-voltage transformers.

(*c*) *Pulse transmission between windings.* The electromagnetic field propagates axially along the high-voltage winding with the moderate velocity v_H, and along the low-voltage winding with the higher velocity v_L, both given by Eqs. (22.16) and much lower than the velocity of light in the insulating space as given by Eq. (22.25). We call v_H and v_L the natural velocities of the windings. The complete electromagnetic state of the transformer is composed of both these slow- and fast-moving

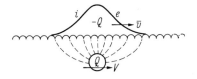

Fig. 22.4

fields. Each of them cuts the helical wires of its own winding with the velocity v given by Eq. (22.25). The wires of the other winding, however, are cut with a velocity considerably different from this. The problem therefore arises of the manner in which a winding responds if a pulse is impressed on it whose velocity does not coincide with its natural velocity.

Let us investigate this performance of windings by means of a method that is easy to visualize. We consider (Fig. 22.4) a coil with a definite natural axial velocity $\bar{v}$, along which an electric charge Q is moved with a given velocity V. Then three different cases are possible.

(1) The velocity of the charge is equal to the natural velocity of the coil: $V = \bar{v}$. In this case the electromagnetic induction of the moving charge produces voltage and current in the coil that move in synchronism with the charge. For an observer moving with the charge, the pulses of current and voltage represented in Fig. 22.4 seem to be at rest. They constitute electromagnetic energy that travels along the coil always in parallel with the charge. This case occurs in transformers with a turns ratio of 1:1, though this is quite rare. The pulses on the two windings here

interact in the same way as pulses on parallel smooth lines, as treated in Chapter 11.

(2) The velocity of the charge is higher than the natural velocity of the coil: $V > \bar{v}$. In this case, represented by Fig. 22.5, the moving charge Q is not able to bind an opposite charge on the coil, for such a charge can move only with a lower natural velocity $\bar{v}$ and thus cannot follow the high velocity V of the charge Q. If no charge can be bound on the coil

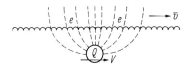

Fig. 22.5

then no current can flow in its turns, and, if no current exists in the wires, the coil acts exactly like an insulator, having no conductivity for waves moving with higher velocity than the natural velocity of the coil.

On account of the low natural velocity, any voltage on the coil produced by a faster-moving charge is therefore merely a statically induced voltage, which cannot flow away from the zone of influence of the charge Q. The coil, therefore, interacts with a fast-moving charge as if it consisted only of

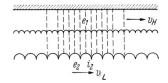

Fig. 22.6

conducting rings not connected with one another, instead of turns of a helical wire; thus it appears transparent for the electromagnetic excitation of a fast-moving charge and interacts with it as a nonconductor.

These results can easily be transferred to two transformer coils, as shown in Fig. 22.6. Any pulse consisting of voltage e_2 and current i_2 on the low-voltage winding and moving with higher velocity v_L than the natural velocity v_H of the high-voltage winding can induce within the latter only a static voltage

e_1 but no current i_1. The high-voltage winding therefore causes merely a capacitive voltage division. This winding does not absorb any energy; instead, the entire energy of the pulse is conducted by the low-voltage winding with its high natural velocity. Under these circumstances there is no reaction from the

Fig. 22.7

high-voltage winding back to the low-voltage winding, the former appears nonexistent from the viewpoint of the latter and its pulses.

(3) The velocity of the charge is lower than the natural velocity of the coil: $V < \bar{v}$. In this case, shown in Fig. 22.7, the charge Q can bind opposite charges on the coil, as it would do even at rest. However, no voltage can be produced opposite the moving charge, since every voltage immediately flows away in both directions with the higher natural velocity $\bar{v}$ of the coil. Thus the winding acts like a metallic layer, conducting in the axial

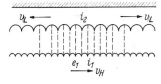

Fig. 22.8

direction. From the viewpoint of the slowly moving charge the coil appears opaque and acts as a perfectly conducting screen. The countercharges moving with the low velocity V of the charge Q produce currents varying in space and in time in the coil. Therefore an intense magnetic interaction exists between the slow-moving charge and the coil.

Figure 22.8 represents this interaction within transformer coils if a pulse propagates with the low velocity v_H along the high-voltage winding. Its voltage e_1 and current i_1 induce only current i_2 at the low-voltage winding but not voltage e_2. The low-voltage winding appears, therefore, from the viewpoint of the high-voltage winding and its pulses, as if it were perfectly conducting in every direction or a complete metallic screen. The induced current i_2 is carried along the low-voltage winding directly opposite the slowly moving high-voltage wave. The whole energy of the electromagnetic field is conducted by the high-voltage winding with its low natural velocity.

In order to determine the consequences of this different behavior of the windings in

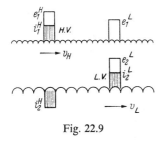

Fig. 22.9

transformers, we go back now to our analytical expressions. Figure 22.9 shows in general, by the example of short rectangular pulses, the voltages and currents that can flow in the windings. The two types of pulse, the fast one with the energy concentrated on the low-voltage winding and the slow one with the energy concentrated on the high-voltage winding, travel independently of each other and form by superposition the entire electromagnetic state of the transformer. We distinguish the two pulse systems by superscripts L and H on all currents and voltages, and consider first the fast low-voltage pulses.

We see from Fig. 22.9 that there is no low-voltage induced current in the transparent high-voltage winding, that is,

$$i_1^L = 0, \qquad (22.28)$$

and thus Eq. (22.5) simplifies to

$$e_2^L = Z_2 i_2^L, \quad e_1^L = \frac{z}{Z_2} e_2^L. \qquad (22.29)$$

So for the low-voltage winding there remains effective only the surge impedance Z_2, while in the high-voltage winding a transmitted voltage appears whose magnitude is determined by the ratio of surge impedances given by the second of Eqs. (22.27). This high-voltage winding thus has no influence on the

performance of the low-voltage pulses; they suffer only a voltage division according to the coupling factor z/Z_2. From the standpoint of the low-voltage winding, the high-voltage winding appears as air.

For the slow high-voltage waves, on the other hand, we have, as in Fig. 22.9, within the opaque low-voltage winding,

$$e_2^H = 0. \qquad (22.30)$$

The second of Eqs. (22.5) therefore gives

$$\frac{i_1^H}{i_2^H} = -\frac{z}{Z_2}, \qquad (22.31)$$

by which the current in the secondary winding is determined. If we insert this in the first of Eqs. (22.5), we obtain the primary voltage

$$e_1^H = Z_1 i_1^H - \frac{z^2}{Z_2} i_1^H = Z_1^H i_1^H. \qquad (22.32)$$

Thus we can define an effective surge impedance of the high-voltage winding,

$$Z_1^H = Z_1 \left(1 - \frac{z^2}{Z_1 Z_2} \right). \qquad (22.33)$$

We see that this is considerably less than Z_1 on account of the reaction of the low-voltage winding with respect to the slow-moving high-voltage pulses.

By introducing into Eq. (22.33) the values of Eqs. (22.26) and (22.27), we can determine

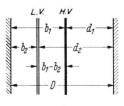

Fig. 22.10

the effective high-voltage surge impedance by the dimensions of the windings as

$$Z_1^H = 4\pi v_0 \sqrt{\frac{\mu}{\varepsilon}} \frac{N_1}{h} \frac{d_1(b_1 - b_2)}{d_2} (\cdot 10^{-9} \, \Omega). \qquad (22.34)$$

If in Fig. 22.10 we examine the distances given in the last fraction of Eq. (22.34) and compare them with Eq. (17.8) and Fig. 17.1, we see that Z_1^H is constituted exactly as if the low-voltage winding were replaced by a ground electrode.

The low-voltage winding therefore reacts intensively on pulses in the high-voltage winding, in screening their rapid electromagnetic field completely. The effective surge impedance of the high-voltage winding is therefore given by the leakage flux of both windings, as indicated already by the form of Eq. (22.33). Any influence of the magnetic steel core disappears here completely.

For verification of these results we consider measurements made by H. Norinder with a 450-kVA oil transformer for 20 kV/112 V with $b_1 - b_2 = 1.4$ cm insulating distance between the windings. The high-voltage winding consisted of 80 sectional coils each with 20 turns, each coil covering an axial length of 0.865 cm. The secondary winding was grounded at the neutral point. The dielectric constant for the insulating spaces can be taken as $\varepsilon = 3.5$, the permeability as $\mu = 1$, and the tank wall as far away.

By Eq. (22.34) we compute an effective high-voltage surge impedance

$$Z_1^H = \frac{120\pi}{\sqrt{3.5}} \cdot \frac{20}{0.865} \cdot 1.4 = 6530\ \Omega.$$

This quantity was measured from cathode-ray oscillograms and was found to be $6600\ \Omega$, a very close agreement indeed. The wave velocity in the wires was measured as 148 m/μs. This is in good agreement with the value given by Eq. (22.25),

$$v = \frac{300}{\sqrt{3.5}} = 160\ \text{m}/\mu s.$$

Since the low-voltage winding had a distance to the core $b_2 = 1.4$ cm, the high-voltage surge impedance, with $b_1 = 2.8$ cm, if it were not reduced by the second winding, would be $Z_1 = 13{,}060\ \Omega$, which is twice as high as the effective as well as the measured value.

The fact that the slow pulse within the high-voltage winding cannot transmit voltage to the low-voltage winding does not mean that no voltage at all appears on that winding. This would be the case only if the low-voltage winding were grounded throughout. In many

actual cases, however, an incident high-voltage surge produces both pulse systems at the terminals of the winding, the slow one and the fast one, in the primary as well as in the secondary winding, and these waves travel further through the windings.

Figure 22.11 shows the performance of a junction in a special but important case, to which we shall confine ourselves for the sake of simplicity. The surge impedance of the incident high-voltage line may be small

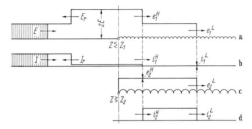

Fig. 22.11

compared with that of the high-voltage winding, while the low-voltage winding may be open at the terminal or connected to a very high-impedance load. We follow only the formation of traveling pulses and again ignore any standing-wave distributions.

The voltage at the high-voltage terminal, doubled by reflection of E as shown in Fig. 22.11a, is composed of both voltages

$$e_1^L + e_1^H = 2E, \qquad (22.35)$$

and the low-voltage terminal current is zero as shown in Fig. 22.11d

$$i_2^L + i_2^H = 0. \qquad (22.36)$$

By using Eq. (22.29) we can transform e_1^L into e_2^L and by successive application of Eqs. (22.32), (22.31), (22.36), and (22.29) we can also transform e_1^H into e_2^H. This gives, together with the use of Eq. (22.33),

$$2E = e_2^L \left(\frac{z}{Z_2} + \frac{Z_1^H}{z} \right) = \frac{Z_1}{z} e_2^L, \quad (22.37)$$

and thus we obtain for the important voltage at the open low-voltage terminal, using Eq. (22.27),

$$e_2^L = \frac{z}{Z_1} 2E = \frac{b_2}{b_1} \sqrt{\frac{N_2}{N_1}} \cdot 2E. \quad (22.38)$$

We see that this voltage, transmitted by traveling-pulse interaction of the winding, is given by twice the value of the surge coupling ratio and therefore is determined by the square root of the turns ratio and by the ratio of the insulating distances between windings and core. For actual transformers, both distances b_1 and b_2 may be corrected with respect to the nonuniform distribution of ε and μ in accordance with the method of Chapter 17, Sec. (*b*).

Because of the importance of the secondary voltage pulses, which may occur on account of incident lightning discharges or terminal flashover of the primary winding, many investigations have been published on transmitted voltages that can easily destroy the insulation of the low-voltage winding. However, only a few of them are made under such well-determined conditions that they can be compared with Eq. (22.38). Table 22.1 gives the results of two such experiments in which a step-pulse front incident on the high-voltage winding was transmitted to an open low-voltage winding; primary and secondary pulses were measured from cathode-ray oscillograms. The characteristics of the transformers are given in the table, so far as they have been published, and estimated values for the distances are added, appropriate to power and voltage. The measured secondary surge voltages are all only slightly smaller than the theoretical values computed by Eq. (22.38).

This discrepancy is due to many neglected minor factors; however, in the main the agreement is such as to verify the main concepts underlying our calculations.

In a similar manner it is possible to examine a voltage pulse transferred to the high-voltage winding by a voltage step pulse E incident upon the low-voltage winding. The result is of the same form as Eq. (22.38), except that the turns ratio is now inverted to $\sqrt{N_1/N_2}$.

Thus, for example, for a turns ratio of 50:1 the pulse voltage transferred from the high-voltage to the low-voltage winding is reduced by $\sqrt{50} \cong 7$ times; this would considerably exceed the steady-state low voltage, which is only 1/50 of the high voltage, and thus would be dangerous for the regular insulation. In contrast to this, a voltage pulse transferred from the low-voltage to the high-voltage winding is increased 7 times, which is not dangerous when we remember that the regular high-voltage winding insulation has to handle 50 times the voltage in the low-voltage winding. This is confirmed by the fact that in actual practice such difficulties are experienced only in the direction of transfer from high-voltage to low-voltage windings.

The double pulses, as shown in Fig. 22.11, travel with their two velocities along the whole length of the windings and are reflected at the opposite end. This can lead to a later voltage rise at the entrance terminals if the

TABLE 22.1. Comparison of computed and measured secondary pulse voltages in transformers.

Investigator	REBHAN	ELSNER
Connection diagram at test	*(connection diagram, input E, output e)*	*(connection diagram, input E, output ε)*
Rated high voltage	15 kV	110 kV
Rated three-phase power	400 kVA	15000 kVA
Turns ratio N_2/N_1	1:37·5	1:5·33
Distance ratio b_2/b_1	1:2	1:3
Transmitted voltage e/E (percent)		
computed	16·3	28·8
measured	15	26

far end is open-circuited or, for a three-phase winding, if the star point is isolated from ground. On the other hand, the secondary voltage will be lower than in Fig. 22.11 if the low-voltage winding is connected to a relatively low-impedance load such as a generator with a single-layer winding, several cables in parallel, or a protective condenser of appropriate size.

Although it would be necessary to use a treatment of the coupling between windings, such as is developed in detail in Eqs. (22.35) to (22.38), to be able to examine such reflections and transmissions properly, it is often sufficient to consider that such repeated reflections at both ends excite the natural oscillations of both windings. These will be slow in the high-voltage winding and rapid in the low-

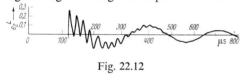

Fig. 22.12

voltage winding. They can be easily recognized in oscillograms of the secondary voltage, such as are shown for example in Fig. 22.12 as observed by R. Elsner. They constitute the gradual transition to the steady-state voltage transfer. The inital step pulse, shown as e_2^L in Fig. 22.11, is at first quite faithfully reproduced. However, after a short time the discontinuities of the sharp oscillations appear, rounded and damped by the effects of internal capacitance and other high-frequency dissipation phenomena in both windings, so that finally the oscillograms become more or less sinusoidal in shape.

The capacitive and the inductive transfer of voltage are treated simultaneously by means of coupling between pulses and thus Eq. (22.38) gives the total voltage shortly after arrival of the incident pulse from the line. If the incident pulse front is so flat that only subcritical frequency components are present, these penetrate into the windings just as shown in Fig. 22.11. However, if steeper pulse fronts are encountered, their supercritical frequency components separate in the terminal coils owing to internal capacitance effects, as discussed in Chapters 19 and 20, and give rise to a higher insulation stress in these entrance coils. This applies equally to the secondary windings, which can thus experience voltages in their end coils that exceed those in their middle coils by a factor of 2.57. This does not alter the voltage with respect to ground.

If protective coils or condensers are connected in front of the transformer, to flatten the incident pulses to such an extent that their front rise and back decay overlap, the transformer will exhibit quasi-steady-state behavior. It can then be replaced by an equivalent LC circuit, which, while it reflects the external behavior of the transformer with respect to its connecting lines, cannot, of course, in any way represent or give information on the transfer processes between the transformer windings.

The interaction of internal and external capacitances with the self-inductance (Chapters 19 and 20) and the subdivision into separate elements (Chapter 21) reduce the natural velocities of propagation in both windings somewhat. However, since the absolute magnitude of these velocities does not affect the development of the magnitude of the voltages, this does not alter the transfer mechanism examined here.

PART V

SPARKS, LIGHTNING, AND INSULATION

CHAPTER 23

SPARK DISCHARGES

Processes connected with the electric spark have two important influences on transient and fault phenomena in electric power networks. On the one hand, the spark represents the most dangerous discharge of excess voltages because it ruptures the air, oil, or solid insulation materials in a manner leading to temporary or permanent damage of the equipment. On the other hand, it is involved in almost all switching phenomena and thus determines the onset, timing, and initial value of all traveling-pulse phenomena. This is particularly true for the intentional or unintentional switching of very high voltages, extending even to naturally occurring lightning. Often the spark is used as a measuring tool to study switching processes of all kinds and at the same time attempts are made to understand the basic processes in electric sparks as much as possible, even though they occupy only very short time intervals.

(a) *Spark lag.* The insulation between two electrodes is broken down by a spark, which thus initiates contact if either the voltage between them is high enough or the distance between them is low enough. In air and other gases the current is then conducted by electrically excited or ionized molecules. Because the production of these charge carriers in the required quantities depends on the influence of the electric field, which is not instantaneous, a certain time—very short, to be sure—elapses between the onset of the high voltage and the passage of the discharge current. This time lag is called the delay of the discharge or the spark lag. The magnitude of the spark lag is a function of many extrinsic and intrinsic factors. In general, it lies between the extreme limits of 10^{-3} and 10^{-8} sec and thus may vary substantially in different cases.

In homogeneous electric fields the ionization of the gas molecules can take place simultaneously along the entire spark path. Thus here the spark lag is relatively small. However, in inhomogeneous fields the ionization starts at places of high field strength and propagates from these places because the ionization will begin to act as a partial conductor to transfer the field concentration to adjacent stretches of the gap. For electrodes that have regions where the radius of curvature is small compared with the electrode separation, particularly for needle and knife-edge gaps, the field inhomogeneity causes a considerable increase in the spark lag.

In oil or other liquid insulators the energy required for ionization is much greater and thus for comparable arrangements the spark lag is much longer than in air. The same is true for solid insulators, such as porcelain, cardboard, and mica. For asymmetrical electrodes, which we encounter frequently in line insulators, there is a strong effect due to polarity. In general, effects at the negative electrode are dominant because it emits most of the electrons that are accelerated in the field to produce ionization.

The current is conducted by the spark gap as follows. A large number of free electrons are accelerated by the field and collide with neutral atoms or molecules. If they carry enough energy the atom or molecule is ionized, releasing more electrons. The positive ions and the electrons can now travel to the proper electrode under the influence of the electric field and thus conduct current. For the initiation of a spark discharge at least one electron must be present in the region of high field strength between the electrodes to permit ionization by collision. Without it even the highest field strength could not cause such ionization. The accelerated electrons ionize by collision and the ionized particles in turn are accelerated; this process is repeated avalanche fashion until in a very short time there is available a large number of ionized particles to support a current in a channel established by the spark.

We can follow the distinct phases of a spark breakdown according to the time sequence given in Table 23.1. The first electrons in the channel must be caused by ionization by an outside source, such as irradiation by light, ultraviolet, or x-rays, radioactive or cosmic-ray particles, or electron field emission from the electrodes at localized high-field-strength concentrations on sharp protuberances. From these initial effects an ion avalanche can build

electrons from the surfaces of the electrodes and thereby reduce the time lag of the spark initiation by several orders of magnitude. Because the sensitivity of the surfaces to the effect of light is very important and depends strongly on the character of the surface layer, it is possible, on the one hand, by the use of clean surfaces and light-sensitive materials such as magnesium or aluminum to produce a very significant reduction in the spark lag. On the other hand, by oxidation or other means of surface desensitization it is possible to increase the spark lag considerably. The

TABLE 23.1. Time sequence of spark-discharge processes.

Spark lag			Discharge current	
Separate ionization	Self-ionization by avalanche	Streamer formation	Shock waves in spark channel	Discharge of connected circuit
Build-up of the discharge		Foredischarge	Main discharge	

up rapidly, from microamperes to milliamperes and finally to amperes. These processes determine the build-up of the discharge. If at some local spot on the electrodes or in the channel the discharge builds up prior to its completion along its entire path, it grows from these spots by streamer or foredischarge. As soon as a complete conductive channel is formed between the electrodes by ionized gas, the main discharge occurs. It usually first discharges the capacitive energy that is stored mainly in the electrodes; this causes a series of shock-wave pulses to be reflected repeatedly within the gap from electrode to electrode. These very rapid pulses already are the start of the main discharge of the circuit connected to the electrodes. The main discharge now follows the laws discussed in previous chapters.

The space available for initiation and formation of the discharge by free electrons is much larger in the case of sphere or flat-plate gaps and similar homogeneous-field arrangements than in the case of sharp needle or knife-edge gaps with highly localized field-strength concentration. Thus the spark lag is much shorter in the homogeneous field than in the inhomogeneous field. By external irradiation with visible or ultraviolet light, x-rays, or radioactive particles it is possible to eject

lowest possible spark lag, however, can be produced by making the negative electrode extremely sharp so that the field strength at the surface is large enough to cause field emission of electrons. This is the case for a field strength of about 10^6 V/cm.

To a very large extent, of course, the spark lag depends on the magnitude of the voltage

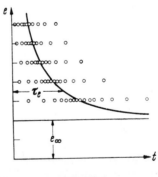

Fig. 23.1

between the electrodes. In Fig. 23.1 we let e_∞ be the sustained breakdown voltage, that is, it is essentially a lower limiting voltage which would lead to breakdown after some rather long time under steady-state d.c. or a.c. peak-amplitude voltage stress. For any higher voltage e applied to the electrodes we then obtain increasingly shorter time lags for spark breakdown because the ion avalanche builds up much faster with increased field strength. In every case, however, the appearance of the first electron in the discharge channel is a random event governed by many extraneous factors; hence it is not possible to predict the time lag at a given voltage level with any accuracy. Instead, for any given excess voltage $e - e_\infty$, repeated observations give a range of time delays (shown by small circles in Fig. 23.1) with a mean value τ_e, as

for which the spark discharge sometimes can take place only during the voltage-decay part of the pulse.

For any such impulse of known rise and decay shape, the ratio of the lowest impulse-voltage peak e_{im} for which breakdown occurs to the steady-state breakdown voltage e_∞ is called the impulse ratio. As shown in Fig. 23.2c, the value of e_{im} may be larger than the instantaneous voltage at the instant of breakdown. For exponential impulses with very short rise and decay times, of the order of 5–50 μs, some observed values of the impulse

TABLE 23.2. Impulse ratio for insulation breakdown.

Air			Transformer oil	Solid insulators
Large spheres	Needles	Chain-suspension insulators		
1.0–1.1	1.8–2.2	1.6–2.0	2–4	1.5–3.0

shown in Fig. 23.1, which is large for a small excess voltage and small for large excess voltages. Thus we must recognize the random probabilistic nature of the spark lag, which implies that large deviations from the mean value will occur.

It is possible to measure the spark lag for different behavior of the voltage as a function of time. Figure 23.2 shows (a) rectangular

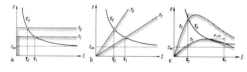

Fig. 23.2

step pulses, (b) two pulses with uniformly rising voltage, and (c) two impulses with an exponential rise and decay. For different voltage amplitudes one can obtain from many measurements the mean time-lag curve, shown as a heavy hyperbolic line in all three figures. Note that these lines will be different for the three different pulse shapes. The shortest lag occurs for step pulses; the lag is longer for the linear-rise pulses, which can begin significantly to initiate ionization only after the instantaneous voltage exceeds e_∞; and the lag is longest for impulses with flat heads

ratio are given in Table 23.2, wherein the larger values apply to impulses with the shorter decay times.

Figure 23.3 shows a number of curves for the mean spark lag τ_e for several shapes of

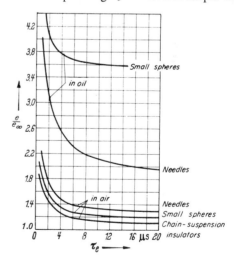

Fig. 23.3

copper electrodes and for impulses with a very short rise time, of about $1\,\mu$s, and a gradual decay of about 250 μs. The very high values for oil insulation are evidence of the fact that transformers have great strength with respect to voltage impulse stresses. The very low values for chain-suspension insulators are mainly due to the initiation of surface discharges on the porcelain insulators. Such low values can also occur for other insulators unless special shielding provisions are made. Even for shorter pulses, with 5–50-μs decay times, the impulse ratio may not be relied upon to be greater than 1.3.

It is remarkable that the impulse characteristics of porcelain insulators are rather independent of the ambient moisture conditions.

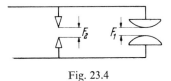

Fig. 23.4

For very short impulses, of 2-μs decay time, the same impulse breakdown is obtained in dry weather as in rain. For longer impulses, however, the insulation strength is reduced under rain or fog conditions by 5–15 percent and for dirty insulators by as much as 20 percent.

These values for gases and liquids are independent of the number of sequential discharges because air and oil can reestablish or regenerate their insulating properties. However, for solid insulating materials this is not the case, so that in them each impulse that exceeds the steady-state breakdown voltage e_∞ weakens the insulation and thus assists the eventual breakdown. Thus frequently repeated impulses can lead in such cases to breakdown at values of peak voltage as much as 25 percent lower than that for a single impulse.

Two spark gaps with large and small spark lags, such as a needle and a large-sphere gap, are shown connected in parallel in Fig. 23.4. If they are now adjusted so that for a steady-state voltage the needle gap F_2 breaks down at somewhat lower voltage than the sphere gap F_1, then for a gradual voltage rise the needle gap will initiate the discharge while an incident step or other steep-rise pulse will initiate a discharge in the sphere gap because it has a shorter spark lag. Thus to afford

protection against fast transients all protective spark gaps should have low spark-lag times and should therefore be constructed as far as possible with homogeneous field distribution and suitable electrode materials.

Owing to the finite velocity of propagation of traveling pulses, it is possible to represent the time lag in breakdown as a spatial delay. If, for example, the spark lag of the needle gap of Fig. 23.4 for a given excess voltage is 3.3 μs, the voltage pulse travels a distance of about 1000 m during this time. A line of this length then must be inserted between two spark gaps, the first with and the second without spark lag, if they are to break down at the same time upon arrival of an impulse at the first. We thus see that spark gaps with appreciable spark lags are not suitable for the protection of the insulation of lines or windings.

Figure 23.5 shows an example of a steep-rise traveling impulse incident upon an installation protected by a spark-gap arrester A having a short spark lag. We see that after τ_e at least the back portion of the impulse is suppressed by the discharge in A. If, owing to sharp corners, for example, the bus bars S and the terminal bushings D of the transformer winding have a longer spark lag than A, the traveling impulse shortened by A cannot cause breakdown. Even inside the transformer T where the pulse is mostly amplified on

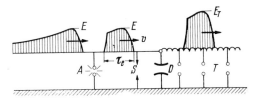

Fig. 23.5

account of transmission, as a consequence of the higher surge impedance, it cannot produce breakdown if the spark lag is increased by the oil a little more than the voltage. The pulse reflected at the transformer entrance terminal can cause breakdown in the bushing D only if its front is so steep that its voltage upon return to A has risen sufficiently, even though a multiple but partial reflection has occurred at the bus bar S on the way. Even at the free star point of the transformer, the again reflected and doubled voltage usually cannot overcome the long spark lag of the oil. However, if the star point of the transformer is brought to a terminal point in air, instead of being left inside the oil-filled tank, then flash over is apt to happen at this terminal on account of the short spark-lag time in air.

(b) *Velocity of the discharge front.* For long sparks the field just prior to breakdown is generally inhomogeneous. This can be the result of the shape of the electrodes or of the mechanism of spark formation or of space

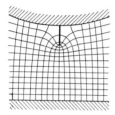

Fig. 23.6

charge in the electric field. Even if the electrodes have large radii of curvature in comparison with their separation and the field in the gap should be homogeneous, a concentration of the field strength may occur on account of some irregularity in the space, as at the upper electrode in Fig. 23.6. This then gives rise to a local breakdown. In the example the breakdown leads to the sharp penetration of the upper-electrode potential into the space, so that an intense field is formed at its tip by a compression of the equipotentials.

This in turn leads to further early breakdown at the tip, which essentially results in propagation of the spark further into the space until it reaches the opposite electrode. The conducting path so established can now support the main discharge.

Thus while the breakdown in a homogeneous field occurs instantaneously, that is, simultaneously along the entire channel, in an inhomogeneous field the spark-discharge front propagates with a finite velocity. The various portions of the air gap are broken down in sequence and the high breakdown field strength need exist only immediately ahead of the

propagating discharge front. As the breakdown field strength is thus transferred gradually, even though the total time is very short, from one electrode to the other, the total breakdown voltage across such an inhomogeneously initiated gap is much lower than if the breakdown field strength had to be established across the entire gap as in the homogeneous case. Thus the total breakdown voltage is less than proportional to the gap length and the breakdown field strength averaged over the gap is much lower for long spark gaps than for short ones.

In practice there occur with every spark discharge substantial space charges, since the spark is originated by free positive and negative charges. However, we shall here consider the electromagnetic relations only and assume that the insulating material breaks down when a given field strength is exceeded. The magnitude of this field strength is of

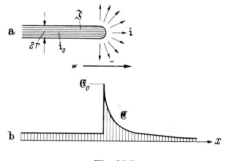

Fig. 23.7

course given by the ionization mechanism in the insulator.

Along the discharge channel of a spark as shown in Fig. 23.6 there are thus entirely different electrical conditions behind and ahead of the traveling front; these are shown schematically in Fig. 23.7. In the spark itself and thus behind the front there is a conduction or convection current of density i_0, so that the total spark current in the circular channel of radius r is

$$\mathfrak{I} = i_0 r^2 \pi. \tag{23.1}$$

Ahead of the discharge front, however, there is a strong displacement current, which forms a closed system with the conductive current, completing a circuit, so to speak. Neglecting space charge, its current density is

$$i = \varepsilon \frac{d\mathfrak{E}}{dt}, \qquad (23.2)$$

where ε is the dielectric constant and $\mathfrak{E}$ is the electric field strength in the vicinity of the front.

If the discharge front maintains its shape while propagating with the velocity w, the field lines in its vicinity will change only very little and in essence everything will move along with the velocity w. We can then change the differentiation with respect to time into one with respect to space by reference to Eq. (23.2) and Fig. 23.7 and can put

$$i = \varepsilon \frac{\partial \mathfrak{E}}{\partial x} \frac{dx}{\partial t} = \varepsilon w \frac{d\mathfrak{E}}{dx}. \qquad (23.3)$$

The equipotentials in the vicinity of the front, as seen in Fig. 23.6, are almost hemispherical, so that the displacement current shown in Fig. 23.7 flows almost radially and only into the right-hand hemisphere as shown. The displacement-current density as a function of the distance x from the tip of the front is given approximately by

$$i = \frac{\mathfrak{J}}{2\pi x^2}. \qquad (23.4)$$

Of course this distribution over the hemisphere is not so uniform but the value given will be a sufficiently representative average for our purposes.

If we insert Eq. (23.4) into Eq. (23.3) we obtain

$$\frac{d\mathfrak{E}}{dx} = \frac{\mathfrak{J}}{2\pi\varepsilon w x^2}. \qquad (23.5)$$

This gives the spatial distribution of the electric-field strength in the vincinity of the spark-discharge front. By integration we obtain

$$\mathfrak{E} = \frac{\mathfrak{J}}{2\pi\varepsilon w} \int \frac{dx}{x^2} = \frac{\mathfrak{J}}{2\pi\varepsilon w} \left[\frac{-1}{x} \right]. \qquad (23.6)$$

To determine the limits of integration we assume that the spark-discharge front is approximately hemispherical and of the same radius r as the discharge channel. Thus the lower limit of integration can be put at $x = r$. For the upper limit we may use infinity because larger distances produce increasingly smaller contributions since the square of the distance appears in the denominator of the integral. We then obtain for the field strength in the vicinity of the discharge front

$$\mathfrak{E}_0 = \frac{\mathfrak{J}}{2\pi\varepsilon w} \left[\frac{-1}{x} \right]_r^\infty = \frac{\mathfrak{J}}{2\pi r\varepsilon w}, \qquad (23.7)$$

and this is of course the breakdown field strength of the insulation in which the spark front is propagating.

For the velocity of progress of the head of the spark-discharge front we obtain

$$w = \frac{\mathfrak{J}}{2\pi r\varepsilon \mathfrak{E}_0}. \qquad (23.8)$$

Thus we can calculate it if we measure the current and the radius of the discharge because the dielectric constant and the breakdown field strength are known in general.

If we measure $\mathfrak{J}$ in amperes, $\mathfrak{E}_0$ in volts per centimeter, and ε as the specific ratio referred to a vacuum, and if we want to obtain w not in centimeters per second but in kilometers per second, we must multiply Eq. (23.8) by a factor

$$4\pi v^2 \cdot 10^{-9} \cdot 10^{-5} = 2\pi \cdot 18 \cdot 10^6,$$

which then gives us

$$w = \frac{18\mathfrak{J} \cdot 10^6}{r\varepsilon \mathfrak{E}_0} \text{ km/sec.} \qquad (23.9)$$

If we substitute for the total current $\mathfrak{J}$ the current density i_0 in the spark as given by Eq. (23.1), we obtain

$$w = 18\pi \frac{r i_0}{\varepsilon \mathfrak{E}_0} 10^6 \text{ km/sec.} \qquad (23.10)$$

From these equations we see that the velocity of propagation of the spark front is lower the greater the dielectric constant and strength but is higher the larger the discharge diameter and current density.

If we take as an example a breakdown field strength in air of 50,000 V/cm, then according to Eq. (23.9) a spark-discharge front of diameter 3 mm that carries a current of 10 A has a velocity of propagation

$$w = \frac{18 \cdot 10 \cdot 10^6}{0.15 \cdot 1 \cdot 50,000} = 24,000 \text{ km/sec;}$$

this is about $\frac{1}{12}$ of the velocity of light. Surface sparks or blue corona brushes have considerably lower current density and therefore by Eq. (23.10) also have proportionately lower velocities of frontal propagation.

Very little is known about the relation between diameter and current in the front of a spark discharge. Similarly, it is not possible to be definite about the breakdown field strength in the region immediately ahead of the propagating front. In the example just calculated a number was used that corresponds to breakdown distances of about 1 mm, but it is known that for smaller distances it becomes much larger. The frontal propagation velocity will then become much lower and can possibly become as low as 1000 km/sec. In the mechanism treated here it appears likely that very high field strengths are found immediately in front of the discharge, even though the average field strength between the electrodes is only moderate.

On account of this high field strength ahead of the discharge front, this particular spot appears especially bright because the energy density is so high. The portions of the discharge farther back should appear less bright. Photographs of spark and lightning discharge fronts taken with high-speed (Boyes) cameras confirm this.

J. J. Torok has succeeded in producing spark discharges, as shown in Fig. 23.8, that were interrupted during formation, that is, before the front had propagated all the way across the gap. He did this by using very short impulses to feed the discharges. The spark fronts taken at about 500 kV show a number of spark streamers, partly uniform and partly branched, whose individual current was just about 100 A and whose diameter was a good 1 cm. This gives about the same order of magnitude for the velocity of propagation as was obtained in the calculated example.

This mechanism of initiation of the spark discharge also explains the well-known jagged form of very long spark discharges. Because

the breakdown field strength in the hemisphere ahead of the discharge front is fairly uniform, as shown in Fig. 23.7, very small accidents of ionization can rather randomly determine in exactly which direction the front will continue to propagate. It need not move exactly in the direction determined by the undisturbed lines

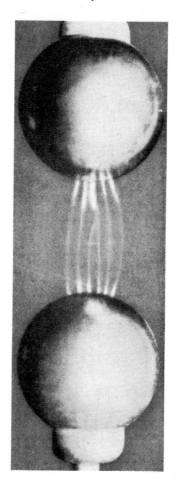

Fig. 23.8

of force of the field, as given in Fig. 23.6, but may instead frequently make small sideways excursions, which in the case of large parallel electrodes far apart may lead to a very appreciable lateral shift.

Natural lightning also propagates such a discharge front, called a leader stroke or precursor, which has a finite velocity of propagation. Since here there are no clearly defined electrodes, we must imagine that the discharge begins because, at an arbitrary point in the electric field extending over the large space, the electric breakdown field strength of air is reached. From this point the fronts of the discharge propagate in both directions until they either encounter the conducting ground or are taken up by space charges in the thunderclouds. Figure 23.9 shows this finite rate of growth of

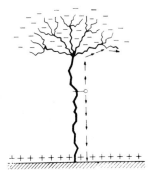

Fig. 23.9

lightning leader strokes from a central point, which may however also lie at the cloud edge.

The time of formation of such a spark discharge propagated over a length a is

$$\tau_a = \frac{a}{w}. \qquad (23.11)$$

For example, lightning of length 1 km initiated at a speed 0.1 that of light has a time of formation

$$\tau_a = \frac{1}{3 \cdot 10^4} = 3.3 \cdot 10^{-5} \text{ sec} = 33 \ \mu s$$

if propagated in only one direction along this whole length. During this time the entire voltage giving rise to the lightning stroke is applied closer and closer to the space at the electrode toward which the front is traveling, as shown in Fig. 23.6.

When the spark-discharge front has reached the opposite electrode, a complete conducting channel is formed through which the main discharge can now take place. Only at this time can the voltage between the electrodes or space-charged clouds decrease significantly. The current in the discharge at the same time

increases to an amount that is determined by the external impedance or other constants of the feeding network.

(c) *Measurement of traveling pulses.* If a spark discharge is used to excite a traveling pulse, the front rise obtained is very steep, as shown in Fig. 23.10. If the discharge occurs in air the maximum voltage rise as a fraction

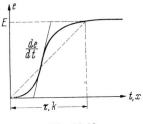

Fig. 23.10

of time and distance upon breakdown will be about

$$\frac{de}{dt} = 2000\text{—}4000\text{—}6000 \text{ kV}/\mu s$$

or $\qquad\qquad (23.12)$

$$\frac{de}{dx} = 6.7\text{—}13.3\text{—}20 \text{ kV/m},$$

and the average rise, shown by the broken line in Fig. 23.10, will be about

$$\frac{E}{\tau} = 500\text{—}1000\text{—}1500 \text{ kV}/\mu s$$

or $\qquad\qquad (23.13)$

$$\frac{E}{k} = 1.7\text{—}3.3\text{—}5 \text{ kV/m}.$$

These values are the result of a number of representative measurements and have very little dependence on the actual voltage used.

Thus the rise τ or the front length k of the pulse is very much shorter for low voltages and short gaps, which are therefore very useful for a variety of measurements. For example, a voltage $E = 15$ kV gives a front rise of

$$\tau = \frac{15}{1000} = \frac{1}{67} \ \mu s \quad \text{or} \quad k = \frac{15}{3.3} = 4.5 \text{ m}.$$

For large-amplitude pulses used in the testing of high-voltage installations, on the other hand, the discharge times are correspondingly higher; for example, a breakdown voltage $E = 500$ kV gives a front rise of about

$$\tau = \frac{500}{1500} = \frac{1}{3}\,\mu s \quad \text{or} \quad k = \frac{500}{5} = 100 \text{ m.}$$

This is still a sharp front rise, particularly if we note, as shown in Eq. (23.12), that the maximum slope is about four times this average value. If the breakdown occurs in oil, the

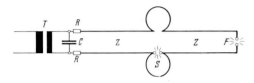

Fig. 23.11

discharge times and thus also the front-rise lengths are only a fraction of their values in air. We see that traveling pulses produced by spark discharges, while not discontinuous as a step are extremely steep and thus have short front-rise lengths, of only a few meters, while the rise time is between 1 and 0.1 μs. For many applications we can thus regard them as discontinuous step pulses.

It is possible to use such spark discharges to charge or discharge lines, cables, condensers, or other apparatus for which an examination

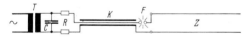

Fig. 23.12

of the behavior of traveling pulses is desired. As an example, Fig. 23.11 shows an experimental arrangement in which a cable or line Z is fed by a high-voltage transformer T. In the line is a delay loop S of variable length whose ends are connected by an adjustable spark gap. A short circuit is initiated at the end of the main line by gradual closure of the spark gap F. The discharge sends a steep-front pulse backward from F along the line Z. Then the spark gap across S will break down at various voltage settings, depending on the length of S. As S is increased beyond a few meters, the final peak voltage of the pulse is

reached. By a series of such measurements the values given in Eqs. (23.12) and (23.13) can be obtained. The resistances R and the condenser C suppress the reflection of the pulses back into Z from this end. At the same

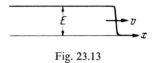

Fig. 23.13

time they protect the transformer against step pulses and short circuits. Figure 23.12 shows a similar arrangement for the observation of charging pulses traveling along an open line Z, produced at the junction with a cable K by a spark gap F.

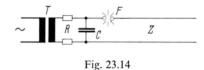

Fig. 23.14

The arrangements of both Figs. 23.11 and 23.12 produce very steeply rising pulses with flat backs, initially constant, as shown in Fig. 23.13. To produce impulses it is customary to use a condenser discharge, as shown in Fig. 23.14. Here the voltage drops back to

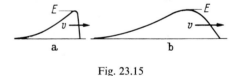

Fig. 23.15

zero after its initial rise (Fig. 23.15a). Owing to the physical size and self-inductance of the condenser, such impulses, as shown in Fig. 23.15, usually have a somewhat flatter front than those produced by the discharge

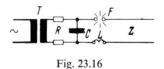

Fig. 23.16

of a cable. To further flatten the front (Fig. 23.15b), and to make it completely independent of the spark characteristics, we can introduce an inductance (Fig. 23.16). If it is made sufficiently large, an oscillatory pulse train is produced (Fig. 23.17). If the pulses in this train are required to have a rectangular step shape (Fig. 23.18), the discharge must come from a cable (Fig. 23.12) with an open

Fig. 23.17

end effectively produced by making the value of the resistances R very large.

If the circuits for producing the pulses as shown in Figs. 23.12, 23.14, and 23.16 are excited by alternating current, a suitable adjustment of the spark gap will produce a

Fig. 23.18

discharge during each half-period of the supply frequency. Thus a rapid sequence of pulses is produced on the line to be observed and the pulse frequency is twice that of the alternating-current supply. If only a few pulses are desired, the condenser must be charged by means of direct current I by a high-voltage rectifier V, shown in Fig. 23.19 as a

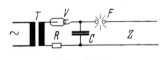

Fig. 23.19

vacuum tube. If the charging current in the rectifier circuit is small compared with the capacitance of the condenser, the voltage across the condenser ($de/dt = i/C$) rises only slowly with time and a discharge occurs only after the breakdown voltage of the spark gap is reached. This was shown in Fig. 6, p. 564, of Ref. 1. Thus the frequency of the pulses can be easily regulated by a choice of voltage, resistance R, and capacitance C.

It is very economical to use a resonant transformer to feed the rectifier circuit. This

will also gradually charge the condenser after each discharge but avoid resistance losses and still allow frequent pulses.

To obtain very high voltages from moderate charging sources it is possible to charge a condenser bank in parallel, using very large resistances, and discharge it in series, using

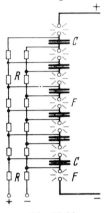

Fig. 23.20

spark gaps. This arrangement, due to E. Marx, is shown in Fig. 23.20. It has been used to produce impulses of several million volts amplitude.

If the aim is to examine pulse behavior between a line and ground, the arrangements discussed so far are adequate because the spark gap can be connected in nonsymmetric fashion between the ungrounded side of the charging device and the end of the line. However, if the pulse must be generated between lines it becomes necessary to obtain

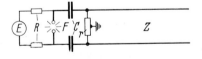

Fig. 23.21

a symmetric arrangement of the circuit. Two spark gaps cannot be used, since there would be no assurance that they would break down at the same instant. In fact, the various circuit shunt capacitances would allow pulse formation in one line only temporarily due to the breakdown of one spark gap. It is thus necessary to use an arrangement such as is shown in Fig. 23.21, which has a single spark gap, but the capacitance C and, if necessary, an inductance L are each split into two equal portions. Further, the test line is balanced with respect to ground by means of a very high resistance r with a center tap, which assures symmetry for the initial pulse.

Because the experimental line cannot be made arbitrarily long, reflections at the ends

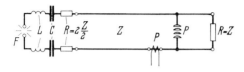

Fig. 23.22

may cause great interference in the observed results. These can be completely avoided, as shown in Fig. 23.22, if at both ends of the line an ohmic resistance $R = Z$ is connected. At the charging end of the line this is split into two equal portions, $R = Z/2$. Because the line current will produce a voltage drop across R equal to that across the surge

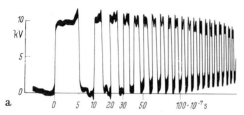

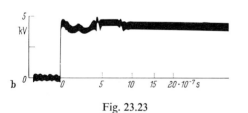

Fig. 23.23

impedance Z, the charging arrangement must now produce twice the voltage required across the line.

Figure 23.23 shows a step pulse produced in such an arrangement (a) without and (b) with reflection-suppressing resistances R.

If it is desired to observe the behavior of a piece of electrical apparatus such as an insulator chain, a current or voltage transformer, or a voltage limiter while it is in use

on the line, it can be connected at P (Fig. 23.22), either in series or in parallel. The tested element P is thus exposed to the pulse and its own behavior as well as its effect on the behavior of the line can be examined. Frequently only the behavior of the tested element under a given voltage impulse is to be examined. Then the surge impedance of the line could be kept so small as to introduce negligible interaction. However, it is still simpler to use no line at all but merely an ohmic resistance R in the discharge circuit. Figure 23.24 shows two equivalent arrangements for this purpose,

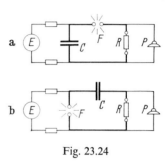

Fig. 23.24

differing only in the relative positions of the condenser C and the spark gap F.

Because the detailed behavior of the voltage breakdown depends on many uncontrolled influences, it is not possible to produce a clearly defined and repeatable pulse front length. To remove these uncertainties it is possible to introduce a small self-inductance L, as shown in Fig. 23.22, into the circuit. The current, and thus the voltage on the line or across the discharge resistor, can then only increase with the time constant L/R, which then governs and defines the pulse rise time beyond the uncertainties. The self-inductance can be very small; for high-voltage test arrangements usually a few turns, or even the inherent self-inductance of a large condenser, are sufficient. Because any such self-inductance contains some shunt capacitance, independent oscillations of very high frequency are produced in this oscillatory circuit, which are superimposed on the pulse front as seen in Fig. 23.23a. They can cause very serious distortion of the pulse shape and introduce very undesirable, though short-duration, high-voltage spikes.

With the voltage-multiplier arrangement of Fig. 23.20 this phenomenon becomes very pronounced. This is due to the fact that the condensers at the upper end of the chain reach their full voltage only upon breakdown of the series spark gaps and thus the voltage pulse is suddenly applied to the entire circuit. As shown in Fig. 23.25, this gives rise to an oscillation in the circuit formed by L, C, and the shunt capacitance c that develops

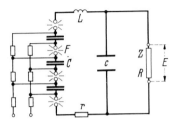

Fig. 23.25

as seen in Ref. 1, Fig. 3, p. 60. The frequency is mainly governed by L and c because C is very large and thus has a low impedance at the high frequency.

For an impulse generator of 3×10^6 V, $L = 50 \ \mu$H self-inductance and $c = 800 \times 10^{-12}$ F shunt capacitance, the natural frequency is

$$\nu = \frac{1}{2\pi\sqrt{50 \cdot 10^{-6} \cdot 800 \cdot 10^{-12}}}$$
$$= 8 \cdot 10^5 \text{ cy/sec,}$$

and thus a quarter period is 0.31 μs. This is of the order of magnitude of the desired pulse-rise time and thus will cause very serious spike overshoots in voltage to twice the desired value. It is necessary to suppress these oscillations by a damping resistance

$$r \geq 2\sqrt{\frac{L}{c}} \qquad (23.14)$$

in series with L and c, which in the example used here will be at least

$$r = 2\sqrt{\frac{50 \cdot 10^{-6}}{800 \cdot 10^{-12}}} = 500 \ \Omega.$$

In its place a parallel resistance of one-quarter of this value could be used. If the impulse generator is used to produce pulses across a

line of moderate surge impedance, this is usually sufficient to produce damping. However, if a very large discharge resistance R is used to obtain a very flat pulse back, a special damping resistance r is needed and is best distributed uniformly around the circuit to avoid disturbances due to local discontinuities. All resistances used in impulse-generating circuits must, of course, have low self-inductance and capacitance to avoid parasitic oscillations of their own. Thus their physical dimensions must be appropriately restricted; for example, they must be short. Carbon resistors are particularly suitable.

The shape of the voltage rise across the high-voltage terminals of Fig. 23.25 for an overdamped or aperiodic circuit with a high load resistance R is, according to Ref. 1, Eq. (33), p. 228,

$$\frac{e}{E} = 1 - \frac{T_1\varepsilon^{-t/T_1} - T_2\varepsilon^{-t/T_2}}{T_1 - T_2}, \qquad (23.15)$$

where

$$T_1 = cr \quad \text{and} \quad T_2 = L/r \qquad (23.16)$$

are the time constants of the overdamped circuit. The load resistance R or Z in the main circuit and its time constant $T = CR$ will now govern the shape of the pulse decay, while the overdamped circuit governs the

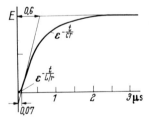

Fig. 23.26

front-rise time and shape of the pulse. In the example used here an adequate damping resistance of 750 Ω produces

$$T_1 = 800 \cdot 10^{-12} \cdot 750 = 0.6 \ \mu\text{s,}$$

$$T_2 = \frac{50 \cdot 10^{-6}}{750} = 0.07 \ \mu\text{s.}$$

Figure 23.26 shows this voltage rise, which agrees well with experimental observations.

In all these arrangements the pulses or impulses were produced in a circuit and then applied to the object of the test. However, it is rather simple to test the windings of alternating-current machines and transformers by means of pulses produced directly in these windings. Figure 23.27 shows an arrangement,

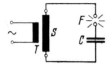

Fig. 23.27

suitable for this purpose, in which the voltage of the winding is connected to a condenser by a spark gap. In Ref. 1, Chaps. 42 and 44, it is shown that the condenser is charged at each spark breakdown to at least the peak value of the feeding voltage. Thus if the spark gap is set to break down at twice the nominal voltage of the winding, at each reversal of voltage the condenser is charged from a voltage $-E$ to $+E$ and vice versa. This always produces a step pulse of voltage amplitude $2E$ that is reflected back into the winding.

The stray inductance S of the winding also forms an oscillatory circuit with the capacitance C of a frequency that in practice is only a

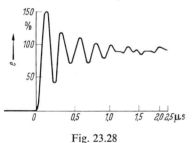

Fig. 23.28

few thousand cycles per second and thus cannot produce noticeable effects on the step pulse. However, it does produce an increased voltage stress in the winding at these frequencies, as seen in Ref. 1, Fig. 4, p. 586. The shunt capacitance of the winding and the self-inductance of the test circuit form another oscillatory circuit whose frequency is very high, of the order of a few megacycles per second, and this gives rise to further impulse-type voltage stresses on the windings. Figure

23.28 shows a cathode-ray oscillogram of these high-frequency oscillations at the terminals of a winding. In Ref. 1, Fig. 12, p. 596 are shown the slower oscillatory phenomena observed by means of a far more slowly moving mirror oscillograph. In Ref. 1 it is possible to recognize clearly the voltage e_C across the condenser and the current i in the spark-gap circuit at each charge reversal.

For three-phase windings it is possible to test the three phases simultaneously with impulses, simply by means of three spark gaps, as shown in Fig. 23.29. Small firing-time differences between the gaps can be neglected

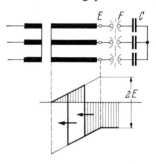

Fig. 23.29

here since the interaction between windings on a very short-time basis is negligible and will in practice also be present owing to nonuniform phase timing of pulses. For 60-cy/sec alternating current a step pulse of twice the nominal operating voltage enters each winding every 1/120 sec, which over a 10-sec period amounts to 1200 such test pulses. As shown in Fig. 23.29, these pulses penetrate into the winding and produce severe voltage stresses, particularly in the insulation of the coils and layers.

The testing of windings with self-produced pulses has a further advantage in that the entire winding is at full operating voltage during the entire test. Any breakdown in insulation due to the pulses is then immediately developed to support high fault currents that will really rupture the weak spots in the insulation, which are thus clearly marked, usually by burns.

Voltage limiters and arresters also must be tested under full operating voltage and conditions. The symmetric arrangement of Fig.

23.30 is particularly suitable, since it allows the testing of two limiters P in series while the operating voltage is supplied in parallel. The choke coils D prevent the pulses from entering the alternating-current source.

Owing to the short duration of traveling pulses, the instruments used to measure their

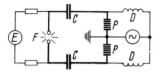

Fig. 23.30

amplitude must have accurate response in very short times. On account of their homogeneous field distribution, sphere gaps are suitable, particularly if their breakdown-initiation time delay is made short and somewhat more regular by irradiation. To measure

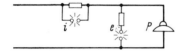

Fig. 23.31

voltage a series resistor is introduced, as shown in Fig. 23.31, to reduce the current drawn and thus effects upon the test circuit. To measure current the sphere gap is connected across a series resistance in the line as shown in Fig. 23.31. This resistance must be low so

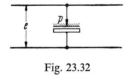

Fig. 23.32

as to interfere as little as possible with the pulses. During tests the gap spacing is gradually reduced in length until a spark occurs. This requires a sequence of pulses and thus the alternating-current arrangements are particularly suitable.

Without this trial-and-error closing of a gap a method due to Lichtenberg can be used. This is shown in Fig. 23.32 and results

in Lichtenberg patterns in a photographic film p if it is exposed to a short pulse between a needle electrode and a plane. Figure 23.33 shows such a pattern after the film is developed. The outer diameter of the pattern can be

Fig. 23.33

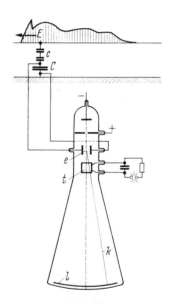

Fig. 23.34

used with some accuracy to estimate the maximum voltage amplitude. In the klydonograph such patterns are produced sequentially on a moving film and thus give a good picture of the rapid transient overvoltages that may occur in a network. Because such patterns are limited to voltages of about

3–20 kV by the thickness of the usual films, the instrument must be connected to lines of higher voltage by means of a capacitive voltage divider.

The best observational tool for fast pulses and transients is the cathode-ray oscillograph, shown in Fig. 23.34. The voltage to be observed is impressed across one set of deflection plates e (usually to produce deflection along the vertical), while an auxiliary voltage is used across a perpendicular set of deflection plates to produce a time base (usually linear) which can be synchronized. This then gives a true continuous account of instantaneous voltage as a function of time t instead of the maximum-value samples obtained with sphere gaps and clydonographs. Only for extremely short pulses encountered in high-frequency microwave regions and very fast computer technology does the standard cathode-ray tube introduce noticeable limitations on account of its various intrinsic time constants. For power-system transients it is completely adequate in time response.

Capacitive voltage dividers, as shown in Fig. 23.34, frequently introduce inaccuracies when used with high voltages because the measuring circuit contains additional resistances and self-inductances that are hard to determine. Instead it is preferable to use carbon resistors having high specific resistance and low temperature and voltage coefficients of resistance. Because they can be short, their self-inductance and capacitance, which could lead to parasitic effects and thus errors in the observations, will be negligible.

CHAPTER 24

LIGHTNING DISCHARGES

The strongest and most dangerous disturbances in open-line networks are caused by natural lightning. The origin and generation of lightning are still not fully understood. However, the actual progress of a discharge can be quite accurately assessed with the aid of the concepts of oscillations and traveling pulses as developed in previous chapters.

In general, the air above the ground surface is slightly charged. This charge and the

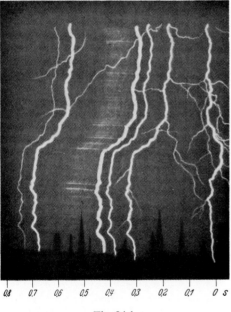

Fig. 24.1

accompanying field strength can increase considerably with changes in meteorological conditions until the breakdown strength of air, which is about 30 kV/cm, is exceeded in a few places, in which case the voltage is equalized by lightning discharges across large distances. The electric-field changes as well as the lightning discharges produce effects on the lines that may break down the insulation along the line as well as equipment in the terminal stations. Figure 24.1 shows a lightning discharge between a thundercloud and ground observed by B. Walter in Hamburg with a rotating camera. It is seen that the actual discharge channels as well as the time interval between successive discharges observed during about 0.8 sec vary somewhat. However, some features persist and others appear to follow earlier incompletely broken-down channels.

(a) *Numerical values.* Measurement of lightning storms and strokes is extremely difficult owing to their transient and dangerous nature; the results are often so variable that numerical values are not easily or well defined. For all observed values it is important to distinguish between values of the field strength in free space and the voltage on a line. Measurements on a line are easier to implement, because only one dimension is involved, that along the line, for which the laws of propagation are well known. In free space, however, the field due to lightning changes rapidly over very large regions, and observations are limited to probe antennas or lightning conductors, which may influence the observation by their presence. In recent years measurements with airborne instruments have also become available. In general, values on lines are somewhat more reliable than those in free space.

The following statements, based on a large number of observations, can be made about the field of a thunderstorm. The voltage between the lightning cloud and the ground is estimated to be between 10 and 100 MV. The field strength, which during normal atmospheric conditions is about 150 V/m, grows during a thunderstorm to several thousand volts per meter. At the time of the discharge it may reach 50–200 and even 300 kV/m near the ground. The spatial distribution of the

248

field strength between ground and cloud is not known. However, the field strength is on the average far below the 3-MV/m breakdown strength of air mentioned earlier. The field strength fluctuates very markedly with time. Changes occur over time periods of 5–20 μs, which is also about the duration of the lightning discharge itself.

There is a wide difference of opinion about the strength of currents in lightning discharges. The estimates begin with a few to 50 kA and reach as high as 100–500 kA. The spatial extent of the high field strength due to a thunderstorm appears never to exceed 2 to 4 km in any direction, while frequently dimensions of clouds and heights apparently involved are limited to 0.2–0.4 km. The lightning discharge is usually overdamped, not oscillatory, though not too much is known about its detailed temporal behavior. Frequently partial discharges occur in the same channel or previously uncompleted branches at intervals of about 0.1 sec, while the total sequence of partial discharges occupies about 1 sec.

On high-voltage lines, cathode-ray oscillograph observations gave lightning voltages which during direct hits amounted to between 500 and 1000 kV and in a few cases even rose to 3000–5000 kV. The voltage pulse usually has a steep front with a rise time of 0.5–2 μs and up to 10–20 μs, while the decay time is of the order of 5–20 or even 100 μs. In addition to such unipolar impulses, oscillatory pulse trains have frequently been observed with a half-period of microseconds to tens of microseconds. The field strength along the line due to the steep rise of the pulse front often is 0.3–3 kV/m and may rise to 15 kV/m. This corresponds to a time rate of voltage rise of about 100–1000 and even 5000 kV/μsec.

The current amplitude in the line can be determined from the voltage and the surge impedance and amounts to 1000–10,000 and even 50,000 A. It can be measured by an oscillograph activated during an actual lightning stroke. Frequently it is measured by the remanent magnetization of special steel rods suitably proportioned, mounted, and calibrated with respect to the line. These have also been mounted at the footing of line towers, though doubts have arisen whether a rod near one of the four feet of the tower really measures one-fourth of the total current. Damping of the pulses during travel along the lines was most pronounced if the corona voltage level was exceeded, in which cases the amplitude was reduced to half its initial value over a distance of 10–20 km. Further, the steep rise and decay are also flattened considerably after the pulse has traversed a few kilometers.

Table 24.1 gives a summary of all results.

As in the case of the spark discharge just treated in Chapter 23, the main discharge in lightning, as described in Table 24.1, takes place only after a discharge channel has been prepared by preliminary discharges. According to measurements by B. F. J. Schonland with high-speed rotating-lens cameras, this preliminary discharge, or leader stroke, propagates not with uniform velocity but in a number of discontinuous jumps each about 50 m long. It is even considered probable that this observable preliminary discharge is prepared by an earlier nonglow discharge that propagates relatively slowly and preionizes the air. This would correspond to the dark-current precursor at the head of the spark discharge.

Because the main discharge of the lightning stroke probably has a larger channel diameter

TABLE 24.1. Measured and estimated lightning parameters.

Parameter	In free space	On line
Voltage (MV)	10–20–100	0.5–1–5
Field strength (kV/m)	50–200–300	0.3–3–15
Time duration (μs): Front rise	5–10–20	0.5–5–20
Back decay		5–20–100
Current strength (kA)	10–100–500	1–10–50
Temporal behavior	Mostly overdamped	Single or multiple impulse
Spatial extent (km)	0.2–2–4	10–30

than the preliminary discharge, the head of the main discharge probably has to use additional energy for ionization. This will have an influence on the development of the discharge by slowing down its propagation along the channel. We shall not consider this effect in the analysis that follows.

The polarity of thunderclouds is most frequently negative and photographs with very high-speed cameras show that usually the lightning stroke proceeds from the cloud to ground. In this case electrons can produce ionization more easily than ionized positive particles could in the reverse case. In the vicinity of the bottom of the cloud initiating regions, like those shown in Fig. 23.9, are probably more numerous and sharper than on the ground surface. However, lightning strokes to the Empire State Building in New York, being initiated by the sharp lightning conductor at the top of the building, were observed to start at the ground and proceed upward into the cloud.

(b) *Oscillations of the main discharge.* The lightning discharge occurs either from cloud to cloud or, as shown in Fig. 24.2, from cloud

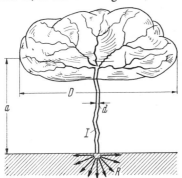

Fig. 24.2

to ground. The discharge channel has a diameter of 1 cm to several decimeters. When the channel is fully developed and visible, it will have a voltage drop along it of about 10 V/cm, corresponding to an electric-arc discharge. This amounts to about 100 kV for each 100 m of the stroke length, which is very modest compared with the effective voltage of the cloud prior to the discharge by the stroke. Thus the discharge

will behave like the spark discharge between two large spheres or flat plates forming a condenser that is in effect shorted out by a central conductor. The behavior with respect to time is mainly determined by the capacitance between the cloud and ground and the self-inductance of the lightning channel.

Thus for the main features of the lightning discharge we can construct the equivalent

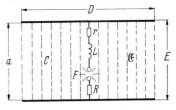

Fig. 24.3

circuit of Fig. 24.3. This contains a circular-plate condenser with electrodes of cloud diameter D and a spacing a equal to the average cloud height above ground. It is charged to the voltage E and is suddenly discharged by the spark gap F. The self-inductance L and resistance R of the discharge channel are also shown. The capacitance of the condenser can easily be calculated; however, the self-inductance of the channel, which is an un-closed circuit, does not have its usual character. Thus it is better to derive both from the energy content of the electromagnetic field, which is a function of the square of the electric-field strength $\mathfrak{E}$ and the magnetic induction $\mathfrak{B}$.

Each unit volume contains electric energy

$$w_e = \frac{1}{8\pi v^2} \, \mathfrak{E}^2 \qquad (24.1)$$

and magnetic energy

$$w_m = \frac{1}{8\pi} \mathfrak{B}^2, \qquad (24.2)$$

where $v = 3 \cdot 10^{10}$ cm/sec is the velocity of light. If the oscillations are so slow that in the space between the electrodes, as shown in Fig. 24.3, the electromagnetic waves have no important phase differences, the electric field strength in this entire space will be uniform and have the value

$$\mathfrak{E} = \frac{e}{a}. \qquad (24.3)$$

We neglect here any distortion of the field lines at the edge of the condenser by assuming the electric lines of force all to be straight, as shown in Fig. 24.3. This is valid when D is much larger than a, and when it is not, an equivalent value for D can be used to include the edge effect. The total electric energy stored is then, by summation over the space between the electrodes,

$$W_e = \frac{1}{8\pi v^2} \frac{e^2}{a^2} \frac{aD^2\pi}{4} = \frac{D^2e^2}{32v^2a}. \quad (24.4)$$

If we equate this to the energy stored in a condenser,

$$W_e = \tfrac{1}{2}Ce^2, \quad (24.5)$$

we obtain for the capacitance of the condenser

$$C = \frac{D^2}{16v^2a}, \quad (24.6)$$

which is the usual expression for a circular-plate condenser.

The magnetic induction $\mathfrak{B}$ is produced by the transient currents. We assume that the

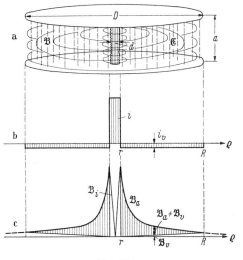

Fig. 24.4

lightning stroke occurs at the center of the plates and has a channel diameter d. This gives rise to a magnetic field having concentric lines of force surrounding it, as shown in Fig. 24.4a. The strength of this external field $\mathfrak{B}_a$ is obtained from the fundamental law that the line integral is equal to 4π times the

enclosed current. With symmetric behavior along a cylindrical path of radius ρ we then obtain

$$2\pi\rho\mathfrak{B}_a = 4\pi i \quad (24.7)$$

or

$$\mathfrak{B}_a = \frac{2i}{\rho}. \quad (24.8)$$

Figure 24.4c shows the field strength, which decreases hyperbolically with increasing ρ.

The magnetic field of the channel current i would extend beyond the edge of the plates to infinity. However, there is a displacement current in the condenser, due to the time change of the electric field, which actually flows to provide a return path for the stroke. Because the electric field $\mathfrak{E}$ is uniformly distributed, this return displacement current is also uniformly distributed. Figure 24.4b shows the main stroke current i as well as the displacement current i_v in the air space. The current density of the main-stroke current is very high, because it is confined to a narrow channel, whereas the current density of the displacement current is very low, because it is distributed over a very large space. The total displacement current is equal to the main-discharge current, but it flows in the opposite direction to complete the circuit. Each circular line of force of radius ρ encloses only the displacement current in its interior that is proportional to the enclosed area, hence to the square of the radius, up to $\rho = R = D/2$. Thus the magnetic field strength or induction $\mathfrak{B}_v$ produced by the displacement current is

$$2\pi\rho\mathfrak{B}_v = -4\pi \left(\frac{\rho}{R}\right)^2 i, \quad (24.9)$$

where R is the radius of the cloud, or

$$\mathfrak{B}_v = -\frac{2i\rho}{R^2}. \quad (24.10)$$

Figure 24.4c shows the magnitude of the magnetic field $\mathfrak{B}_v$ due to the displacement current; it increases linearly with ρ and at the edge of the plates when $\rho = R$ it is of equal magnitude but opposite polarity to $\mathfrak{B}_a$ produced by the main-discharge current.

Thus the main electromagnetic-energy exchange is confined to the space between the plates.

The magnetic-energy density in the space outside the channel is thus, from Eqs. (24.2), (24.8), and (24.10),

$$w_m^{(a+v)} = \frac{1}{8\pi} (\mathfrak{B}_a + \mathfrak{B}_v)^2 = \frac{i^2}{2\pi} \left(\frac{1}{\rho} - \frac{\rho}{R^2} \right)^2.$$
(24.11)

To obtain the total magnetic-energy content we integrate over the entire volume of the condenser, using an elementary cylindrical shell cylinder of height a, circumference $2\pi\rho$, and thickness $d\rho$. Then we obtain

$$W_m^{(a+v)} = a \int_r^R w_m \cdot 2\pi\rho \, d\rho$$

$$= ai^2 \int_r^R \left(\frac{1}{\rho} - \frac{2\rho}{R^2} + \frac{\rho^3}{R^4} \right) d\rho$$

$$= ai^2 \left[\ln \frac{R}{r} - \frac{3}{4} + \left(\frac{r}{R} \right)^2 - \frac{1}{4} \left(\frac{r}{R} \right)^4 \right].$$
(24.12)

If the resistance of the conducting channel were vanishingly small, the high-frequency currents of the discharge would be confined to the surface of the channel. Skin effect would thus make the lightning discharge a hollow tube. In that case Eq. (24.12) would give the total magnetic-energy content. However, the actual channel resistance is probably large enough to avoid skin effects, so that the current density is uniform over the cross section of the channel. In this case there is a magnetic field $\mathfrak{B}_i$ in the interior of the channel that varies linearly with the radius up to r similarly to $\mathfrak{B}_v$ between r and R and is shown in Fig. 24.4c. Substituting r for R in Eq. (24.10) we obtain

$$\mathfrak{B}_i = \frac{2i\rho}{r^2}.$$
(24.13)

This field has the same polarity as $\mathfrak{B}_a$ and is equal to it at $\rho = r$.

The energy content of this internal magnetic field we obtain again by integrating the energy density as given by Eq. (24.2) over the channel of radius r,

$$W_m^i = ai^2 \int_0^r \frac{\rho^3}{r^4} \, d\rho = \frac{ai^2}{4}.$$
(24.14)

If we add this to Eq. (24.12) and neglect the terms in the second and higher powers of the ratio r/R, since $r \ll R$, then only the first two terms remain and we obtain simply

$$W_m = ai^2 \left(\ln \frac{D}{d} - \frac{1}{2} \right),$$
(24.15)

in which the diameters of Fig. 24.4a have been substituted for radii.

If we equate this to the magnetic energy due to the current in an inductance,

$$W_m = \tfrac{1}{2} Li^2,$$
(24.16)

the effective self-inductance of the arrangement is determined finally as

$$L = 2a \left(\ln \frac{D}{d} - \frac{1}{2} \right).$$
(24.17)

The circuit constants C given by Eq. (24.6) and L given by Eq. (24.17) now allow us to calculate the frequency of oscillation of the discharge,

$$\nu = \frac{1}{\sqrt{LC}} = \frac{2\sqrt{2}v}{D\sqrt{\ln D/d - 1/2}} \simeq \frac{2.88}{D} 10^5$$
(24.18)

if D is in kilometers, and the oscillatory impedance,

$$\sqrt{\frac{L}{C}} = 4\sqrt{2} \frac{va}{D} \sqrt{\ln \frac{D}{d} - \frac{1}{2}} \simeq 500 \frac{a}{D} \, \Omega.$$
(24.19)

Both are seen to depend on only a few basic parameters of the arrangement.

The channel diameter d is very hard to determine. The time exposure in Fig. 24.5 leads to an estimate for d of about 10 cm based on the apparent thickness and known distances and objects. B. F. J. Schonland measured it as between 10 and 25 cm for heavy discharges. However, it varies somewhat with the total current and thus with the capacitance and hence the diameter of the cloud, among other factors. Thus its ratio to the cloud diameter will probably not vary very much and, since in Eqs. (24.18) and (24.19) only the square root of the logarithm of this ratio introduces d, its influence must be small. Thus in both equations an approximation is introduced, $d/D = 10^{-4}$, which is in accord with values observed in practice. Then it is seen that the frequency is merely inversely proportional to the cloud diameter D while the impedance is further proportional directly to the cloud height a. This is because as a increases the self-inductance increases in the same ratio as the capacitance decreases.

We made the assumption that the voltage on the electrodes was the same between all

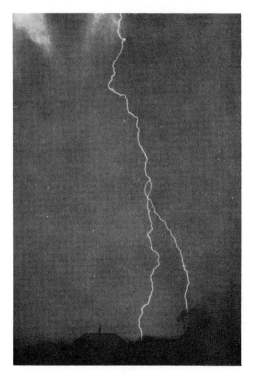

Fig. 24.5

opposite points, which implied that the wavelength of the oscillations was large compared with the dimensions of the condenser. This wavelength is given by the quotient of the velocity of propagation and frequency as

$$\lambda = 2\pi \frac{v}{\nu} = \frac{\pi D}{\sqrt{2}} \sqrt{\ln \frac{D}{d} - \frac{1}{2}} \cong 6.55\, D,$$

(24.20)

which is indeed several times the condenser diameter D and thus justifies the assumption.

We should, however, note that shorter wavelengths can appear in the form of harmonics. To determine their behavior would require a rigorous use of the Maxwell equations in integral form. Their amplitudes will always be small in comparison with the fundamental considered here, and thus in practice they play only a subordinate role.

Thus if a thundercloud as shown in Fig. 24.2 has a diameter $D = 1$ km and is discharged by a stroke in a channel of diameter $d = 10$ cm, the frequency is, by Eq. (24.18),

$$\nu = \frac{2 \cdot \sqrt{2} \cdot 3 \cdot 10^5}{1 \cdot \sqrt{\ln 1000/0.1 - 1/2}} = \frac{8.5 \cdot 10^5}{2.95}$$

$$= 2.88 \cdot 10^5 \text{ cy}/2\pi \text{ sec} = 45{,}800 \text{ cy/sec.}$$

The impedance for a cloud height $a = 1$ km is, by Eq. (24.19),

$$\sqrt{\frac{L}{C}} = 4\sqrt{2} \frac{3 \cdot 10^{10} \cdot 1}{1} 2.95 \cdot 10^{-9} = 500\ \Omega.$$

The amplitude of the lightning current is given by the same relations:

$$I = E \sqrt{\frac{C}{L}} = \frac{D}{4\sqrt{2}v\sqrt{\ln D/d - 1/2}} \frac{E}{a}$$

$$\cong \frac{D}{500} \frac{E}{a}, \quad (24.21)$$

with D in kilometers. Here also the approximation $d/D = 10^{-4}$ is introduced. To obtain E in volts and I in amperes it is necessary to multiply by the dimensional factor 10^9.

The quantity E/a represents the average breakdown-voltage field strength in the air. Thus we see that if we again neglect the small influence of d/D as we did before, the current strength is almost entirely a function of the cloud diameter D and is directly proportional to it. Hence the current strength and frequency of the lightning discharge are independent of the cloud height and are mainly a directly proportional function of the planar extent of the cloud. For $D = 1$ km and an average breakdown field strength $E/a = 100$ kV/m, which corresponds to 1000 V/cm, we obtain a lightning current of

$$I = \frac{1 \cdot 1000 \cdot 10^9}{4\sqrt{2} \cdot 3 \cdot 10^5 \cdot 2.95} = 200{,}000 \text{ A,}$$

which is a very high value. The discharge leads

to radiation into the surrounding space with a wavelength of

$$\lambda = \frac{\pi}{\sqrt{2}} 1000 \cdot 2.95 = 6550 \text{ m},$$

which draws a considerable amount of energy from the discharge. This radiation can be observed at very great distances, and can be used to locate the occurrence of lightning. At the same time it constitutes a large fraction of the background noise or interference in the lower part of the radio-frequency spectrum.

For a range of cloud diameters likely to occur in nature, Table 24.2 gives the frequency and also the current strength for several values of breakdown-field strengths. The frequency is given in kilocycles per second and a quarter-period rise time in microseconds is also given. The current is given in kilo-amperes. We see that the frequency is in the range of 10–500 kc/sec and the rise time in the range of 0.5–30 μs, which latter agrees closely with the observed values listed in Table 24.1. The current strength for small clouds and low breakdown strength is as low as 10 kA and rises to well above 500 kA for very large clouds and high breakdown voltage. Because a number of factors can give rise to strong damping, the currents listed represent only possible maximum values. It is open to question whether clouds of 5-km diameter are discharged by a single stroke so that the values for the clouds of largest diameter in the table may not occur in nature.

The relations developed for a parallel-plate condenser apply quite well to an umbrella type of antenna used to propagate radio-frequency transmissions as shown in Fig. 24.6. The natural frequency is given approximately by Eq. (24.18) and the wavelength by Eq. (24.20) if D is taken as the diameter at the base of the umbrella and d the diameter of the central supporting mast. For self-inductance and capacitance Eqs. (24.6) and (24.17) give somewhat less reliable

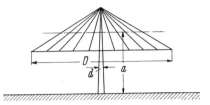

Fig. 24.6

estimates, because the average height a is hard to estimate and because the discrete, spaced wires of the antenna have lower C and higher L than a conducting plate would have. However, all this cancels for the natural frequency. An umbrella of diameter $D = 200$ m supported by a mast of diameter $d = 2$ m has, as given by Eq. (24.20), a natural wavelength

$$\lambda = \pi 200 \sqrt{\frac{\ln 100 - 0.5}{2}} = 900 \text{ m}.$$

Electromagnetic oscillations of this wavelength, which corresponds to a frequency of 333 kc/sec, will be radiated into space if this antenna is excited by a sharp impulse.

For many experimental investigations it is appropriate and convenient to use smaller

TABLE 24.2. Lightning frequency and current.

	Cloud diameter, D (m)							
	100	200	500	800	1000	2000	3000	5000
Frequency, $\nu/2\pi$ (kc/sec)	460	230	92	58	46	23	15.2	9.2
Rise time, $\mathfrak{T}/4$ (μs)	0.55	1.1	2.8	4.4	5.5	11	16.5	27.5
Breakdown field strength, E/a (kV/m)	Current, I (kA)							
50	10	20	50	80	100	200	300	500
100	20	40	100	160	200	400	600	1000
200	40	80	200	320	400	800	1200	2000
300	60	120	300	480	600	1200	1800	3000

condensers with parallel plates. For plate diameters of 1 m and spark-channel diameters of 1 mm, Eq. (24.18) gives a natural frequency of

$$\nu = \frac{2\sqrt{2} \cdot 3 \cdot 10^{10}}{100\sqrt{\ln 1000/1 - 1/2}} = \frac{8.5 \cdot 10^8}{2.54}$$

$$= 3.35 \cdot 10^8 \text{ cy}/2\pi \text{ sec} = 53.3 \cdot 10^6 \text{ cy/sec}.$$

The current strength in the spark discharge for a breakdown-field strength of 10 kV/cm is, from Eq. (24.21),

$$I = \frac{100 \cdot 10^9}{4\sqrt{2} \cdot 3 \cdot 10^{10} \cdot 2.54} \cdot 10 \cdot 10^3 = 2400 \text{ A},$$

which is a remarkably high value.

Measurements made on circular-plate condensers with mica insulation and with diameter $D = 4$ cm and rupture-channel diameter $d = 0.15$ mm at the center gave a wavelength $\lambda = 55$ cm, while from Eq. (24.20) we obtain 57 cm. For $D = 2.9$ cm, a wavelength of 42 cm was observed, while the calculated value was 39 cm. Thus the mechanism of these discharges is accurately accounted for in our analysis.

(c) *Ground penetration.* Now that we have analyzed the general development of the lightning discharge, we can examine in more detail the initial events connected with it.

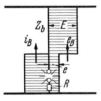

Fig. 24.7

Because the head of the preliminary discharge in lightning propagates with only a fraction of the velocity of light, this initial process occurs very slowly in comparison with times required for the propagation of traveling pulses. Thus, if we imagine, as shown in Fig. 24.7, that the lightning stroke propagates slowly from a charged cloud down to the ground, we may regard the entire field prior to completion of the stroke to the ground

as electrostatic. Only after the preliminary discharge has prepared a conductive channel can the main discharge or stroke occur.

Because the lightning current, as shown in Fig. 24.2, is distributed uniformly in all directions at the point of entry, its ground-penetration resistance, by Ref. 1, Eq. (12), p. 318, is approximately

$$R = \frac{s}{\pi d}. \tag{24.22}$$

Here s is the specific ground resistance, which for average moist ground is about $10^4 \Omega$ cm. Thus for a channel diameter $d = 10$ cm the ground-penetration resistance is

$$R = \frac{10^4}{\pi \cdot 10} = 318 \ \Omega.$$

Depending on ground resistivity and channel diameter, the ground-penetration resistance can be anywhere in the range between 30 and 3000 Ω.

The discharge now begins, as shown in Fig. 24.7, as if a charged line were grounded by an ohmic resistance as in Fig. 7.5. Here the surge impedance of the lightning channel comes into play; it can be calculated from the self-inductance per unit length, by Eq. (24.17), and is

$$Z_B = \frac{L}{a} v = 2v \left(\ln \frac{D}{d} - \frac{1}{2} \right). \tag{24.23}$$

For the diameter ratio $D/d = 10^4$ we obtain

$$Z_B = 2 \cdot 3 \cdot 10^{10} \cdot 8.7 \cdot 10^{-9} = 520 \ \Omega.$$

For a range of diameters this parameter is close to this value, so that for lightning channels we may also assume an average surge impedance of 500 Ω, which is equal to the value representative of open lines. It is independent of the length of the stroke.

The first step of the pulse stroke to ground will have a current amplitude, as given by Eq. (7.7), of

$$i_B = \frac{E}{Z_B + R}, \tag{24.24}$$

and a discharge pulse of voltage

$$e_B = - \frac{Z_B}{Z_B + R} E \tag{24.25}$$

travels up the stroke channel, as shown in Fig. 24.7. The voltage at the ground point hit by the stroke is

$$e = E - e_B = \frac{R}{Z_B + R} E. \quad (24.26)$$

For a channel impedance $Z_B = 500 \, \Omega$ and a ground resistance $R = 300 \, \Omega$, a cloud voltage of 20,000 kV gives rise to an initial stroke-current step pulse of

$$i_B = \frac{20,000 \cdot 10^3}{500 + 300} = 25,000 \text{ A.}$$

This discharge pulse reduces the lightning voltage in the cloud by

$$\frac{e_B}{E} = \frac{500}{500 + 300} = 63 \text{ percent,}$$

while at the ground location it produces a voltage that is

$$\frac{e}{E} = \frac{300}{500 + 300} = 37 \text{ percent}$$

of the cloud voltage and thus amounts to 7400 kV.

The discharge pulse travels along the channel up to the cloud, where it is reflected back to the ground. This process is periodic, as shown in Eq. (7.13), but very heavily damped because the ground resistance is smaller than the surge impedance. It is even more strongly damped for very moist ground, a good lightning conductor, a tree with highly conducting roots, or bodies of water of low resistivity, down to $s = 10^2 \, \Omega$ cm, all of which will have even lower ground resistance. However, for very dry ground, when R may be a multiple of $500 \, \Omega$, the pulses will behave aperiodically along the channel.

We shall not examine the repeated travel along the channel because the reflection at the cloud is not well defined on account of the many branches and even bundles of branches that constitute the termination of the stroke channel at the cloud end, as seen in Fig. 24.2. However, the return of portions of the discharge pulse requires at least a time equal to twice that required to traverse the channel length a, hence

$$\tau_B = \frac{2a}{v}. \quad (24.27)$$

For a cloud height $a = 200$ m this is

$$\tau_B = \frac{2 \cdot 200}{3 \cdot 10^8} = 1.33 \ \mu s,$$

and becomes 6.67 μs for a cloud height of 1000 m.

Thus the main stroke or discharge of lightning begins with such step pulses. However, the blurred reflection at the cloud and other damping tend to smooth out these steps and the discharge soon settles down to the quasi-stationary oscillations of the lumped equivalent circuit shown in Fig. 24.3 and analyzed earlier. This sinusoidal oscillation gives the mean values about which the detailed

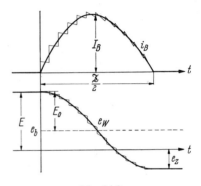

Fig. 24.8

steps represent small excursions, as shown in Fig. 24.8. We note that these steps are quite pronounced at the beginning of the sinusoid.

The time duration of a half-period, by Eq. (24.18), is

$$\mathfrak{T}/2 = \frac{\pi}{v} = \frac{\pi}{2.88} D \cdot 10^{-5} \quad (24.28)$$

(D in kilometers), and for a cloud of diameter $D = 1$ km,

$$\mathfrak{T}/2 = \frac{\pi \cdot 1 \cdot 10^{-5}}{2.88} = 11 \ \mu s.$$

Thus we see that a number of steps can occur during a half-period as shown in Fig. 24.8. The actual number of steps is given by Eqs. (24.27) and (24.28):

$$\frac{\mathfrak{T}/2}{\tau_B} = 1.64 \frac{D}{a}. \quad (24.29)$$

We see that our analysis, using the equivalent lumped circuit, is fairly valid for lightning strokes of length up to the cloud diameter.

Let us examine the damping of these lightning oscillations. This is caused primarily by the ground resistance, derived in Eq. (24.22), which for average ground conditions is about 300 Ω, while the surge impedance of the lightning channel, given by Eq. (24.19), is about 500 Ω for $a = D$. Such a discharge is periodic and the ratio of successive half-period peaks is

$$\varepsilon^{-\pi R/2\sqrt{L/C}} = \varepsilon^{-\pi 300/2\cdot 500} = \varepsilon^{-0.94} = 0.39.$$

The discharge becomes aperiodic, according to Eq. (16), p. 50, Ref. 1, only when, from Eqs. (24.19) and (24.22),

$$\frac{R}{\sqrt{LC}} = \frac{s}{\pi \cdot 100d} \frac{D}{500a} = \frac{10^{-4}}{5\pi} \frac{sD}{ad} = 2,$$

$$(24.30)$$

where d is here expressed in meters. For ground of resistivity $s = 10^4 \, \Omega$ cm and $D/d = 10^4$ this gives a limiting value of cloud height,

$$\bar{a} = \frac{10^{-5}}{\pi} \frac{sD}{d} = \frac{10^{-5}}{\pi} 10^4 \cdot 10^4 = 320 \text{ m}.$$

For lower clouds the discharge will be aperiodic.

For longer lightning strokes or lower ground resistance we must also consider that the oscillation will be damped by radiation of energy. The radiation resistance, by Eq. (24.20), is

$$R_s = 160\pi^2 \left(\frac{a}{\lambda}\right)^2 = 160 \left(\frac{\pi a}{6.55D}\right)^2 = 36.8 \left(\frac{a}{D}\right)^2.$$

$$(24.31)$$

For the limiting cloud height $a = 320$ m and diameter $D = 1$ km, this is only

$$R_s = 36.8 \left(\frac{320}{1000}\right)^2 = 3.8 \, \Omega.$$

However, for long strokes from small clouds much larger values will occur. For example, for $a/D = 2$,

$$R_s = 36.8 \cdot 2^2 = 147 \, \Omega,$$

which represents a very considerable addition to the ground resistance and thus further supports the tendency of long lightning strokes to be aperiodic.

How much influence the branching of the channel in the cloud, as shown in Fig. 24.2, has upon the damping is hard to estimate. One may assume that it has an influence similar to the voltage drop in the channel, which acts essentially as an arc discharge to cause arithmetic damping, as treated in Ref. 1, Chapter 43, p. 577. The arc burning voltage e_b and even more so the reignition voltage e_z in the lightning channel cause an early interruption of the current in its damped harmonic oscillation. For values of these that are small compared with the initial voltage E, reignition of even the second half-period discharge may be impossible. This then leads to a single unipolar lightning stroke.

Such a single-pulse lightning stroke occurs when the pertinent parameters are related by the equation

$$\frac{R}{\sqrt{L/C}} = \frac{2}{\pi} \ln \frac{E - e_b}{e_z + e_b}, \qquad (24.32)$$

which was derived in Ref. 1 as Eq. (23), p. 583. If we let the burning voltage in a lightning stroke be twice that of arcs at $e_b = 20$ V/cm $= 2$ kV/m on account of the high radiation cooling and then let the reignition voltage be 25 times this value, that is, $e_z = 50$ kV/m, then for an initial electric field strength $E = 100$ kV/m the limiting ratio of resistance to surge impedance for which only a single pulse stroke occurs becomes

$$\frac{R}{\sqrt{L/C}} = \frac{2}{\pi} \ln \frac{100 - 2}{50 + 2} = \frac{2}{\pi} \ln 1.88 = 0.4,$$

which is only one-fifth the value given by Eq. (24.30). On the other hand, the limiting cloud height with the same ground resistance is increased fivefold, to

$$\bar{a} = 320 \cdot 5 = 1600 \text{ m}.$$

Because thunderclouds are usually at lower altitudes, unipolar single strokes are by far the most common.

Even for the highest values of field strength $E = 300$ kV/m, this limiting height is $\bar{a} = 575$ m from the values of parameters generally used previously. Only when it strikes a highly conducting ground or a well-grounded object can a lightning stroke develop into a set of several impulses in a damped oscillatory train. Figure 24.8 shows the current i_B and the cloud voltage e_W for the single-impulse interrupted lightning stroke. Both are terminated by the voltage drop required for maintenance and reignition of the discharge in the channel. In this way the cloud retains a remainder charge of opposite polarity until atmospheric charging processes change it.

We must distinguish between the two regions for which the ohmic resistance is

$$R \lessgtr 2\sqrt{\frac{L}{C}} \qquad (24.33)$$

in order to calculate the lightning currents and voltages. The main component of the

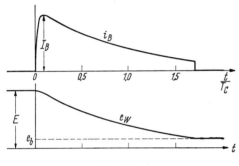

Fig. 24.9

effective resistance will be the ground resistance. For smaller R we obtain the periodic interrupted discharge just analyzed and shown in Fig. 24.8. For larger R an overdamped exponentially decaying impulse discharge, as shown in Fig. 24.9, will occur.

In the case of the sinusoidal discharge the maximum value of current for both single- and multiple-pulse trains is

$$i_B = \frac{E}{\sqrt{L/C}}\,\varepsilon^{-\rho t}\sin \nu t. \qquad (24.34)$$

We have neglected here the burning voltage e_b, since it is very small compared with the initial field strength. The peak value of the lightning current after a quarter-period is then at most

$$I_B = \frac{E}{\sqrt{L/C}}\,\varepsilon^{-\pi R/4\sqrt{L/C}}$$

$$\cong \frac{D}{500}\frac{E}{a}\,\varepsilon^{-\pi RD/4\cdot 500\cdot a}. \qquad (24.35)$$

For a field strength $E/a = 100$ kV/m, a cloud diameter $D = 1$ km and height $a = 1$ km, and a ground resistance $R = 300\ \Omega$, this gives

$$I_B = \frac{1000}{500}\,100\varepsilon^{-(\pi/4)\,300/500} = 200\cdot 0.625\ \text{kA}$$

$$= 125{,}000\ \text{A}.$$

Evidently damping by the ground resistance reduces the lightning current very considerably. The current values in Table 24.2 may thus be very much reduced by damping.

The highest possible voltage difference between the point at which the lightning strikes and the surrounding ground can now be calculated; it is

$$E_B = RI_B = \frac{R}{\sqrt{L/C}}\,E\varepsilon^{-\pi R/4\sqrt{L/C}}$$

$$\cong \frac{R}{500}\frac{D}{a}\,E\varepsilon^{-\pi RD/4\cdot 500\cdot a}. \qquad (24.36)$$

For values just used this gives a voltage

$$E_B = 300\cdot 125 = 37{,}500\ \text{kV}.$$

Thus this highest possible ground voltage is considerably less than the cloud voltage, which at 1-km height was here 100,000 kV. However, in the immediate vicinity of the ground object hit by the stroke it produces voltage gradients of enormous magnitude, which are extremely dangerous for people and animals.

For larger R and hence, according to Eq. (24.33), an exponentially damped discharge, the lightning current is

$$i_B = \frac{E}{R}\frac{\varepsilon^{-t/T_C} - \varepsilon^{-t/T_L}}{1 - T_L/T_C}, \qquad (24.37)$$

as developed in Ref. 1, Eq. (34), p. 228. Here, too, we have neglected the small burning voltage in the channel. The shape of this type of discharge is shown in Fig. 24.9. The rise-time constant is inductive and

$$T_L = \frac{L}{R} \cong 17.4 \frac{a}{R} 10^{-4}, \quad (24.38)$$

with R in ohms and a in kilometers and the approximate diameter ratio as used before. We see that it is directly proportional to the cloud height. Similarly, the decay-time constant is capacitive and

$$T_C = CR \cong \frac{RD^2}{144a} 10^{-6}, \quad (24.39)$$

where again dimensions are in kilometers. It decreases with increasing cloud height but increases with the square of the cloud diameter.

With our usual cloud of diameter $D = 1$ km and height $a = 1$ km, a ground resistance $R = 3000 \, \Omega$ in a poorly conducting area will give a rise-time constant

$$T_L = 17.4 \frac{1}{3000} 10^{-4} = 0.58 \, \mu s$$

and a decay-time constant

$$T_C = \frac{3000 \cdot 1^2}{144 \cdot 1} 10^{-6} = 20.8 \, \mu s.$$

Thus the current rises very rapidly but decays slowly, with a resulting broad back pulse. When the cloud voltage has been reduced by this discharge to the burning voltage e_b in the channel, the discharge can no longer be sustained and is interrupted, as shown in Fig. 24.9.

Because the rise-time constant T_L is very small compared with the decay-time constant T_C when the ground resistance is very high, it can be neglected in Eq. (24.37) for this case and the maximum lightning current is approximately

$$I_B = \frac{E}{R}. \quad (24.40)$$

In our example this is

$$I_B = \frac{100,000}{3000} = 33.3 \text{ kA}.$$

We see that the high ground resistance has led to a very large reduction. On the other hand, the ground voltage becomes almost equal to the cloud voltage,

$$E_B = E, \quad (24.41)$$

because the inductive voltage drop in the channel is very small.

Thus we see that the shape and strength of lightning strokes as well as the effect at the ground can vary very significantly with the ground resistance of the object struck.

If the stroke hits a building via a good lightning rod, which may have an average ground resistance of $5 \, \Omega$, then even for very low clouds the discharge will be periodic.

For a cloud of diameter $D = 1$ km and height $a = 200$ m, a voltage gradient of 100 kV/m gives a cloud voltage of 20,000 kV, which according to Eq. (24.26) gives a first voltage step-pulse discharge of

$$e = \frac{5}{505} 20,000 = 198 \text{ kV}.$$

For the ensuing lightning current there is very little damping and thus the voltage at

Fig. 24.10

the point of ground entry will rise, according to Eq. (24.36), almost to

$$E_B = \frac{5}{500} \frac{1000}{200} 20,000 = 1000 \text{ kV}.$$

The entire lightning-rod installation with all its conductive branches, parts, and connected supports will be raised to this high voltage, which will be even higher when the ground resistance or electric field is higher.

If now, as shown in Fig. 24.10, the building contains other grounded conductors, such as water or gas pipes or electric conduit, that approach the lightning-rod system at some points, a breakdown of the intervening insulation may easily occur at these points. Because such extended pipe systems have extremely low ground resistances, of the order of $r = 0.5\ \Omega$, the major portion of the lightning current may now transfer to these conductors. While the lower ground resistance reduces the voltage rise, the arc or spark at the flashover point can cause considerable damage and in particular may start a fire. Thus such places at which flashover can occur must be avoided, particularly if a fire might be started. It is better to connect all such conductive installations properly with the system of lightning-rod conductors as high up in the building as possible.

(d) *Induced charges on lines.* Figure 24.11 shows a thundercloud discharged via an open

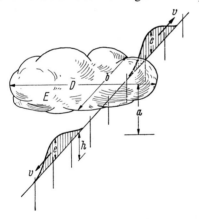

Fig. 24.11

line either to ground or to a neighboring cloud (not shown). Discharge pulses are produced on the line whose voltage was derived in Eq. (3.30):

$$e = \frac{bh}{2v}\frac{d\mathfrak{E}}{dt}. \qquad (24.42)$$

Apart from the height of the line h and the

width of the cloud b it is mainly determined by the change with respect to time of the electric-field strength $\mathfrak{E}$ and is propagated with the velocity of light away from the location of the cloud. The spatial distribution along the line is determined by the temporal distribution of the field below the cloud.

Now the lightning-discharge current of the cloud is given by

$$i = C\frac{de_C}{dt} = Ca\frac{d\mathfrak{E}}{dt}. \qquad (24.43)$$

We thus see that the shape of the charge-induced voltage on the line is an image of the time variation or shape of the current in the lightning stroke irrespective of whether this current varies in sinusoidal, exponential, periodic, aperiodic, or any other complicated form as a function of time. We always have

$$e = \frac{bh}{2vaC}i = 8v\frac{bh}{D^2}i, \qquad (24.44)$$

where C is the capacitance of the cloud as derived in Eq. (24.6). Thus if the diameter of the cloud D and the width of its zone of influence over the line b, as shown in Fig. 24.11, can be estimated, it is possible to measure the strength and time variation of lightning currents with some accuracy by observation of voltage on the line.

For a sinusoidal behavior of the current we can introduce the maximum possible value of Eq. (24.21) into the first version of Eq. (24.44) and obtain a maximum possible induced voltage

$$e_0 = \frac{bh}{2v}v\mathfrak{E}_0 \cong 0.48\frac{b}{D}h\mathfrak{E}_0, \qquad (24.45)$$

where the field strength $\mathfrak{E}_0$ has been used in place of E/a and the frequency v in place of $\sqrt{LC}$. In the last version we have used the approximation of Eq. (24.18). For a field strength $\mathfrak{E}_0 = 100\ \text{kV/m}$, an average line height $h = 10\ \text{m}$, and a line-to-cloud geometry that makes $b = D$, we obtain an induced voltage on the line of

$$e_0 = 0.48 \cdot 10 \cdot 100 = 480\ \text{kV},$$

which is a very considerable voltage. Depending on the magnitude of the ground resistance, it will be reduced by damping, which in the earlier numerical example would bring it down to 300 kV. Furthermore, the cloud will only rarely be centered above the line, so that these charge-induced voltage pulses will usually be less than these upper-limit values.

If we examine a section of line open at both ends, centered below the cloud, so that the traveling pulses cannot arise, then the electric-field changes of the undamped discharge oscillation would induce a voltage of $2\ h\mathfrak{E}_0$. This is about four times the value given in Eq. (24.45) and in the numerical example about 2000 kV.

For an exponential shape of the current we can substitute the highest possible value of Eq. (24.40) into Eq. (24.44) to obtain the highest possible value of the induced traveling voltage pulse,

$$e_0 = \frac{bh}{2v}\frac{\mathfrak{E}_0}{T_C}. \qquad (24.46)$$

Here T_C is the time constant of Eq. (24.39), which replaces the product of capacitance and resistance and determines the shape of the back of the pulse shown in Fig. 24.9. The higher the resistance encountered by the stroke, the lower is the current and the higher is the discharge-time constant and consequently the lower is the induced voltage. With the values of our earlier numerical example we now obtain for $T_C = 20.8\ \mu s$ an induced voltage pulse of

$$e_0 = \frac{1 \cdot 10}{2 \cdot 3 \cdot 10^5} \frac{100}{20.8 \cdot 10^{-6}} = 80 \text{ kV}.$$

A voltage of this magnitude is usually of no danger to high-voltage transmission lines.

We see from these considerations that lightning strokes to a dry ground produce tolerable voltages by induction in neighboring lines, especially if, according to Eq. (24.39), the thundercloud has a large diameter and is low over the ground. In contrast, lightning that strikes objects in wet ground or with low ground resistance for any other reason may induce dangerous voltages on a line, which are further governed by the extent of the zone of influence of the cloud over the line.

If the voltage on the line due to the released charges rises so high that there is a flashover across the line insulators, these sudden discharges of the line to ground will in turn produce step pulses, which will be propagated in the network along with those due to the atmospheric discharge. Figure 24.12 shows this process. The voltage on the line rises gradually and propagates in both directions until finally at a critical voltage E an insulator breaks down and shorts the line effectively to ground, so that the pulse, which has a gently rising front, has a very steep and abrupt steplike back. If the field strength above the line changes further, a whole train of such charge and discharge pulses can travel along the line away from the region of lightning.

Fig. 24.12

Because the breakdown or flashover voltage of insulators is always several times the normal operating voltage, such traveling pulses due to atmospheric effects represent very considerable excess voltages. Thus it would seem practical to keep the insulator breakdown voltage only slightly above the normal operating voltage. In practice, however, an exactly opposite trend is evident in that the breakdown voltage of insulators is set at at least three times the operating voltage. This is done so that the total number of breakdowns due to such atmospheric-charge-induced voltages is kept low and so that in turn very few moderate excess voltages can produce the dangerous step-terminated pulses in the network.

CHAPTER 25

LINES STRUCK BY LIGHTNING

If lightning is in the process of formation above a power line, its actual path and point of striking are determined by the naturally occurring ionization of the air and the distribution of the electric field in the vicinity of the ground. In Ref. 1, Chapter 28, it was shown that three-phase power lines in a rapidly changing electric field in air behave as if they were well insulated. Even though for much slower changes they act as a grounded

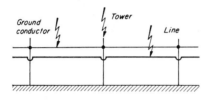

Fig. 25.1

conductor, the large effective diameter applicable in this mode increases the field strength at the ground but little. Thus the head of the lightning stroke has only a slight tendency to strike the actual line conductors.

However, as Fig. 25.1 shows, within the vicinity of the line, the tower is very liable to lightning strokes because its pointed needle-like shape causes a high concentration of electric-field strength at its tip, which is also the highest point of the ground surface. Very high field-strength concentrations also exist at ground conductors if they are strung from tower to tower above the line and thus they are next most likely to be struck by lightning. The field strength at the line conductors is by far the lowest and thus they are least likely to be struck.

(a) *Stroke to a line or ground conductor*. Figure 25.2 shows a lightning stroke that has just entered a line or ground conductor of surge impedance Z. We must take note of the fact that its head approached the line with

a velocity of propagation much lower than that of light. Thus the charges created by field concentration and ionization ahead of the stroke head are discharged with greater speed in both directions along the conductor. Hence large current and voltage effects arise only at the instant when the head of the lightning stroke actually reaches the conductor. At that instant the cloud voltage E, which up to then was almost static, is switched to the line formed by the conductor. This is true even if the stroke travels entirely or partly from the ground upward instead of in the opposite direction previously considered.

At the instant of the stroke a lightning current develops that divides into two charging pulses traveling in both directions into the

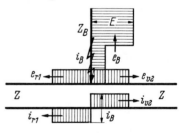

Fig. 25.2

line. The lightning voltage E drops to the line voltage e, which causes a reflected pulse of amplitude e_B to travel up the lightning channel. This causes a step increase in the lightning current i_B if the cloud is heavily charged to sustain it. As shown in Fig. 25.2, the voltage pulses at the strike point give

$$E - e_B = e_{r1} = e_{v2} = e, \qquad (25.1)$$

and for the current pulses,

$$i_B = i_{v2} - i_{r1} = 2i, \qquad (25.2)$$

where e and i designate pulses on the continuous line which are equal in magnitude. The fundamental voltage-current relation for the conductor or lightning-stroke channel,

$$e = \pm Zi, \qquad (25.3)$$

then gives the voltage on the line, by Eq. (25.2),

$$e = \frac{Z}{2} i_B, \qquad (25.4)$$

which is proportional to the lightning current and to one-half the surge impedance. If we insert this in Eq. (25.1) and call the surge impedance of the lightning channel again Z_B, we have

$$E - Z_B i_B = \frac{Z}{2} i_B, \qquad (25.5)$$

which gives for the lightning current

$$i_B = \frac{E}{\frac{1}{2}Z + Z_B}. \qquad (25.6)$$

The lightning voltage on the conductor line is

$$e = \frac{E}{1 + 2Z_B/Z}, \qquad (25.7)$$

and the discharge pulse propagated into the lightning channel is

$$e_B = \frac{E}{1 + Z/2Z_B}. \qquad (25.8)$$

As the surge impedance of the lightning channel and the line are about equal, the lightning-stroke voltage on the line is in practice about one-third of the cloud voltage.

For a moderate cloud voltage $E = 10,000 \, \text{kV}$, Eq. (25.6) gives a lightning-stroke current

$$i_B = \frac{2}{3} \frac{E}{Z} = \frac{2}{3} \frac{10,000 \cdot 10^3}{500} = 13,300 \, \text{A},$$

and Eq. (25.7) a conductor line voltage

$$e = \frac{E}{3} = \frac{10,000}{3} = 3333 \, \text{kV},$$

while the discharge pulse traveling up the stroke channel has a pulse-front amplitude $e_B = 6666 \, \text{kV}$, as given by Eq. (25.8). Line insulators cannot cope with voltage pulses of such high amplitude and will flash over or break down.

In the absence of direct measurements of the cloud voltage E, it is possible only to infer it indirectly from measurements of the line voltage e. The highest cathode-ray oscillograms recorded showed line voltages of about 5000 kV. However, as these observations were obtained at some distance from the strike point, the damping experienced makes it probable that the values of cloud voltage that occur are considerably more than three times this value. Thus we must consider it likely that cloud voltages can be as high as 20,000 or even 100,000 kV. There is as yet no very clear understanding of the generating processes leading to such high values.

If the line has a ground conductor, a voltage e is induced in it if the line conductor is struck by lightning. As Fig. 25.3 shows, the

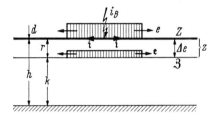

Fig. 25.3

ground-conductor currents i traveling in both directions should have opposite polarity owing to their opposite direction of travel. However, they must also be equal, since no current can flow into the ground conductor at the strike locality. These two conditions require that these currents are zero. Thus the ground conductor acts purely in an electrostatic fashion and causes no inductive coupling to affect the lightning currents and voltages on the struck line. These remain as given in Eqs. (25.6) and (25.7). The ground conductor merely has a voltage induced on it electrostatically, which, by Eq. (11.17), is

$$\mathfrak{e} = \frac{z}{Z} e = \frac{\ln \left[(h + k)/r \right]}{\ln (4h/d)} e, \qquad (25.9)$$

and whose value, as determined earlier by numerical examples, will usually be in the range of 20–40 percent of e.

If the lines are carried by highly insulating wooden poles and the steel crossbars supporting the insulators are connected by a ground wire that is grounded only at a few widely spaced points, only the difference between these two voltages,

$$\Delta e = e - \mathfrak{e} = \left(1 - \frac{z}{Z}\right) e, \quad (25.10)$$

acts across the insulators to cause stress. Thus a stroke on such a line causes a voltage impulse stress on the insulators of 60–80 percent of the line voltage to ground. The balance of the voltage causes a stress along the wooden pole, which it can usually withstand for the short time involved.

If the stroke enters the ground conductor, instead of one of the line conductors, the conditions are the same since it is grounded only at infrequent intervals. The value e now applies to the ground conductor and $\mathfrak{e}$ is induced in the line conductor. The stress on the insulators is given by Eq. (25.10) as before. However, the full voltage e is now a stress on the wooden poles that frequently splinters them.

Actually a line always has several conductors. In every case those not actually struck experience only the electrostatically induced voltage, provided their insulators do not break down.

(b) *Stroke to a tower.* Figure 25.4 shows a steel tower, connected by a ground conductor

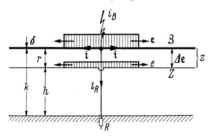

Fig. 25.4

of surge impedance $\mathfrak{Z}$ to its neighbors, hit by a lightning stroke. It shows that the lightning current can flow in part directly to ground via the ground resistance R. If the surge impedance of the tower is small compared with $\mathfrak{Z}$, the lightning voltages are considerably reduced and the lightning current now is

$$i_B = 2i + i_R = 2\frac{\mathfrak{e}}{\mathfrak{Z}} + \frac{\mathfrak{e}}{R}. \quad (25.11)$$

Thus the traveling voltage pulse on the ground conductor becomes

$$\mathfrak{e} = \frac{3 i_B}{2 + 3/R}. \quad (25.12)$$

If we insert this in Eq. (25.1), which is still valid, we obtain

$$E - Z_B i_B = \frac{i_B}{1/R + 2/\mathfrak{Z}}, \quad (25.13)$$

and thus the lightning current is

$$i_B = \frac{E}{Z_B + (1/R + 2/\mathfrak{Z})^{-1}}. \quad (25.14)$$

According to Eq. (25.12), the ground-conductor voltage is now

$$\mathfrak{e} = \frac{E}{1 + 2Z_B/\mathfrak{Z} + Z_B/R}. \quad (25.15)$$

If we again let the surge impedance of the ground conductor be equal to that of the lightning channel, we obtain the simple expression

$$\mathfrak{e} = \frac{E}{3 + Z_B/R} = \frac{R}{Z_B + 3R} E \quad (25.16)$$

for the voltage at the top of the tower and on the ground conductor.

If the tower ground resistance R is much smaller than the surge impedance Z_B of the lightning channel, it mainly determines the voltage according to Eq. (25.15), and the surge impedance of the ground conductor has little influence. However, if the tower ground resistance is high, the voltage can be reduced by the use of several ground conductors in parallel, which gives a considerably lower surge impedance $\mathfrak{Z}$.

For our numerical example of a cloud voltage $E = 10,000$ kV, a tower ground resistance $R = 20 \, \Omega$ gives a lightning current, according to Eq. (25.14),

$$i_B = \frac{10,000 \cdot 10^3}{500 + (1/20 + 2/500)^{-1}} = 19,300 \text{ A},$$

and a voltage, according to Eq. (25.16),

$$\mathfrak{e} = \frac{10,000}{3 + 500/20} = 357 \text{ kV}.$$

While here the lightning current has increased

by about 50 percent, the voltage has been reduced very considerably, to about 10 percent of the former value, which makes it quite tolerable for many high-voltage lines. As we have seen, the cloud voltage has peak values of at least 20 MV. Table 25.1 gives values of e, calculated from Eq. (25.15), for a tower struck from a cloud of this voltage by lightning of channel impedance $Z_B = 500 \, \Omega$, for a range of ground resistances and for either 0, 1, 2, or 3 ground conductors. It shows the great value of a low ground resistance toward obtaining low voltage stresses due to lightning. For even more severe thunderstorms, with cloud voltages E up to 100 MV, the values in Table 25.1 increase by a factor of 5. In the line conductors, shown in Fig. 25.4 below the ground conductor, there may again be no traveling pulse current. Thus no inductive back emf interferes with the lightning voltage and only a voltage

$$e = \frac{z}{3} e \qquad (25.17)$$

is induced electrostatically on the line conductor.

The voltage difference between the tower and the line is thus

$$\Delta e = e - e = \left(1 - \frac{z}{3}\right) e, \quad (25.18)$$

which represents a short impulse stress on the insulators.

As we saw in Chapter 11, the ratio of the surge impedances depends strongly on the number of ground conductors and is in the range of 0.1–0.15 for 1 conductor to 0.3–0.4 for 2 or 3 conductors. Thus the voltage stress on insulators due to lightning is only about 60–90 percent of the values given in Table 25.1 when ground conductors are used. If owing

to this voltage Δe an insulator flashes over from the tower to the line, the line is also raised to the full voltage e. The traveling pulses on the line and ground conductors now travel together, unless there has been a delay in the flashover. The total voltage is then reduced a little, according to Eq. (25.15), because the effective surge impedance 3 has become smaller owing to an extra conductor.

(c) *Effect of adjacent towers.* Pulses travel in both directions along the line from the

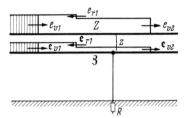

Fig. 25.5

point struck, whether it is at the tower or in the ground or line conductor. As shown in Fig. 25.5, we shall designate the forward-traveling voltage pulse on the line conductor by e_{v1} and that on the ground conductor by e_{v1}. Because the ground conductor is grounded through the ground resistance R of the tower, the voltage pulse traveling along it will be transmitted at the tower. According to Eqs. (6.35) and (6.36), it is transmitted into the next sector with an amplitude reduced to

$$e_{v2} = \frac{e_{v1}}{1 + 3/2R}, \qquad (25.19)$$

while a discharge pulse of amplitude

$$e_{r1} = \frac{-e_{v1}}{1 + 2R/3} \qquad (25.20)$$

TABLE 25.1. Step voltage (kV) at a tower due to lightning stroke from a cloud at 20 MV.

Ground conductors	Tower ground resistance, R (Ω)											
	2	3	5	10	20	30	50	100	200	300	500	1000
0	80	119	198	392	768	1132	1818	3333	5710	7470	10000	13333
1, $3 = 500 \, \Omega$	79	118	194	377	714	1016	1539	2500	3636	4280	5000	5710
2, $3 = 300 \, \Omega$	79	117	192	368	681	952	1394	2142	2922	3333	3750	4140
3, $3 = 200 \, \Omega$	78	116	189	357	645	881	1250	1818	2352	2604	2836	3075

is reflected in the opposite direction. These secondary pulses caused by the tower current cannot produce current pulses in the main line on account of symmetry. Thus the transmission at the tower of the pulses on the ground conductors is not affected by the pulses on the main line conductors.

However, the voltage pulses e_{v2} and e_{r1} on the ground conductor are coupled electrostatically to the main-line conductors by the mutual surge impedance z and thus produce additional voltage pulses on the line. The rearward-traveling pulse on the main line, according to Eq. (25.17), is

$$e_{r1} = \frac{z}{3} e_{r1} = -\frac{z/3}{1 + 2R/3} e_{v1}, \quad (25.21)$$

and the forward-traveling pulse on the main line is reduced by the same amount to

$$e_{v2} = e_{v1} - \frac{z/3}{1 + 2R/3} e_{v1}. \quad (25.22)$$

Equations (25.19), (25.20), (25.21), and (25.22) are the laws of reflection and transmission at any tower for pulses traveling along the line and ground conductors. The pulses e on the ground conductor will always be reduced for low ground resistance R of the tower. The reduction of the line-conductor pulse e depends in addition on the degree of coupling.

If the stroke hits the tower or ground conductor, then, as shown in Fig. 25.4, the main pulse travels along the latter. The degree of reduction at the first tower encountered can be calculated from Eq. (25.19); for example, at a tower of 20-Ω ground resistance it is

$$\frac{e_{v2}}{e_{v1}} = \frac{1}{1 + 500/2 \cdot 20} = 7.4 \text{ percent.}$$

The same process is repeated at the next tower, where in the example the pulse will be reduced to less than 1 percent of its original value. We thus see that the ground-conductor pulses are heavily attenuated and can traverse only a few intertower distances, depending on the value of the tower ground resistance.

The original incident pulse on the main line conductor is given in this case by Eq. (25.17) with the subscript v_1. If we substitute e_{v1} from this into Eq. (25.22) we obtain for

the transmitted voltage pulse on the line conductor

$$\frac{e_{v2}}{e_{v1}} = 1 - \frac{z/3}{1 + 2R/3} \frac{3}{z} = \frac{1}{1 + 3/2R}. \quad (25.23)$$

Thus, by comparison with Eq. (25.19), we see that it is reduced in the same measure as the ground-conductor voltage pulse. Hence the voltage pulses induced on the line conductors also cannot be propagated beyond a

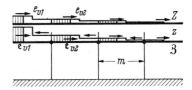

Fig. 25.6

few towers. Figure 25.6 shows the behavior diagrammatically.

If the line conductor is struck, the main pulses are produced in it, as shown in Fig. 25.3. The pulses in the adjacent ground conductor are now given by Eq. (25.9) and are usually between 20 and 40 percent of the main pulses. The transmission at the first tower encountered is again given by Eq. (25.19) and thus the pulses on the ground conductor are also quickly reduced to zero here.

There is a back effect on the main pulse. If Eq. (25.9) is inserted into Eq. (25.22), the transmission at the tower becomes

$$\frac{e_{v2}}{e_{v1}} = 1 - \frac{z^2/Z3}{1 + 2R/3}, \quad (25.24)$$

and thus the reduction depends to a large extent on the mutual surge impedance z. For an average value of the coupling factors of 30 percent, the lightning pulses on the line conductors are reduced at each tower to

$$\frac{e_{v2}}{e_{v1}} = 1 - \frac{0.3 \cdot 0.3}{1 + 2 \cdot 20/500} = 1 - 0.083,$$

which is a very small reduction. Even for zero ground resistance, $R = 0$, the reduction amounts to only 9 percent. A corresponding reduction of the main pulse occurs at each tower encountered and this forms a geometric series down to a final value. The total reduction is

$$\frac{z^2/Z3}{1 + 2R/3} + \left(\frac{z^2/Z3}{1 + 2R/3}\right)^2$$

$$+ \cdots = \frac{1}{(Z3/z^2)(1 + 2R/3) - 1}, \quad (25.25)$$

and in our example amounts to

$$\frac{1}{(100/8.3) - 1} = 9.1 \text{ percent,}$$

which is thus seen to be not much more than the reduction at the first tower. Figure 25.7

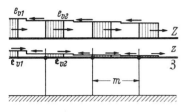

Fig. 25.7

shows the behavior again in diagrammatic form.

The lower reduction at the farther towers is due to the fact that there the ground-conductor currents have disappeared almost completely. Thus the traveling pulse on the line is propagated beyond the farther towers. Instead of the induced voltage in the first tower span adjacent to the hit, a counter current without voltage induced in the ground conductor in the farther spans acts to reduce by induction the main voltage pulse on the line conductor, as just calculated.

The pulses reflected at the first and subsequent towers travel back to the point struck, as shown in Figs. 25.5 and 25.7, to reduce the voltage at that point. For a strike on the actual line, Eq. (25.25) represents the total reduction again and it is always only a few percent.

However, when the tower or ground conductor is struck, the voltage, as shown in Fig. 25.6 and Eq. (25.19), is reduced very considerably after only a few steps. After repeated to-and-fro travel of the pulses, there remains only a voltage determined by the steady-state ground resistance of all towers in the vicinity of the strike. From the resis-

tances of the ground conductors r, the resulting effective resistance to ground is, as given in Ref. 1, Eq. (36), p. 362,

$$\Sigma R, r = \tfrac{1}{2}\sqrt{Rr}. \quad (25.26)$$

Now the discharge of a long line over a pure resistance, by Eqs. (7.7) and (7.9), gives a lightning current that rises to the value

$$i_B = \frac{E}{Z_B + \Sigma R, r}, \quad (25.27)$$

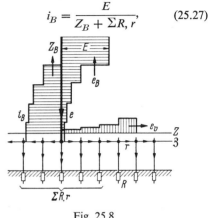

Fig. 25.8

as shown in Fig. 25.8, while the voltage at the tower drops to

$$e = i_B \Sigma R, r = \frac{E}{1 + Z_B/\Sigma R, r}. \quad (25.28)$$

For a ground-conductor resistance $r = 1.25\,\Omega$ between towers and a tower grounding resistance $R = 20\,\Omega$, the combined effective resistance is $2.5\,\Omega$. Thus the lightning current in the earlier numerical example now rises to

$$i_B = \frac{10,000 \cdot 10^3}{500 + 2.5} = 39,500 \text{ A.}$$

At the same time the voltage at the point struck drops to the low value of

$$e = \frac{10,000}{1 + 500/2.5} = 49.7 \text{ kV,}$$

which is of hardly any danger to the insulators of high-voltage power lines.

The ground-conductor resistance used is that of a usual steel ground wire. Owing to skin effects it has a higher effective resistance for traveling pulses. However, the lightning voltage pulses on the line in the example given will stay below dangerous levels.

The duration of these equalizing effects is determined by the propagation time of the pulses over twice the distance m between adjacent towers. For towers 250 m apart this is only

$$\tau_m = \frac{2m}{v} = \frac{2 \cdot 250}{3 \cdot 10^8} = 1.67 \cdot 10^{-6} \text{ sec. (25.29)}$$

Thus, because the ground-conductor pulses are almost completely extinguished at the first tower, the duration of the transition shown in Fig. 25.8 is usually only a few microseconds. Hence in the tower and the ground conductors the high initial voltages produced by a lightning stroke are reduced by good grounding to very small remainders over a short interval of time as well as of distance.

However, on the line conductors no such reduction is encountered and the pulses propagate outward from the point struck as shown in Fig. 25.7. If the voltage at the tower struck stays so low that the voltage across the insulator given by Eq. (25.18) does not lead to flashover, only the transmitted voltage pulses, as given by Eq. (25.17), are propagated outward. However, if an insulator flashover occurs at the tower struck, the full tower voltage is impressed on the line. The propagated pulse now has the same shape, with stepwise-dropping back, though the magnitude of the voltage is greater. Because the ground and the line conductor are now connected at the tower, they have pulses of equal amplitude in the first span. At the next tower, however, the ground-conductor voltage pulse is sharply reduced while that on the line conductor is reduced very little. Thus a second insulator flashover may occur at this adjacent tower, which will tend to equalize the voltage pulses on the ground and line conductors, thereby lowering the latter pulse considerably.

(d) *Development of the lightning current.* The initial discharge steps after the lightning stroke grow into a total lightning current that varies in amplitude as a function of time. The shape of this growth was shown in Chapter 24 to be mainly determined by the grounding resistance R in relation to the surge impedance $\sqrt{L/C}$ associated with the lightning channel. We saw that for

$$\frac{R}{\sqrt{L/C}} \begin{cases} >2 \text{ for exponential behavior,} \\ <2 \text{ for sinusoidal behavior,} \end{cases} \tag{25.30}$$

and that in the latter case

$$\frac{R}{\sqrt{L/C}} \gtrless \frac{2}{\pi} \ln \frac{E - e_b}{e_z + e_b} \tag{25.31}$$

according as the lightning stroke is unipolar or multipolar, respectively.

When the lightning strikes the line, the effective value of R becomes, in the case of a stroke to a line conductor, according to Eq. (25.4), one-half of its surge impedance, and in the case of a stroke to a tower or a ground conductor the effective resulting ground resistance as given by Eq. (25.26). The development of the current after the initial stroke will, as in the case of a stroke to ground, be composed of a sequence of pulses that will travel to and fro along the lightning channel.

A stroke to a line supported by insulating towers can be illustrated by Fig. 25.9. For the

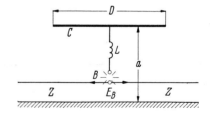

Fig. 25.9

criterion of Eq. (25.30) we obtain, using Eq. (24.19),

$$\frac{Z}{2} \gtrless 2 \cdot 500 \frac{a}{D} \tag{25.32}$$

or

$$\frac{a}{D} \gtrless \frac{Z}{2000} \cong \frac{500}{2000} = \frac{1}{4}. \tag{25.33}$$

Thus thunderclouds with heights greater than one-fourth their diameter lead to strokes of sinusoidal shape, while lower clouds lead to strokes of exponential shape. Figure 25.10 shows the exponential shape including the detailed structure of the initial step pulses. Pulses of this shape are propagated along the line in both directions away from the point struck.

If we take for an example a large and low thundercloud of diameter $D = 1200$ m and height $a = 200$ m, the surge impedance is

$$\sqrt{\frac{L}{C}} = 500 \frac{200}{1200} = 83.3 \ \Omega,$$

while the surge impedance of the line is $Z/2 = 250 \ \Omega$. The rise-time constant of the

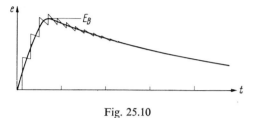

Fig. 25.10

lightning pulse on the line is then, according to Eq. (24.38),

$$T_L = 17.4 \frac{a}{Z/2} 10^{-4} = 17.4 \frac{0.2}{250} 10^{-4} = 1.4 \ \mu s,$$

and depends mainly on the cloud height a. The decay-time constant is, according to Eq. (24.39),

$$T_C = \frac{Z/2 \cdot D^2}{144a} 10^{-6} = \frac{250 \cdot 1.2^2}{144 \cdot 0.2} 10^{-6}$$

$$= 12.5 \ \mu s,$$

and depends in addition on the cloud diameter D.

In the course of this exponential behavior the peak of the voltage pulse in the line approaches closely the initial voltage of the cloud E, as shown in Eq. (24.41). The peak value of the lightning current is then, by Eq. (24.40),

$$I_B = \frac{E}{Z/2}. \quad (25.34)$$

This current divides into two equal halves which flow into the line in both directions. For the example chosen and a cloud voltage $E = 10,000$ kV,

$$I_B = \frac{10,000}{250} 10^3 = 40,000 \ A.$$

which is about three times the amount in the first pulse step as given by Eq. (25.6).

Figure 25.11 is the cathode-ray oscillogram of a direct lightning stroke to a line supported by wooden poles and observed at a distance of 6.5 km from the point struck. It is seen to have almost exactly the time constants just calculated and thus we may suppose that the thundercloud had the dimensions of the example, namely a diameter of 1.2 km and a height of 200 m. Such observations then will

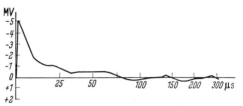

Fig. 25.11

usually permit an estimate of the height and diameter of the thundercloud.

A stroke to a steel tower without a ground conductor is essentially the same as a stroke to ground, only the ground resistance is always much less because of the large footing diameter of the tower. Thus the cloud height below which the shape is exponential is considerably reduced. For example, in the case of a cloud of diameter $D = 1$ km, a uniform tower diameter $d = 160$ cm, and a ground resistivity $s = 10^4 \ \Omega$ cm, the tower ground resistance is 20 Ω and the cloud height for exponential behavior, according to Eq. (24.30), has been reduced to

$$\bar{a} = \frac{10^{-5} \ sD}{\pi \ d} = \frac{10^{-5} \ 10^4 \cdot 1000}{\pi} = 20 \ m.$$

Thus such strokes will usually exhibit a sinusoidal behavior. Whether they consist of a single sinusoidal impulse or a train of impulses depends, according to Eq. (25.31), mainly on the ground resistance and the reignition (or restrike) voltage of the lightning channel.

The voltage at the tower can be determined from Eq. (24.36). For the given example of a cloud of diameter $D = 1$ km and height $a = 200$ m, an electric-field strength of

100 kV/m, and a tower ground resistance $R = 20\,\Omega$, it is

$$E_B = \frac{20}{500} \cdot \frac{1000}{200} \cdot 20{,}000\ \varepsilon^{-\frac{\pi}{4}\frac{20}{500}\frac{1000}{200}}$$

$$= 4000 \cdot 0.855 = 3416\ \text{kV},$$

which will always lead to insulator flashover.

Figure 25.12 shows a stroke to a tower connected to its neighbors by a ground conductor. In this case the criterion of Eq. (25.30) gives a sinusoidal behavior even more frequently because the ground resistance, now

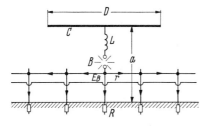

Fig. 25.12

given by Eq. (25.26), will almost always be less than the surge impedance. Because the developing lightning current can now flow into the low ground resistance of a large number of adjacent towers, this low effective ground resistance will lead to multiple-pulse trains according to the criterion in Eq. (25.31) unless the total effective resistance is increased by additional loss processes due to radiation from the channel or resistance in the cloud channels.

Whether the stroke enters at the tower or somewhere along the ground conductor is of no consequence for the subsequent current developments. Even a stroke to a line conductor will lead to the same lightning current if there is a flashover at the adjacent towers, which is usually the case for high lightning voltages. If the lightning voltage is so low that no flashover occurs, the pulses on the line develop exactly as shown above for a line supported by wooden poles.

As shown in Fig. 25.8, the lightning current develops as a sequence of discharge step pulses that travel to and fro between the towers and take a time τ_m to do so, which was calculated from Eq. (25.29) to be about 1–2 μs. The thundercloud also discharges, according

to Eq. (24.27), in a series of step pulses spaced at intervals τ_B of about equal duration. The ratio

$$\frac{\tau_B}{\tau_m} = \frac{a}{m} \tag{25.35}$$

depends only on the cloud height a and the spacing between towers m. For a cloud height $a = 1000$ m and tower spacing $m = 250$ m, for example, four ground-conductor steps will occur during one lightning-channel step. Because the lightning-channel steps are in general of short duration compared with the subsequent stroke-current development according to Eq. (24.29), we see that the different intermingled steps determine the shape of the initial rise of the stroke but soon become merely small ripples superimposed on the general shape of the discharge. Thus Fig. 25.13 shows the general

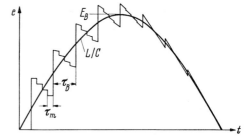

Fig. 25.13

shape of the voltage and current behavior of the stroke at the point struck and in its immediate vicinity.

If we further regard the tower with its finite height as constituting a line having a surge impedance and terminated in the ground resistance, we obtain an additional series of very short steps of about one-tenth the duration of the ground-conductor steps. This will be examined somewhat further in the next chapter.

The peak value of the developing lightning current is given by Eq. (24.35), with the effective ground resistance of the towers from Eq. (25.26):

$$I_B = \frac{E}{\sqrt{L/C}}\ \varepsilon^{-\frac{\pi}{8}\sqrt{\frac{Rr}{L/C}}} \cong 2D\mathfrak{E}_0\varepsilon^{-\frac{\pi}{8}\frac{\sqrt{Rr}}{500a/D}}, \tag{25.36}$$

with D in kilometers and $\mathfrak{E}_0$ in kilovolts per meter. In the example here for a cloud diameter $D = 1$ km, height $a = 200$ m, and $\mathfrak{E}_0 = 100$ kV/m, this is

$$I_B = 2 \cdot 1 \cdot 100 \cdot \varepsilon^{-\frac{\pi}{8}\frac{\sqrt{20 \cdot 1.25}}{500 \cdot 1/5}}$$

$$= 200 \cdot \varepsilon^{-0.02} = 196 \text{ kA}.$$

The damping due to the exponent is very small because the ground resistance is low. The very high lightning current flows to ground over about 12 tower footings in both directions adjacent to the point struck.

The lightning voltage at the tower struck is the product of current and ground resistance

$$E_B = \frac{\frac{1}{2}\sqrt{Rr}}{\sqrt{L/C}} \varepsilon^{-\frac{\pi}{8}\sqrt{\frac{Rr}{L/C}}} E$$

$$\cong \sqrt{Rr} D \mathfrak{E}_0 \varepsilon^{-\frac{8}{\pi}\frac{\sqrt{Rr}}{500a/D}}, \quad (25.37)$$

which in the numerical example becomes

$$E_B = \sqrt{20 \cdot 1.25} \cdot 1 \cdot 100 \cdot 0.98 = 490 \text{ kV}.$$

This is sufficiently low that the insulators of many high-voltage lines can withstand it for the short impulse times involved.

Thus use of the ground conductor has reduced the peak voltage due to the fully developed stroke to exactly one-seventh of the value without it.

Table 25.2 gives the amplitude of the fully developed voltage pulse at a tower over a range of ground resistances for 0, 1, 2, or 3 ground conductors when struck by lightning from a cloud of the parameters used in the example above. Depending on conditions, it can be larger or smaller than the amplitude of the first step-voltage pulse as given in Table 25.1. Again these voltages will increase

severalfold if a more highly charged lightning cloud is involved.

The skin effects due to the traveling pulses will also again diminish somewhat the beneficial reduction due to the ground conductors.

We learn from Eq. (25.37) that to obtain low lightning voltages in the case of direct strokes to a tower or ground conductor the grounding resistance R of the towers must be kept as low as possible and that in addition it is necessary to use ground conductors with low resistance r between the towers. All this then helps to protect the line from damage due to lightning.

If the lightning voltage flashes over the insulator from the tower to the line conductor, it is propagated as a voltage pulse along the

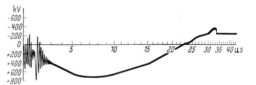

Fig. 25.14

line conductor in both directions. The flashover reduces it very little because the surge impedance of the line conductors is always very large compared with the effective total impedance of the multiple tower and ground-conductor network. As such sinusoidal impulses, shown in Fig. 25.13, travel along the line, the front rise is flattened, the back is distorted into a long tail, and the initial steps are strongly dispersed, so that the pulse shape approaches that of Fig. 25.10. Figure 25.14 is the cathode-ray oscillogram of a lightning pulse taken 5 km from the point struck on a 220-kV line with steel towers and ground conductors. It shows very clearly the initial

TABLE 25.2. Peak voltage (kV) at a tower struck from a cloud of diameter 1 km at 20 MV.

Ground conductors	Tower ground resistance, R (Ω)											
	2	3	5	10	20	30	50	100	200	300	500	1000
0	394	586	961	1848	3416	4740	6750	9120	12600	15400	17800	19000
1, $r = 1.25$ Ω	157	192	248	348	490	598	766	1070	1488	1793	2262	3072
2, $r = 0.625$ Ω	111	136	176	248	348	426	546	766	1070	1300	1650	2262
3, $r = 0.417$ Ω	91	111	144	203	286	348	448	629	880	1070	1366	1883

spikes of the stroke and the insulator flashover after which it develops as a damped sinusoid of at least two half-periods and then returns to the operating voltage. If the time of 8 μs to the first voltage peak is taken as a quarter-period, the lightning-stroke frequency was about

$$f = \frac{10^6}{4 \cdot 8} = 31{,}300 \text{ cy/sec},$$

which is well within the range of values corresponding to our numerical examples and which, according to Eq. (24.18) implies a cloud diameter

$$D = \frac{2.88 \cdot 10^5}{2\pi \cdot 31.300} = 1.46 \text{ km}.$$

We note that this is close to the value indicated by the oscillogram in Fig. 25.11.

CHAPTER 26

IMPULSE CHARACTERISTICS OF TOWERS AND GROUND RODS

Up to now we have regarded the ground resistance of a tower struck by lightning as a constant and have neglected effects due to the height of the tower. Bad experiences with very high towers, especially on poorly conducting ground, indicate that this may not always be satisfactory.

Owing to the strong preionization of the space ahead of the slowly descending head

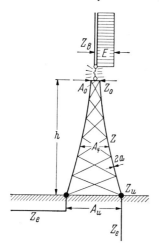

Fig. 26.1

of the lightning stroke, the actual stroke at the top of the tower occurs in a very short time, of the order of small fractions of a microsecond. The rise time of the first voltage and current pulse is thus extremely short, perhaps only 0.05 μs or less, which corresponds to a pulse-front length of 15 m or shorter. This is only a fraction of the height of a high-voltage tower, say 50 m, or of the length of a ground rod, say 100 m.

For such short times and hence sharp pulses, one can no longer use the ground resistance of the tower. Instead it is necessary to regard the actual pulse propagation as governed by

the dimensions in order to estimate correctly the effects of lightning impulses.

(a) *Impulse characteristics of ground rods.* To obtain a good ground for line towers or other objects to be protected against lightning, it is customary to drive rods into the ground vertically or horizontally in the radial direction as a counterpoise to the line conductors, as shown in Fig. 26.1. The ground resistances for both arrangements were fully analyzed in Ref. 1, Chapter 25 (*b*), pp. 328ff and

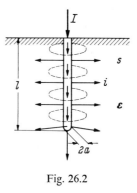

Fig. 26.2

Chapter 27 (*b*), pp. 363ff, respectively. Here we must examine to what extent a current impulse in the electrode and the adjacent ground can follow the rapidly changing lightning voltage because rod or wire grounds have inductive and capacitive properties.

The best criteria are the electric time constants and the natural frequency of the electrodes. Figure 26.2 shows the current I as it enters a vertical ground rod and is propagated into the ground which in addition to its specific resistance s has also a dielectric constant ε. Thus in parallel with the conductive current into the ground there will be a capacitive displacement current if the electrode voltage changes as a function of time. The

273

displacement current follows exactly the lines of propagation of the conductive current. The ground resistance for a rod of radius a and length l was derived in Ref. 1, Eq. (20), p. 329:

$$R = \frac{s}{2\pi l} \cdot \ln\left(\frac{2l}{a}\right) \ \Omega. \qquad (26.1)$$

By reciprocal analogy the ground rod then can be shown to have a capacitance

$$C = \frac{\varepsilon l}{2 \cdot \ln(2l/a)} \cdot \frac{10^{-9}}{9} \ \text{F}, \qquad (26.2)$$

where ε is the dielectric constant relative to a vacuum.

The current in the rod and ground gives rise also to a magnetic field, which is shown by the broken circles in Fig. 26.2. This field is strongest near the rod, where the current density is highest. Thus the self-inductance of the rod due to ground currents is mainly determined by the current distribution in the rod. The current has its full value at the top of the rod and drops to zero at the bottom. The self-inductance of such a rod by analogy to the capacitance is

$$L = 2l\mu \ln\left(\frac{2l}{a}\right) \cdot 10^{-7} \ \text{H}, \qquad (26.3)$$

where the magnetic permeability μ of the space can usually be put equal to 1 even for steel rods.

The capacitive time constant of any circuit element is given by the product of capacitance and resistance and thus for a vertical ground rod is

$$\tau_C = CR = \frac{s\varepsilon}{4\pi \cdot 9 \cdot 10^9}. \qquad (26.4)$$

It is purely a function of the specific resistance s and the dielectric constant ε. The dielectric constant for the usual ground is about 9, on account of its high value for water (80). Thus for a moist ground having $s = 10^2 \ \Omega\text{m}$,

$$\tau_C = \frac{10^2 \cdot 9}{4\pi \cdot 9 \cdot 10^9} = 8 \cdot 10^{-9} \ \text{sec},$$

which is a very small value. For a rock-outcrop ground and $s = 10^4 \ \Omega\text{m}$, τ_C is about $8 \cdot 10^{-7}$ sec.

The inductive time constant is given by the quotient of self-inductance and resistance,

$$\tau_L = \frac{L}{R} = 4\pi \frac{l^2}{s} \cdot 10^{-7}. \qquad (26.5)$$

It is proportional to the square of the length l of the rod. For a rod of length $l = 6$ m in a moist ground it is

$$\tau_L = 4\pi \frac{6^2}{10^2} \cdot 10^{-7} = 4.5 \cdot 10^{-7} \ \text{sec}.$$

For the same rod in a rocky ground it will be only $4.5 \ 10^{-9}$ sec. Both of these values are again very small.

Because all these times are smaller than the rise time usually assumed for the start of a lightning impulse striking a tower, the current and voltage in the rod will follow such slower impressed impulses, with rise times of the order of a microsecond, without any significant time delay and the rod will thus behave as in the steady state. Thus the capacitance and inductance of ground rods of moderate length have no significant influence, even with quite rapid lightning strokes.

The natural oscillations of such a rod have a period

$$T = 2\pi\sqrt{LC} = \frac{2\pi\sqrt{\varepsilon}}{3 \cdot 10^8} l, \qquad (26.6)$$

which is a function of the dielectric constant and rod length only. For the previous example this period is

$$T = \frac{2\pi\sqrt{9}}{3 \cdot 10^8} \cdot 6 = 3.8 \cdot 10^{-7} \ \text{sec}.$$

This is again shorter than that assumed for most lightning strokes and also indicates the frequency limit below which the rod behaves as an ohmic resistance.

The damping of the free natural oscillations is given by the relation of the quotient of ohmic resistance and surge impedance to the number $\frac{1}{2}$, as is seen from Eq. (3.19) based on Ref. 1, Eq. (16), p. 50. The criterion for behavior is thus

$$\frac{R}{\sqrt{L/C}} = \frac{s\sqrt{\varepsilon}}{120\pi l} \begin{cases} > \frac{1}{2} \ \text{periodic}, \\ < \frac{1}{2} \ \text{aperiodic}. \end{cases} \qquad (26.7)$$

For the same rod as above in moist ground,

$$\frac{R}{\sqrt{L/C}} = \frac{10^2\sqrt{9}}{120\pi \cdot 6} = \frac{1}{7.5},$$

so that any oscillations will be overdamped. However, for rocky ground the quotient is 13, which shows that now the rod can oscillate periodically.

An equivalent circuit is often convenient to represent the properties of a circuit element in simplified form. In view of the foregoing numerical values, there is no doubt that for

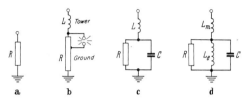

Fig. 26.3

low frequencies a vertical ground rod and all derived ground-electrode arrangements can be represented by a single lumped resistance, as shown in Fig. 26.3a. This remains true for electrodes of moderate length up to high and even very high frequencies. However, it should be noted that a very high tower or other lead connecting the rod to the lightning discharge can have appreciable self-inductance, as shown in Fig. 26.3b.

If the voltage at the top of the ground rod is very high, as is frequently the case during lightning strokes, the field strength adjacent to the rod becomes large enough to break down the ground, and this results in a lowering of the ground resistance. In Fig. 26.3b this is shown by a spark gap across a portion of the ground resistance.

If the discharge to the ground electrode is extremely rapid or the equivalent frequency is extremely high, above the order of megacycles per second, a considerable displacement current in the capacitance of the ground will flow in parallel with the conductive resistance, as shown in Fig. 26.3c. The time constants, natural frequencies, and damping of the ground rod given by Eqs. (26.4)–(26.7) now represent the effects of these circuit elements.

In the most general case, finally, for extremely rapid current changes and steep pulse fronts,

all three elements L, C, and R of the ground rod act together, as shown in Fig. 26.3d.

If the specific resistance of the ground is very high, as for example in rocky terrain, self-inductance and capacitance effects predominate and the resistance has only a minor influence. Thus the surge impedance

$$Z = \sqrt{\frac{L}{C}} = 60\sqrt{\frac{\mu}{\varepsilon}} \cdot \ln\left(\frac{2l}{a}\right) \, \Omega \quad (26.8)$$

becomes the governing factor in the ability of the ground electrode to admit the lightning current. Again for the example of length $l = 6$ m and radius $a = 2.5$ cm, this is, for $\varepsilon = 1$,

$$Z = 60 \cdot \ln\left(\frac{2 \cdot 6}{2.5 \cdot 10^{-2}}\right) = 375 \, \Omega.$$

If there is water in the rocky ground, ε will of course be greater than 1 and the surge impedance can become considerably smaller. Even if several such rods, say four as shown in Fig. 26.1, are used in parallel and their interaction is neglected, $Z/4$ gives $Z_4 = 94 \, \Omega$, which is still very high compared with the few ohms desirable for good lightning protection.

The ohmic resistance of such a rod in the rocky ground of $s = 10^4 \, \Omega$m is

$$R = \frac{10^4}{2\pi \cdot 6} \cdot \ln\left(\frac{2 \cdot 6}{2.5 \cdot 10^{-2}}\right) = 1650 \, \Omega,$$

which is much larger than the surge impedance. If we now consider again the four ground rods in parallel, neglecting interaction, the total effective parallel impedance according to Fig. 26.3d is

$$Z//R = \frac{1/4}{1/375 + 1/1650} = \frac{1}{4} \cdot 305 = 76 \, \Omega,$$

which is still too high a value.

For ground of low specific resistance, lengthening of the ground rods or buried wires cannot indefinitely reduce the effective resistance for rapid lightning strokes. While the capacitance now plays a minor role compared with the resistance, the self-inductance of the longer wires becomes appreciable because the inductive time constant given by Eq. (26.5) increases with the square of the length. The capacitive time constant given by

Eq. (26.4) does not change and the effects may now be neglected owing to the low ground resistance. Since wires buried in the ground usually have an appreciable cross-sectional area, we may also ignore their resistance in comparison with the self-inductance. Figure 26.4a shows the remaining important parameter magnitudes for this example, namely the current I (A) in the wire, the self-inductance L' (H/m) of the wire, the ground resistance

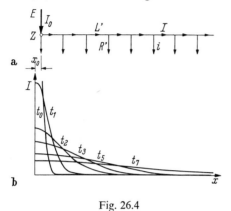

Fig. 26.4

R' (Ω m), and the ground current i (A/m), the last three referred to unit length.

At each point along the wire the sum of the current entering the ground and the change in current along the wire must balance and thus

$$i + \frac{dI}{dx} = 0. \qquad (26.9)$$

On the other hand, the voltage along a small rectangle formed by a line element of self-inductance $L' \, dx$ and the two adjacent resistances R' must be zero, which leads to

$$R' \frac{di}{dx} + L' \frac{dI}{dt} = 0. \qquad (26.10)$$

Equations (26.9) and (26.10) give the differential equation for the current I in the wire,

$$\frac{d^2I}{dx^2} - \frac{L'}{R'} \frac{dI}{dt} = 0, \qquad (26.11)$$

and a corresponding equation for the ground current i.

Equation (26.11) is the well-known equation of heat diffusion. A solution for a short impulse is

$$I = \frac{K}{\sqrt{t}} \varepsilon^{-L'x^2/R't}, \qquad (26.12)$$

where the amplitude is still free as determined by the constant of integration K, while the exponent depends, in addition to L' and R', on the square of the distance x and the time t. This solution is plotted as a function of distance in Fig. 26.4b for various times and as a function of time in Fig. 26.5a for various

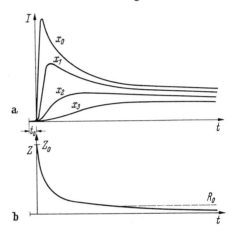

Fig. 26.5

distances. Equation (26.12) represents an impulse curve along the time axis that increases at first slowly, then steeply, reaches a maximum, and decreases gradually to zero. Such a curve might well be used to approximate the time characteristics of a lightning discharge to ground as observed in many oscillograms.

Figure 26.4b shows, according to Eq. (26.12), the manner in which the current in the ground wire, from an initial high value over a small region, gradually spreads over the entire length with a lower value. It is not necessary to identify the origins of x and t in Eq. (26.12) with the origin of the lightning current. Figures 26.4 and 26.5 show that small displacements in space x_0 and in time t_0 can be used for constants of the solution in Eq. (26.12), and by proper selection of these constants every actual case can be closely approximated.

The points at which the spatial curves in Fig. 26.4b are steepest determine the front of the damped propagating current pulse. Their location can be derived from Eq. (26.12) and is given by

$$\frac{L'}{R'}\frac{x^2}{t} = \frac{1}{2}. \qquad (26.13)$$

These points propagate with a velocity

$$v = \frac{dx}{dt} = \frac{1}{4}\frac{R'/L'}{x} = \frac{10^7}{4\pi}\frac{s}{x} \qquad (26.14)$$

along x, where R'/L' is inserted from Eq. (26.5) and both are referred to unit length. Thus the velocity of the pulse front is not constant but decreases rapidly along x.

Long buried wires are often used as counterpoises, radiating from the feet of the towers as shown in Fig. 26.1, when the ground has such high specific resistance that low ground resistances cannot be obtained in other ways. In such wires, buried in ground of resistance $s = 10^3\ \Omega m$, the velocity of the current pulse at a distance of 10 m from the foot of the tower will be about

$$v = \frac{10^7}{4\pi}\frac{10^3}{10} = 0.8 \cdot 10^8\ \mathrm{m/sec}.$$

This is about a quarter of the velocity of light which applies to current pulses on open lines in air. Thus we see that the effect of resistance and self-inductance in ground wires of appreciable length gives a low velocity of pulse propagation, which is reduced even further over short distances. This reduces the ability of such ground electrodes to absorb rapidly rising lightning currents.

The voltage at any point along the wire can be obtained from the ohmic voltage drop in the ground resistance R' due to the ground current i. If we use Eq. (26.9) for i and insert Eq. (26.12) differentiated, we obtain for the voltage

$$E = R'i = -R'\frac{dI}{dx} = 2L'\frac{x}{t}I, \qquad (26.15)$$

where now the change of E for any distance and time is reduced in form to that of the current as given by Eq. (26.12). Thus E is proportional to I according to the quotient of distance and time. Then the surge impedance becomes

$$Z = \frac{E}{I} = 2L'\frac{x}{t}, \qquad (26.16)$$

which makes it proportional to L', the self-inductance per unit length of the wire. For a given point x along the wire, Z is inversely proportional to the time t.

In Fig. 26.5a, t_0 denotes the short time preceding the sharp rise of the current. The corresponding distance x_0 in Fig. 26.4 is about one-half the value from Eq. (26.13), as can be derived from the form of the current shape given by Eq. (26.12). Thus

$$\frac{x_0^2}{t_0} = \frac{1}{4}\frac{R'}{L'}. \qquad (26.17)$$

If we insert this value of x in Eq. (26.16) and use values of R and L given in Eqs. (26.1) and (26.3) for a finite wire, but here referred to the unit length, we obtain for the surge impedance at the beginning of the buried conductor the expression

$$Z_0 = \frac{t_0}{t}\sqrt{\frac{R'L'}{t_0}} = \frac{t_0}{t}\sqrt{\frac{2s \cdot 10^{-7}}{\pi t_0}}\ln\left(\frac{2l}{a}\right)\Omega. \qquad (26.18)$$

Because such wires are usually buried close to the surface, the value of R' must be increased by a factor 2 over that for a vertically embedded wire, as is done in the final term of Eq. (26.18).

As an example we take a ground wire of radius $a = 0.5\ cm$ and length $l = 75\ m$ buried in ground of moderately high specific resistance $s = 10^3\ \Omega m$. We estimate the time delay to the rapid current rise to be $t_0 = 0.2 \cdot 10^{-6}\ sec$. Then the surge impedance at the start of appreciable current at $t = t_0$ is

$$Z_0 = \sqrt{\frac{2 \cdot 10^3 \cdot 10^{-7}}{\pi \cdot 0.2 \cdot 10^{-6}}}\ln\left(\frac{2 \cdot 75}{0.5 \cdot 10^{-2}}\right) = 184\ \Omega,$$

and this value decreases hyperbolically with increasing time, as shown in Fig. 26.5b. For a counterpoise of four such conductors at the foot of a tower we thus have an effective surge impedance $Z_e = 46\ \Omega$.

According to Eq. (26.18) this surge impedance gradually decreases to zero owing to the neglect in this equation of the ohmic resistance of the long wire. Actually it cannot drop below the steady-state ground resistance of the finite wire. In our example this has twice the value given by Eq. (26.1), again owing to its closeness to the surface, and is

$$R_0 = \frac{10^3}{\pi \cdot 75} \ln (3 \cdot 10^4) = 44 \ \Omega$$

thus it represents an asymptote to the decrease of Z with time as shown in Fig. 26.5b by a broken line.

(b) *Step pulses on towers.* If lightning strikes the top of a line-supporting tower, as shown in Fig. 26.1, the quasi-statically charged lightning channel discharges itself into the tower and the front of the lightning surge travels down the tower as a step pulse with the velocity of light, if the magnetic effects due to the steel in the tower are neglected. The magnitude of the current in the tower and the fraction of the cloud voltage impressed on it are determined by the surge impedance of the tower in comparison with that of the lightning channel. The latter was found in Eq. (24.23) to be derived from the usual equations for long lines and so $Z_B = 500 \ \Omega$ represents a good average value.

For the tower we must use the surge impedance of a conductor of height h that rises into the air space above the ground plane. This is mathematically equivalent to the determination of the self-inductance L and capacitance C in Eqs. (26.2) and (26.3) for a long electrode in the ground. Thus we may use here Eq. (26.8) for the surge impedance of the tower if we substitute the tower height h for the length l of the ground rod.

However, we must also replace the radius a of the ground rod by an equivalent radius A to take account of the width of the tower, which varies somewhat on account of its construction, as shown in Fig. 26.1. Then by analogy with Eq. (26.8) the surge impedance of the tower is

$$Z = 60 \ \sqrt{\frac{\mu}{\varepsilon}} \ln \left(\frac{2h}{A}\right) \Omega. \quad (26.19)$$

Table 26.1, taken from Ref. 1, p. 335, gives the equivalent radius A for towers with $n = 2$, 3, and 4 legs each of radius a. Here a may again be an equivalent radius if the legs are not of solid construction; A_n is the actual distance between legs in the different arrangements and the radius a may usually be about 0.1 of this distance. Thus for a four-legged tower the numerical factor in the table becomes 0.61 and

$$A = 0.61 \ A_4, \quad (26.20)$$

where A_4 is the actual center distance between the four legs. Because of the fourth root this factor is fairly independent of the actual magnitude of a. For two-legged towers the factor is only 0.32.

As shown in Fig. 26.1, the spacing A_n of the legs of a high tower varies considerably from top to bottom and may be five times as large at the base. Thus we see that such a tower presents a varying surge impedance

TABLE 26.1. Geometric mean distance of parallel rods.

n	Arrangement	Geometric mean distance	For $a = 0.1A_n$
2		$A = \sqrt{aA_2}$	$A \approx 0.32A_2$
3		$A = \sqrt[3]{aA_3^2}$	$A \approx 0.56A_3$
4		$A = \sqrt[4]{\sqrt{2}aA_4^3}$	$A \approx 0.61A_4$

which decreases considerably from top to bottom.

Figure 26.6 shows in detail a tower for a very high-voltage line that has a height $h = 45$ m and has a width A_4 of 2.35 m at the top and 10 m at the base. For $\varepsilon = \mu = 1$ this gives a surge impedance at the top

$$Z_0 = 60 \ln \left(\frac{2 \cdot 45}{0.61 \cdot 2.35} \right) = 60 \ln 62.5 = 248 \ \Omega,$$

and at the base

$$Z_u = 60 \ln \left(\frac{2 \cdot 45}{0.61 \cdot 10} \right) = 60 \ln 14.7 = 162 \ \Omega.$$

Thus the value at the base is only about $\frac{2}{3}$ of that at the top. The shape of the tower in Fig. 26.6 is similar to an exponential shape

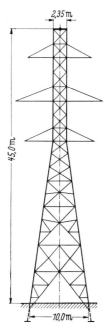

Fig. 26.6

treated in Chapter 10, Sec. (c) for continuously varying line diameter. From Eq. (10.37) we can then here calculate the difference in surge impedance between the top and bottom:

$$\Delta Z = 60 \ln \left(\frac{A_u}{A_0} \right) = 60 \ln 4.25 = 86 \ \Omega.$$

Now let a lightning stroke of voltage E, as in Chapter 24, Sec. (c) or Chapter 25,

Sec. (b), strike the top of the tower, which at first we shall consider not to be joined to others by a ground conductor. The voltage is divided, according to Eqs. (9.27) and (9.28), as shown in Fig. 26.7, into a charge pulse traveling down the tower of amplitude

$$e_{v0} = \frac{Z_0}{Z_B + Z_0} E = \beta E \qquad (26.21)$$

and a reflected discharge pulse that travels up the lightning channel,

$$e_{rB} = \frac{Z_B}{Z_B + Z_0} E = \rho E. \qquad (26.22)$$

Thus in the numerical example the voltage at the top of the tower will be

$$e_{v0} = \frac{248}{500 + 248} = 0.32 \ E,$$

and as the pulse travels down the tower, assuming a smooth transition, the results of

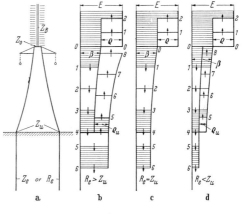

Fig. 26.7

Chapter 10 indicate that it will be reduced at the base to

$$e_{vu} = \sqrt{\frac{Z_u}{Z_0}} \ e_{v0}, \qquad (26.23)$$

when the surge impedance has reached the lower base value Z_u. In the numerical example this becomes

$$e_{vu} = \sqrt{\frac{162}{248}} \ e_{v0} = 0.81 \ e_{v0}.$$

The travel time down the tower is

$$\tau = \frac{h}{v} = \frac{45 \text{ m}}{300 \text{ m}/\mu s} = 0.15 \ \mu s,$$

so that the reflection of the pulse at the base can again reach the top of the tower 0.3 μs later.

If the base of the tower is in moist ground, the ground resistance will be small compared with Z_u. The pulse will be considerably reduced upon reflection and thus upon reaching the top a time 2τ after the original stroke its voltage will be considerably less. On the other hand, it was seen earlier that a tower based in rocky soil can have a ground resistance of several hundred ohms, which is large compared with Z_u. In that case the pulse will be reflected at the somewhat open-circuited base with almost double amplitude and will then, after a time 2τ, very appreciably increase the voltage at the top of the tower. Thus we see that the reflection at the ground has a very important influence on the development of the peak voltage at the top of the tower.

We saw in Sec. 26(a) that even the addition of four ground rods was able to reduce the ground resistance to the order of 100 Ω in soil of high specific resistance, which is close to the surge impedance at the base of the tower. Under these conditions the decrease in surge impedance of the tower from top to bottom is a disadvantage that cannot be avoided unless the width of the tower can be reduced at the base.

The amplitude of the pulse reflected back up the tower at the ground base is given by the surge impedances, from Eq. (6.8),

$$e_{ru} = \frac{Z_e - Z_u}{Z_e + Z_u} e_{vu} = \rho_u e_{vu}, \quad (26.24)$$

where the ground surge impedance Z_e may be replaced by the ohmic ground resistance R_e. This gives the simple relation

$$\frac{Z_e \text{ or } R_e}{Z_u} \begin{cases} < 1 \text{ lowers} \\ = 1 \text{ maintains} \\ > 1 \text{ increases} \end{cases} \text{ the voltage on the tower.}$$

$$(26.25)$$

Thus, depending on the ground resistance at the base, reflection gives rise to either a discharge or a charge pulse that travels up the tower. The amplitude of this pulse is changed by the changing width and hence surge

impedance of the tower, until at the top it is increased in the ratio $\sqrt{Z_0/Z_u}$. In the most favorable case of a very good ground, this pulse reduces the voltage at the top of the tower very considerably and thus reduces the period during which the voltage is high to 2τ. This is thus the minimum time that insulators at the top must be able to withstand high voltages even in the best cases. In the worst cases of high ground resistance the voltage at the top of the tower is increased further, which in practice almost always leads to flashover.

A further disadvantage arises from the fact that the line insulators usually are suspended from the tips of the crossarms. At the point on the tower where the crossarm joins it, the pulse voltage is a little reduced by the additional loading. However, once the pulse enters the crossarm it is doubled by reflection at the tip if the pulse front is short compared with the length of the arm. If the length of the pulse front is greater than that of the crossarm, this effect is somewhat less pronounced.

The pulse reflected at the base and transformed by the changing surge impedance reaches the top of the tower after a time 2τ with the voltage

$$e_{r0} = \sqrt{\frac{Z_0}{Z_u}} e_{ru} = \rho_u e_{v0}, \quad (26.26)$$

where the changing surge impedance has been canceled by the down-and-up traversal. It can be positive or negative depending on the sign of the reflection factor ρ_u from Eq. (26.24). This pulse is again reflected at the top, owing to the junction of the tower and the lightning channel,

$$e'_{v0} = \frac{Z_B - Z_0}{Z_B + Z_0} e_{r0} = \rho_0 e_{r0} \quad (26.27)$$

down the tower. Because Z_0 is probably always smaller than Z_B, ρ_0 will always have a positive value and thus lead to a further voltage increase.

Together with the initial voltage e_{v0} at the top of the tower this gives a second step pulse

$$e''_{v0} = \beta(1 + \rho_u \rho_0)E. \quad (26.28)$$

With the calculated values of the numerical example for the tower top and lightning channel, β is, according to Eq. (26.21), a fraction of about $\frac{1}{3}$ and according to Eq. (26.27) ρ_0 is also a positive fraction of about $\frac{1}{3}$. Because these surge impedances cannot be changed significantly in practice, we see how important it is, with regard to Eqs. (26.24) and (26.25), to make the ground resistance as small as possible compared with the base impedance Z_u of the tower. This will keep ρ_u small, according to Eq. (26.24), and thus keep the tower-top voltage as low as possible after time 2τ.

Figure 26.7 shows the progress of the pulses along the tower for the three impedance ratios of Eq. (26.25). It is clear that the change in surge impedance along the tower has no influence on the top voltage directly but merely through the very bad effect of the lower impedance at the base compared with the ground.

Voltage conditions at the tower and thus for the insulators are considerably improved when the tops of all towers are joined by ground conductors of surge impedance $\mathcal{Z}$. As shown in Sec. 25(b), the lightning current now splits into three portions, one into the tower and two equal ones into the ground conductors in opposite directions. The voltage at the top then becomes, according to Eq. (25.15),

$$e = \frac{E}{1 + 2Z_B/\mathcal{Z} + Z_B/Z_0}. \quad (26.29)$$

The use of two ground conductors instead of one reduces this even further, since $\mathcal{Z}$ will then be only about half as large.

Further, the pulse that returns to the top of the tower from the base after time 2τ encounters a junction of several lines and thus leads to lower values than those in Eqs. (26.27) and (26.28) because it gives rise to three, five, or more transmitted pulses, one along the lightning channel and one each along one, two, or more ground conductors in both directions. This reduces the voltage increase at the top of the tower to negligible proportions even when the ground resistance at the base is considerable. Even though for two or more ground conductors the effective surge impedance is not reduced in inverse ratio to the number, it is possible to reduce the voltage at the top of the tower considerably in more practical cases.

Finally, an effect that is beneficial for the line insulators is introduced by the ground

conductors owing to the fact that by capacitive coupling they raise the voltage of the line conductors toward that of the tower. In Sec. 25(b) we saw that this reduced the voltage across the insulator by 10–15 percent for one ground conductor and by 30–40 percent for two or three ground conductors.

These two beneficial effects of ground conductors are usually sufficient to protect line insulators on high towers against flashover even under the most severe lightning conditions encountered.

Figure 26.8 shows the development of the voltage in a tower for a 345-kV line, shown in

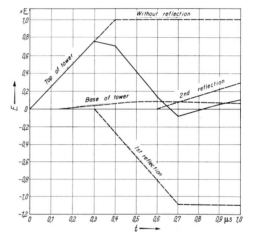

Fig. 26.8

Fig. 26.6, as calculated by C. J. Miller. The lightning stroke is assumed to lead to a linear rise to full voltage in 0.4 μs. The tower had only one ground conductor and frequent flashover of the insulators was experienced; the addition of a second ground conductor considerably reduced the frequency of this flashover.

(*c*) *Low-resistance ground grids.* Because the surge impedance Z_u at the base of a high tower is small, while it is liable to be struck by lightning owing to its great height, it is necessary to keep the ground resistance lower than Z_u as shown by Eq. (26.25) to reduce the danger of lightning strokes. To achieve this in soil of high specific resistance it is necessary to install extensive grounding systems. This leads to a large number of parallel conductors in the ground so as to obtain a low effective ground resistance in spite of the high resistivity s and at the same time keep the lengths short so as to avoid bad effects due to late reflection at the ends. Figure 26.9 shows two such crossed

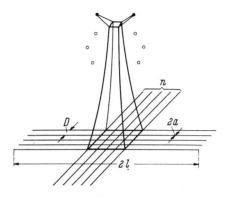

Fig. 26.9

ground grids at the base of a tower, which represents a good solution.

To determine the ground resistance of a grid containing n parallel wires of length $2l$, spaced a distance D between wires and near the surface, as shown in Fig. 26.10, we start

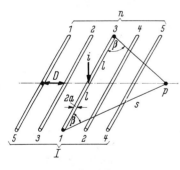

Fig. 26.10

with the potential due to a single conductor, which was derived in Ref. 1, Eq. (16), p. 328:

$$p = \frac{si}{2\pi l} \ln \left(ctn \frac{\beta}{2} \right). \qquad (26.30)$$

Here i is the current in each wire and β is the angle subtended by the wire at the point at which the potential p is observed.

This equation for the potential would be strictly correct if the ground current from the wire to ground had a uniform distribution. The factor $\frac{1}{2}$ in the equation is valid whether the wire is only one-half buried in the ground or completely buried slightly below the surface.

Because potentials are additive, the total potential due to all wires, as shown in Fig. 26.10, is simply

$$p_1 = \frac{s}{2\pi} \left[\frac{i_1}{l_1} \ln \left(ctn \frac{\beta_{11}}{2} \right) \right.$$
$$\left. + \frac{i_2}{l_2} \ln \left(ctn \frac{\beta_{12}}{2} \right) + \cdots \right] \qquad (26.31)$$

at the first wire taken as an example. Similar values obtain for the other wires. Here the length of the wires can be allowed to vary. The angle β can be easily measured relative to two wires between the ends of one and the center of the other. Thus we obtain as many equations as there are unknown currents i.

The arithmetic solution of these equations, up to perhaps ten wires in the grid, is not too difficult. With the aid of a computer it can be easily performed for a very large number of wires. Thus the common potential or the voltage E of the grid wires as a function of the sum I of all wire currents i can be worked out, from which the ground resistance of the entire grid is found. We also obtain at the same time the nonuniform current distribution, which varies from the center conductors to those at the edge of the grid.

However, if we neglect this edge effect and assume a uniform current distribution as well as equal lengths of wires, we can reduce the ground resistance to a simple form as long as $nD < 2l$, that is, if the grid is long compared with its width. Using the radius a of the wires and the spacing D between them we can approximate the arguments of the logarithms in Eq. (26.31) more simply as

$$ctn \frac{\beta}{2} = \frac{l}{a/2}; \ \frac{l}{D/2}; \ \frac{l}{D/2}; \ \frac{l}{2D/2}; \ \frac{l}{2D/2}; \cdots$$
$$(26.32)$$

starting with the center wire in Fig. 26.10. Then instead of the sum of the logarithms we may use the product of all these values up to those of the edge wires as a single logarithm to obtain the total grid voltage,

$$E = \frac{si}{2\pi l} \ln \left[\left(\frac{l}{a/2}\right) \cdot \left(\frac{l}{D/2}\right)^2 \right.$$

$$\left. \cdot \left(\frac{l}{2D/2}\right)^2 \cdots \left(\frac{l}{\frac{n-1}{2}D/2}\right)^2 \right].$$

$$(26.33)$$

If we take n, the number of wires, to be odd and move it outside of the logarithm, we can give the product a factorial form to obtain

$$E = \frac{\sin}{2\pi l} \ln \left(\frac{2l}{\sqrt[n]{aD^{n-1}\left[\left(\frac{n-1}{2}\right)!\right]^2}} \right), \quad (26.34)$$

where

$$in = I \qquad (26.35)$$

is the total grid current and

$$D_n = D \sqrt[n]{\frac{a}{D}\left[\left(\frac{n-1}{2}\right)!\right]^2} \quad (26.36)$$

is the mean geometric distance of the wires referred to the central wire. The ground resistance of the entire grid then is given by

$$R_g = \frac{E}{I} = \frac{s}{2\pi l} \ln \left(\frac{2l}{D_n}\right), \quad (26.37)$$

which is a simple closed expression that can be calculated and discussed.

The number of wires n in Eqs. (26.34) and (26.36) increases the square of the factorial but is in turn under the nth-root sign and a logarithm, so that the actual number reduces the ground resistance only slightly with increasing n. Thus within a given width of the grid it is not necessary to introduce a large number of wires to obtain a low ground resistance. However, the length l of the grid and the specific ground resistance s are of great importance. This is because the length or extent and not the area of the grid determines the ground resistance.

Figure 26.11 compares the effective ground resistance of several different numbers of conductors all of wire 1 cm in diameter or ribbon 2 cm wide. All are buried near the surface in poorly conducting ground of specific resistance 10^3 Ω m and all are 100 m long and thus extend 50 m on each side of the tower. For 9 parallel wires spaced 3 m apart, the ground resistance is reduced to about $\frac{1}{3}$ that for a single wire owing to the mutual effect. Note also that an infinite number of wires produces little additional reduction in ground resistance, while three wires spread over more

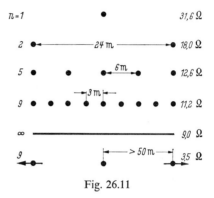

Fig. 26.11

than four times the width again reduce the ground resistance by a factor of 3. Thus the width and hence area of the grid does have an effect.

When the ground conditions are particularly bad, two crossed grids of the kind shown in Fig. 26.10 can be used as shown in Fig. 26.9. Owing to mutual interference it is not possible in this case simply to take one-half the value of Eq. (26.37). However, as shown in Ref. 1, Eqs. (20), p. 329, and (31), p. 332, it is possible to compare a crossed ground wire with a single one near the surface. By analogy we can obtain the ground resistance for two perpendicular grids at the base of a tower, as shown in Fig. 26.9:

$$R_g^+ = \frac{s}{4\pi l} \left[\ln \left(\frac{4l}{D_n}\right) + 1 \right], \quad (26.38)$$

which is a little more than half the value given by Eq. (26.37).

To obtain values for the surge impedance of such grids, which becomes important when the ground conditions are very bad, we can compare Eq. (26.1) for the resistance R and Eq. (26.8) for the surge impedance Z. We notice that they differ only by the factor in front of the logarithm. Thus by analogy with

Eq. (26.37), we may put for the surge impedance of the single grid

$$Z_g = 60 \sqrt{\frac{\mu}{\varepsilon}} \ln \left(\frac{2l}{D_n} \right). \quad (26.39)$$

The factor 60 is further reduced to 30 when the grid is excited at the center, as is usual, because the two halves now act in parallel.

For the surge impedance of the crossed grid, as shown in Fig. 26.9, we obtain by analogy with Eq. (26.38) and connection at the center, the value

$$Z_g^+ = 15 \sqrt{\frac{\mu}{\varepsilon}} \left[\ln \left(\frac{4l}{D_n} \right) + 1 \right], \quad (26.40)$$

which is a little larger than one-fourth the value of Eq. (26.39).

If in Fig. 26.9 the empty corners of the crossed grids were filled in by additional wires, then in particularly difficult cases this could give a further reduction of the ground resistance and surge impedance.

To avoid any nonuniformity in surge impedance, which could lead to additional reflections, the wires of the grid must be directly connected at the rim of the footing of the tower and thus at its full width. In this way, any reductions in the width of the power base are avoided, which is important.

CHAPTER 27

IMPULSE STRENGTH OF INSULATION

High-voltage lines are supported by insulators that vary in design with the different arrangements. All, however, have in common at least one surface exposed to air, which governs the performance of the line with respect to voltage flashover.

It has been demonstrated that the voltage gradient or electric-field strength is not uniformly distributed on a long insulator surface. This nonuniformity is marked during

long distances are usually suspended at the towers as shown in Fig. 27.1. The rated voltage for such insulator chains, as more elements are added, increases only gradually beyond the value for two or three elements and finally approaches a limiting value. This is due to the fact that the conducting link between elements and its associated conducting surfaces assume a potential that depends not only on the electrostatic capacitance of the

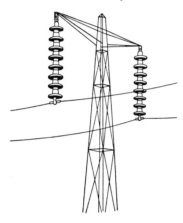

Fig. 27.1

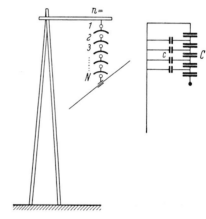

Fig. 27.2

normal operation at the regular voltage by the very slightly conducting surface film that forms in even the cleanest atmospheres and serves to produce a large measure of uniformity during steady-state operation at the low operating frequencies and voltages. However, for very steep step pulses or impulses, caused by switching or lightning, the very high frequency components cause the capacitive impedance of the insulators to predominate in determining instantaneous voltage distribution leading to flashover. This is to be examined here.

(a) *Voltage distribution along insulator chains.* High-voltage lines for distribution over

adjacent elements but also on the capacitance of the link with respect to the tower and ground. This leads to a nonuniform distribution of voltage across the elements, with the largest voltage difference across the element closest to the line conductor and the smallest across the one connected to the tower.

Figure 27.2 shows such an insulator chain with an adjacent equivalent capacitive circuit diagram. Here C is the capacitance between adjacent conducting links and any conducting surfaces connected at the links; c is the capacitance of the links to the ground and grounded tower, which is usually much smaller than C. Of course, such a simplified equivalent circuit gives only an approximate solution for the voltage distribution. It can be slightly improved rather easily if in addition to the link capacitance c to ground we introduce a small capacitance ζ to the line conductor. To obtain a very exact solution it would be necessary to integrate the electric-field equations with due regard to the exact shape of the insulators, which is hardly feasible in practice.

Such capacitance chains form very important functional elements in many insulators, in machinery, and in other equipment used in high-voltage networks. Their behavior during impulsive voltage loading is thus an

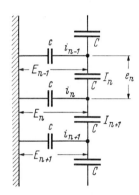

Fig. 27.3

important consideration in the reliability of their operation. The chain type of arrangement is common to many spatially periodic structures in physics and technology. As it is one of the simplest structural forms, it can be analyzed quite easily.

A displacement current I flows through the dielectric of every element of the chain, driven by impulse voltage, and a displacement current i flows from every conductive link through the air to the tower and ground. Let us denote the voltage with respect to ground by E. Figure 27.3 now shows the current distribution in the vicinity of the nth chain element, with $n = 1$ for the element connected to the grounded tower. For the capacitive currents and voltages we then obtain the following equations with respect to ground:

$$i_n = c \frac{dE_n}{dt}, \qquad (27.1)$$

and for the two adjacent elements,

$$I_n = C \frac{d}{dt}(E_n - E_{n-1}),$$

$$I_{n+1} = C \frac{d}{dt}(E_{n+1} - E_n), \qquad (27.2)$$

where the change of voltage with respect to time is given by the time derivative and thus applies to components of any frequency, shape, and duration.

The equilibrium of currents is given by

$$i_n = I_{n+1} - I_n. \qquad (27.3)$$

If Eqs. (27.1) and (27.2) are inserted in Eq. (27.3), the variation with respect to time cancels and we obtain

$$\frac{c}{C} E_n = E_{n+1} - 2E_n + E_{n-1} = \Delta^2 E_n. \qquad (27.4)$$

In Eq. (27.2) we have the first difference ΔE_n as the voltage drop across a single element, while Eq. (27.4) contains $\Delta^2 E_n$ as the second difference, that is, the difference between the voltage drops across two successive elements.

Thus Eq. (27.4) gives the voltage-distribution characteristic of chain insulators as it gives the dependence of the voltage at the nth link on the voltage at the two adjacent links. Equation (27.4) is a difference equation, which would become a differential equation if the differences between successive voltages were very small, as would be the case for a chain with a very large number of elements. However, we shall not examine this limiting case here since the solution of the exact difference equation is actually simpler than that of the approximation using a differential equation.

We are interested in an expression for the voltage E_n as a function of the number of elements n. To obtain this we must use functions E_n of the argument n that satisfy the difference equation (27.4). As can easily be seen, exponential functions form such a solution. Let us use the trial form

$$E_n = A \cdot \varepsilon^{\alpha n}, \qquad (27.5)$$

where A is initially a mere constant and α

will be determined from the difference equation. We have then

$$E_{n+1} = A\varepsilon^{\alpha(n+1)} = \varepsilon^\alpha \cdot A\varepsilon^{\alpha n},$$

$$E_{n-1} = A\varepsilon^{\alpha(n-1)} = \varepsilon^{-\alpha} \cdot A\varepsilon^{\alpha n}. \quad (27.6)$$

Substituting this in Eq. (27.4) and canceling we obtain

$$\frac{c}{C} = \varepsilon^\alpha - 2 + e^{-\alpha} = \left(\varepsilon^{\frac{\alpha}{2}} - \varepsilon^{-\frac{\alpha}{2}}\right)^2$$

$$= \left(2\sinh\frac{\alpha}{2}\right)^2. \quad (27.7)$$

For Eq. (27.5) to be a solution of the difference equation (27.4), α must then be chosen such that

$$\sinh\frac{\alpha}{2} = \frac{1}{2}\sqrt{\frac{c}{C}}. \quad (27.8)$$

Because the capacitance ratio c/C for any shape of insulator members and suspension arrangement can be found by analysis or experiment, α can easily be determined from tables of hyperbolic functions. For a very small capacitance ratio c/C, a rough approximation is

$$\alpha = \sqrt{\frac{c}{C}}. \quad (27.9)$$

Because positive and negative values of α satisfy Eq. (27.7), the complete trial solution of Eq. (27.5) must be expanded to

$$E_n = A\varepsilon^{\alpha n} + B\varepsilon^{-\alpha n}. \quad (27.10)$$

The two arbitrary constants A and B arise from the fact that Eq. (27.4) contains the second difference of E_n. Thus its solution must contain two constants of integration, corresponding to the case of a second-order differential equation.

To obtain the value of these constants, we use the boundary conditions. The zeroth element is grounded at the tower and so, for $n = 0$,

$$E_0 = A + B = 0. \quad (27.11)$$

Thus Eq. (27.10) becomes

$$E_n = A(\varepsilon^{\alpha n} - \varepsilon^{-\alpha n}) = 2A\sinh\alpha n. \quad (27.12)$$

The voltage at the last link is E, that of the line conductor. If there are a total of N insulator elements, we have

$$E_N = 2A\sinh\alpha N = E, \quad (27.13)$$

and the constant is

$$2A = \frac{E}{\sinh\alpha N}. \quad (27.14)$$

So the problem is solved in principle and the voltage at the nth member, from Eq. (27.12), is

$$E_n = E\frac{\sinh\alpha n}{\sinh\alpha N}, \quad (27.15)$$

where E, α, and N are given by the actual construction. Thus the voltage distribution

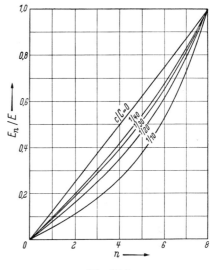

Fig. 27.4

along any insulator chain can be easily determined. Figure 27.4 shows it for a chain of eight elements for different values of the capacitance ratio c/C that cover a range likely to be encountered in practice. A uniform distribution is also indicated by the line $c/C = 0$.

We see that the voltage increases linearly across the first elements of the chain near the tower and then increases more rapidly as the line conductor is approached, according to the hyperbolic sine function. For each value of c/C and thus of α, the curves have a different shape, while the general trend just described is nevertheless present. The distribution becomes more uniform as the ratio of ground capacitance c to the insulator-element capacitance C becomes smaller. Only for very small values of α can the hyperbolic sines in Eq. (27.15) become identical with their arguments αn and αN, in which case the voltage distribution along the chain would be uniform.

In the design of such insulators one should thus attempt to keep the capacitance c of the conducting link and connected conductors to ground and tower as small as possible and

and, with Eq. (27.15),

$$e_n = \frac{E}{\sinh \alpha N} [\sinh \alpha n - \sinh \alpha (n-1)].$$

$$(27.17)$$

For the same range of values of c/C as used in Fig. 27.4 for an eight-element chain, Fig. 27.5 shows this voltage difference as a function of n. It is seen that for c/C large there is a very large nonuniformity.

Of greatest interest, of course, is the highest voltage difference that occurs at the last

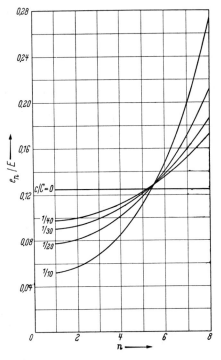

Fig. 27.5

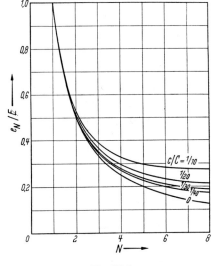

Fig. 27.6

the capacitance C across the individual insulating elements as large as possible. In every case the value of $\sqrt{c/C}$ and thus of α is characteristic for the voltage distribution across the chain and the resultant behavior. In general, it is found that c/C and α increase with the diameter of the insulators.

Equation (27.15) allows the voltage stress across each insulator element in the chain to be calculated. As shown in Fig. 27.3, the voltage difference across an element is

$$e_n = E_n - E_{n-1} = \Delta E_n, \quad (27.16)$$

element, at the line. For $n = N$ Eq. (27.17) in this case gives

$$e_N = E\left[1 - \frac{\sinh \alpha (N-1)}{\sinh \alpha N}\right]. \quad (27.18)$$

Figure 27.6 shows this highest value as a fraction of E, the total voltage across the chain for different numbers of elements N in the chain, and for different values of c/C. Here $c/C = 0$ again shows a uniform distribution for purposes of comparison. As an example we note that for $c/C = 0.05$ an increase of N above five produces only a small decrease in voltage across the last and most stressed insulator.

For chains containing a larger number of elements this highest voltage across the last element approaches a limiting value, which can easily be obtained from Eq. (27.18) if for large N the hyperbolic-sine function is approximated by the exponential function. Then

$$e_{N=\infty} = E(1 - \varepsilon^{-\alpha}). \qquad (27.19)$$

This limiting value depends only on α and thus on the ratio of capacitance c/C of the insulator elements. Because for any given type of insulator element the value of α and the safe voltage e are fixed, we see that we cannot increase the total voltage capability E of a chain by simply increasing the number of elements. Instead, even the longest possible chain of elements can never withstand a limiting line voltage that, by Eq. (27.19), is

$$E_{N=\infty} = \frac{e}{1 - \varepsilon^{-\alpha}}. \qquad (27.20)$$

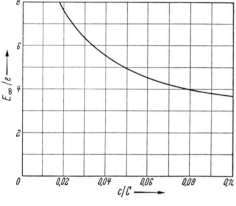

Fig. 27.7

In Fig. 27.7 this is shown as a multiple of the permissible voltage e of each element as a function of the capacitance ratio c/C. We see that for insulator elements with c/C larger than 0.05 even the longest chain can be made to withstand only less than five times the voltage permissible for each element.

The preceding analysis for the voltage distribution across the chain, which was derived from Eqs. (27.1) and (27.2), is also valid in the case of clean and dry insulator surfaces for the normal operating voltages in the case of alternating current. In this case d/dt becomes ω and cancels. For many

insulators in practical use it is found, even for long chains, that about 20 percent of the total voltage is concentrated across the last insulator at the bottom of the suspension chain, at the line conductor. It is also found that this concentration leads to a glow discharge across these lower elements that tends to redistribute the voltage more uniformly and thus may serve to avoid breakdown. However, such permanent discharges produce strong radio-frequency interference, which is objectionable. Thus it is frequently necessary to use other means to obtain a more uniform voltage distribution, particularly in the case of very high-voltage lines.

It is also of interest to determine the total effective capacitance K of the insulator chain. Because the total charging current of the line must flow in the last insulator connected to it and this has a voltage drop e_N across it, we see that we can write

$$\frac{K}{C} = \frac{e_N}{E}, \qquad (27.21)$$

or, using Eq. (27.18),

$$K = C\left[1 - \frac{\sinh \alpha(N - 1)}{\sinh \alpha N}\right]. \qquad (27.22)$$

For large values of N and with Eq. (27.8) this approaches

$$K = \sqrt{Cc}, \qquad (27.23)$$

which is the geometric mean between the capacitance to ground and that between elements.

(b) Insulators with corrugated surfaces. The approach just used to evaluate the unfavorable voltage distribution across a chain insulator due to the shunt capacitance c to ground can be qualitatively applied to other insulating structures, such as insulating mountings and bushings. A smooth insulator mounting, as shown in Fig. 27.8, does not consist of discrete elements with identifiable capacitances. However, each surface element of ring shape has both kinds of capacitance, one to ground and the other between adjacent elements. They are indicated separately in Fig. 27.8 by the electric lines of force, shown dashed.

Their relation governs the ability of the insulator to withstand voltage impulses.

While the respective capacitances C and c vary along the insulator and we can thus not obtain discrete values to use in the equation, the general disposition is the same as for chain insulators. Thus, qualitatively, a similar effect

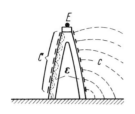

Fig. 27.8

will produce a larger voltage gradient near the insulated object, here at the top, than at the ground, here at the bottom. This effect is accentuated here by the fact that the ratio of capacitance C between ring elements end to end to that to ground c is smaller than in the case of chain insulators.

If such mountings are now provided with corrugated ribs of high dielectric constant ε,

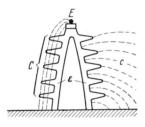

Fig. 27.9

as shown in Fig. 27.9, this increases the series capacitance C considerably while the shunt capacitance c remains about the same. This is because C is essentially proportional to the square of the diameter, while c is only proportional to the logarithm of the diameter. Thus the voltage distribution will become more uniform and the ability to withstand voltage impulses will increase. The same is true for the low-frequency operating voltages and even in the case of direct current. For the latter, the ribs increase the length of the surface path and hence the surface-film resistance.

If insulating rods with or without ribs are used to suspend line conductors, they act of course in a similar manner.

A still more pronounced effect is encountered in the case of feed-through bushings, as shown in Fig. 27.10. Here there is a strong radial

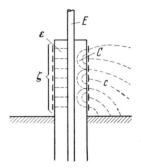

Fig. 27.10

electric field between the outer surface of the insulator and the central conductor, which corresponds to an almost uniformly distributed capacitance ζ. This is much larger in the solid insulator than the external capacitance c in air, which can thus be practically neglected. The concentric surface ring elements have a capacitance C between them just as in the

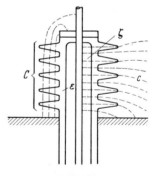

Fig. 27.11

case of the mounting. This arrangement thus corresponds to the capacitive ladder of a chain insulator, as shown in Figs. 27.1 and 27.2, though with the roles of ground and line interchanged so that the highest voltage gradient will now occur near the grounded enclosure in which the bushing is mounted.

The use of corrugated ribs on the outer surface of the bushing is shown in Fig. 27.11.

Here, too, the series capacitance C is increased considerably and thus reduces the ratio ζ/C much more than in the case of the mounting, which had only small shunt capacitance c to start with. Here the shunt capacitance ζ has fairly large values because the distance between the outer surface and the central conductor is small and is occupied mostly by material of high dielectric constant ε. This leads to large changes in the ratio ζ/C and the influence of the ribs, on account of their large series capacitance C, is significant. This is observed in practice where the impulse-voltage strength of feed-through bushing insulators can be increased by a factor of 2 by the introduction of ribs; this is a much larger factor than that found possible for supporting mounts.

If we wish to explain the effect of ribs upon the impulse strength by other simple means, it is possible to imagine the ribs as conductors fixed to the smooth insulating surface. Then the ribs would correspond to the conducting links of a chain considered in the preceding section of this chapter. They would have discrete capacitances to ground and between them. Such conducting ribs have actually often been used in practice. For the distribution of lines of force, which ultimately governs the breakdown considerations, conducting materials can be considered to have $\varepsilon = \infty$. Actually, ε for the ribs is much smaller but still very large in comparison with the surrounding air. Thus the net favorable effect is of considerable magnitude.

To obtain the most nearly uniform voltage distribution on the surface of an insulator, it is always important to keep the series capacitance C as large as possible and the shunt capacitance c or ζ as small as possible. This then will give the highest impulse strength. Insulators made of titanium compounds, with their very large dielectric constant, thus would give a more uniform voltage distribution than the materials, mostly ceramic or plastic, presently used for mounts, rods, chain insulators, and feed-through bushings.

(c) *Equalization of voltage distribution.* Even though an artificial increase in the series capacitance C of insulators produces greater uniformity of voltage distribution, there will

always remain a degree of nonuniformity due to the finite though small value of c/C, as shown in Fig. 27.4 or 27.5. However, it is possible to remove this remaining non-uniformity if the series capacitances C are bridged by two ladders of shunt capacitances, one of c to ground and one of ζ to the insulated conductor. Figure 27.12 shows the

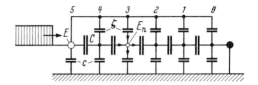

Fig. 27.12

corresponding equivalent circuit, which was already encountered to a degree in the bushings shown in Figs. 27.10 and 27.11.

To be successful and produce complete equalization the relation between the capacitances must be carefully balanced. We shall determine this balance using Fig. 27.13 for the nth link. Here four currents flow, namely I_n and I_{n+1} in the series chain and i_{cn} and $i_{\zeta n}$

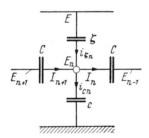

Fig. 27.13

in the shunt chains. Again $n = 0$ at the ground end of the entire chain. The two main currents in their respective capacitances are

$$I_n = C\frac{d}{dt}(E_n - E_{n-1}),$$

$$I_{n+1} = C\frac{d}{dt}(E_{n+1} - E_n), \qquad (27.24)$$

and the two side currents are

$$i_{cn} = c\frac{dE_n}{dt}, \quad i_{\zeta n} = \zeta\frac{d}{dt}(E - E_n). \quad (27.25)$$

The balance at the junction is

$$(i_{\zeta n} - i_{cn}) + (I_{n+1} - I_n) = 0, \quad (27.26)$$

and thus the voltages are related by

$$\zeta(E - E_n) - cE_n + C(E_{n+1} - E_n)$$
$$- C(E_n - E_{n-1}) = 0, \quad (27.27)$$

which can be rearranged as

$$(c + \zeta)E_n - \zeta E = C(E_{n+1} - 2E_n + E_{n-1})$$
$$= C\Delta^2 E_n. \quad (27.28)$$

This again introduces the second voltage difference $\Delta^2 E_n$ along the chain, while the first difference

$$e_n = \Delta E_n = E_n - E_{n-1} \quad (27.29)$$

gives the voltage difference across the nth element and is contained in Eq. (27.24).

Using the assumption of constant capacitances we could now solve the difference

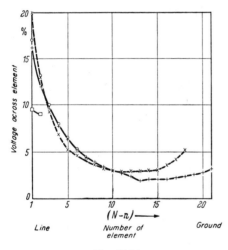

Fig. 27.14

equation (27.28) here and thus obtain a slightly more accurate solution than that of Eq. (27.15). Thus instead of the gradual increase of voltage across elements near the tower, as shown in Fig. 27.5 near $n = 1$, we now obtain at first a decrease in the voltage across elements. Figure 27.14 shows this for 18- and 21-element chains, used for very high voltages, and observed recently by A. S. Denholm. The highest voltage across an element is seen to be here between 16 and 20 percent of the total

chain voltage. This high value for a single element is in good agreement with Fig. 27.5. These observations were made with steady-state 60-cy/sec alternating voltages, rather than with abrupt impulses. However, in the case of clean insulators this is quite applicable and equivalent.

Instead of pursuing this analytical improvement, let us ask whether and under what circumstances a uniform distribution of the voltage among the n elements of the chain can be obtained. The main series capacitance C, which is the coefficient of the voltage terms E_n in Eq. (27.28), can be considered constant, while the side shunt capacitances c and ζ may be regarded as variable.

Uniform voltage distribution across the n elements means that ΔE_n in Eq. (27.29) must be constant, so that the second difference must be

$$\Delta^2 E_n = 0. \quad (27.30)$$

From Eq. (27.28) we then obtain the condition

$$\left(\frac{c}{\zeta} + 1\right) E_n = E. \quad (27.31)$$

However, with e_n constant, the voltage E_n across the chain rises linearly, as shown by Eq. (27.29). If N is the total number of elements in the chain,

$$E_n = \frac{n}{N} E, \quad (27.32)$$

which is the summation of Eq. (27.29). Thus, using Eq. (27.31), we obtain

$$\frac{c}{\zeta} = \frac{N}{n} - 1 \quad \text{or} \quad \frac{\zeta}{c} = \frac{1}{\dfrac{N}{n} - 1}, \quad (27.33)$$

and this is the formal solution of the problem of equalization.

Depending then in a given construction on the given c or ζ, the other shunt chain can be similarly and correspondingly proportioned, according to Eq. (27.33), to produce the equalization aimed for.

We note that the magnitude and influence of the main series capacitance has dropped out completely. The main chain and the two side chains form entirely independent units. The current in the main chain is uniform and the currents in the two side chains are equal at each junction. Thus the conditions of Eq. (27.33) are valid for values of c and ζ that can vary arbitrarily from element to element provided they are balanced in accordance with Eq. (27.33) at each junction. When so balanced, the two side chains send equal currents into the junction, which cancel and allow the voltage distribution to be determined by the uniform main-chain capacitance C. Where one side shunt chain is exposed

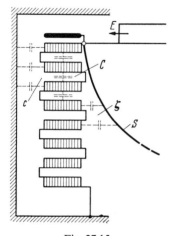

Fig. 27.15

to high voltage, the opposite one has high capacitance, and vice versa. Thus the currents are of equal magnitude and cancel.

Table 27.1 shows the capacitance ratios calculated for a chain insulator having eight elements. These elements may be eight actual members, such as suspension insulators, ribbed insulators, series gaps, voltage-dividing resistors, and the like, or they could be eight circular zone elements on the surface of smooth insulators such as mountings, bushings, slot sheaths in rotating machines, or similar units. The next to last line of Table 27.1 is

for the case of given ζ, while the last one is for the case of given c.

Figure 27.15 shows as an example the use of a conducting shield S to protect the winding of a transformer, consisting of eight pancake coils, against step pulses. We saw in Chapters 19 and 20 that the capacitance to ground c, unlike the constant series capacitance C between coils, produces a very nonuniform distribution of the voltage due to impinging step pulses. This stresses the coils at the entrance terminals particularly heavily. If all these coils are surrounded by a shield S with the shape of a flared horn and its capacitance ζ to each of the coils is in accord with the relations given in Table 27.1, then the voltage concentration in the first coils disappears completely. Because the capacitance is inversely proportional to the distance the shape of the shield in Fig. 27.15 is given by $1/\zeta$. The same equalization can also be produced by conducting rings that surround each of the coils separately and are proportioned to give the proper value of ζ. Still other arrangements are possible.

Apart from such flared or other shields, which protect a large portion of the transformer winding against step pulses, static end shields are often used to protect merely the turns or layers of the first coil, as also shown in Fig. 27.15. Because this first coil contains only a small fraction of the total turns in the winding, the shield needs to be extended over only one or a few of the steps near the line in Table 27.1. Its capacitance ζ must be large and hence its distance small according to Eq. (27.33) and the fact that it must balance the intercoil capacitance C between the first and second coils.

To protect more than only the first pancake coil and its layers, which would transfer the impact farther into the winding, it is possible to use such disk shields between each two

TABLE 27.1. Insulator chain with completely uniform voltage distribution.

Ground	Chain element							Line
$n = 0$	1	2	3	4	5	6	7	8
$N/n = \infty$	8	4	2.67	2	1.60	1.33	1.14	1
$c/\zeta = \infty$	7	3	1.67	1	0.60	0.33	0.14	0
$\zeta/c = 0$	0.14	0.33	0.60	1	1.67	3	7	∞

coils and connect it to the junction between the two coils, as shown in Fig. 27.16. Now the entire system of shields produces a uniform voltage distribution of the step pulse over all layers and coils because a good series capacitance chain is created between the first and

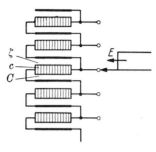

Fig. 27.16

last shields to match the inductive subdivision of the winding coil for coil.

This arrangement is particularly useful for adjustable tap-changing transformers that are in trouble because they have no fixed first coil. As shown in Fig. 27.16, the individual

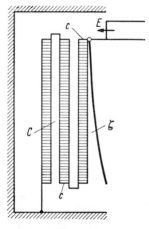

Fig. 27.17

shields now protect each associated coil and the line can be freely changed between taps. Of course the shield occupies a fraction of the important insulating space.

Such capacitive shields can also be used for the windings in slots of generators connected to high-voltage lines. However, this becomes very complicated because they must be wound into the slot insulation so as to increase the capacitance between conductors compared with their capacitance to ground.

For a transformer winding consisting of a few concentric cylindrical layers, as shown in Fig. 27.17, the shield enclosing the top layer is much less flared than that in Fig. 27.15. It can also be slit or otherwise made into an open grid with less surface area as long as the capacitance relations of Eq. (27.33) are met over this first layer, which is only a fraction of the winding.

If in the construction of transformers such shields are used in a region of strong magnetic

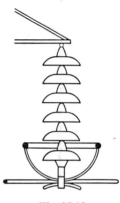

Fig. 27.18

fields, stray or otherwise, they must be made of poorly conducting material or suitably split to keep eddy-current losses low.

Frequently complete equalization is not possible, owing to difficulties in construction, high cost, or many other aspects; then it is necessary to stop short of completion. In most cases shields that cover the first half of the winding are sufficient because some lines of force from the lower end of the flared shield will still produce some effect on the rest of the winding and also the elements in the lower half of the winding experience less stress anyway.

Because Eq. (27.33) and Table 27.1 show that the two shunt capacitances c and ζ should be equal at the center, this gives a dimension for the shield at this point provided the dielectric constants of the two spaces involved are duly taken into account.

Figure 27.18 shows a ring shield at the high-voltage end of an insulator chain. It does not have the exact equalizing value called for by Eq. (27.33), but nevertheless it is frequently used because it increases the capacitance ζ to the line, which is very useful. In Fig. 27.14 the two points for $n = 1$ and 2, shown as squares, indicated that such ring shields were observed during the tests to reduce the voltage across these elements to about one-half of the value without them. Such ring shields for protection against step pulses on the line must be used, as shown, at the line end of the chain, because the grounded end of the chain at the tower is in less danger, as seen in Eq. (27.17) and Fig. 27.5.

However, if the chain is to be protected against the dangerous "backward" flashover, caused by lightning, from the tower to the line, the tower end of the chain is at a very high voltage with respect to the line because, as shown in Chapter 26, the top of the tower is raised temporarily to very high voltages by the lightning stroke. Thus we must now protect these upper elements of the suspension chain insulator by additional capacitance to

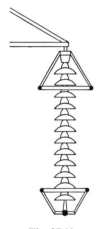

Fig. 27.19

the tower because the normal capacitances, shown in Fig. 27.2, are not of the proper proportion to balance out the very small capacitances to the line as demanded by Eq. (27.33) for proper equalization. Thus a set of stepped ring shields at the tower end of the chain should be used. However, it must be properly designed not to interfere with the ring shield at the lower, line end of the chain.

If, as shown in Fig. 27.19, rings or horns are used at both ends of the chain, they can at the same time act as electrodes for a spark or arc discharge, which may still take place if the voltage becomes so large as to cause a flashover. This discharge will then be kept at some distance from the insulator surfaces, which could easily be damaged by it.

Figure 27.20 shows a flared shield used for bushings on conductors penetrating a grounded

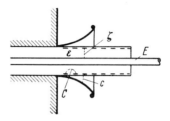

Fig. 27.20

enclosure or emerging from the grounded slots of high-voltage machinery. In both cases the high-voltage conductor is embedded in an insulating sleeve. Here the next to last line of Table 27.1 must be used and the shape of the flared shield is again, as shown in Fig. 27.20, given by $1/c$. It is now grounded. To avoid voltage stresses in the radial direction it is possible to fill the shield with an insulator of high dielectric constant.

Because the different capacitances in the cases of the various shielding arrangements are not always analytically derivable, owing to the complex shape of the lines of force, it is often desirable to obtain their values from measurements with models in an electrolytic tank. This will allow correct design parameters for complex arrangements to be obtained. In the tank the resistance R of an element is measured and the corresponding capacitance C is obtained by the transformation

$$CR = \frac{\varepsilon s}{4\pi v_0^2}, \qquad (27.34)$$

to which must be added a linear scaling factor and in which all dimensions must of course be expressed in the same system of units.

Apart from large series capacitance or the various forms of shields described, the voltage distribution in or across the insulation can also be rendered more uniform in many cases by suitably shaped and distributed conductors embedded in the insulation. Such "condenser-type" feed-through construction is mentioned here merely for the sake of completeness. This principle is used with success for conductors penetrating enclosures or emerging from the slots of machinery. It can even be useful in pancake-coil transformer construction or similar arrangements in resistances or excess-voltage arresters.

CHAPTER 28

PERFORMANCE OF NETWORKS AND WINDINGS

The many separate processes that we have examined in the preceding chapters usually take place in practice at the same time in a network, and accompany any change in the electromagnetic state of the entire network or any of its components. The effects of a disturbance at any point in the network due to a change in load, a lightning stroke, or a switching or other process are propagated as traveling pulses throughout the network and produce observable effects even in its most remote corners. Hence in this chapter we shall examine the behavior of traveling pulses connected with network operation and the precautions necessary in the insulation and protective devices to limit any dangerous effects connected with them.

(a) *Behavior of traveling pulses in the network.* The amplitude of traveling pulses produced by lightning discharges is determined primarily by the properties of the thunderclouds, whereas that of traveling pulses produced by processes within the network is primarily governed by the operating voltage of the system. Every switching-on process results in a step pulse that at its origin may have an amplitude up to the full value of the voltage switched on. The switching-off of alternating current, which usually involves spark ignition and reignition in a circuit containing resistance and self-inductance, may lead to step pulses of amplitude up to twice that of the operating voltage. For circuits containing resistance and capacitance, this may even increase to three times the operating voltage and in the case of oscillatory circuits containing inductance and capacitance it may increase even further.

Pulses that arise anywhere in the network on account of unplanned disturbances are in general more difficult to handle, since they cannot be limited by suitable protective devices at their point of origin, as is possible in the case of planned switching operations. Particularly dangerous is a breakdown to ground in a cable or open line if the network has a high power rating. This is because the ground-fault arc can be reignited every half-period. Furthermore, as seen in Ref. 1, Chapter 44, during the periods between grounding at the fault very high voltages may develop in the lines, which may exceed the high voltages causing the initial breakdown or flashover. Thus a series of large voltage impulses are generated at the fault, which propagate in all directions into the network.

Every network has a large number of oscillatory modes and associated natural frequencies owing to the large number of inductances and capacitances distributed throughout it. All can be excited by the step-pulse charge and discharge processes, which act like hammer blows on a complex mechanical system. The different local sections of the network oscillate with their natural frequencies and change the flat back of the incident step pulse into a periodically shaped form. This disperses the energy of the incident pulse into several fractions in the different frequency regions encountered and so breaks up the incident pulse. The oscillations in the local-circuit portions of the network are propagated also into the network at large and this coupling produces a damping influence upon them. Their frequency can be anywhere in the range of a few thousand to several million cycles per second, depending on the length of line involved or on the size of localized, that is, lumped, capacitances and inductances.

Traveling pulses caused by planned switching operations are usually propagated between two conductors of the system. This is because even in multiphase switching processes closure occurs initially by means of a spark across two contacts, which causes a single loop only to be energized. The distribution of the electric (*e*) and magnetic (*m*) fields in the vicinity of a loop consisting of two conductors due to such a pulse is shown in Fig. 28.1. For the surge impedance applicable here the self-inductance and capacitance of two-wire loops usually leads to values between 600 and 800 Ω. In three-phase circuits the third conductor will

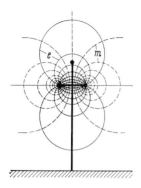

Fig. 28.1

usually be energized about a quarter-period later, when the third contact has also been closed by a spark. Thereupon traveling pulses occur also between this line and the other two, with the field distributions as shown in Fig. 28.2. The surge impedance is smaller than in

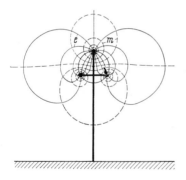

Fig. 28.2

Fig. 28.1 and is between about 500 and 600 Ω.

If a line in the network is shorted to ground the traveling pulses are propagated between the faulty line and ground and the field lines are as shown in Fig. 28.3. The surge impedance here is also about 500 Ω. Such pulse propagation also occurs in planned switching operations if the network has a grounded center point or a large total capacitance, for

then the first switch contact closed by a spark can energize a line whose return to complete a loop is formed by the ground. Because secondary pulses are induced in the other two lines, as shown in Chapter 11, the actual field lines will be slightly changed from those shown in Fig. 28.3.

If, finally, atmospheric discharges in the vicinity of a long three-phase line produce induced voltage pulses on all three lines in common, the three lines are at about the same voltage with respect to ground and the field distribution is as shown in Fig. 28.4. The surge impedance is about 250 Ω. However, direct

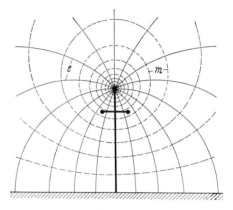

Fig. 28.3

lightning strokes to a line and a flashover from the tower to a line produce traveling pulses whose fields are concentrated, as in Fig. 28.3, only at the faulty line. But a lightning stroke that causes a flashover from the tower to all three lines again has the field distribution shown in Fig. 28.4.

These four different field distributions associated with traveling pulses also occur in three-phase cables between the conductors and between conductors and the sheath. The field distribution for pulses between parallel conductors is identical in shape with those of the steady-state voltages and currents. For different arrangement of the conductors in the cable, however, this is no longer the case and considerable distortion may take place, which will no longer allow us to use fixed values of surge impedance.

Every branching involving equal lines reduces the pulse voltages by refraction and reflection into partial pulses that have a fraction of the incident amplitude. However, a transition to a different line may produce an increase or a decrease, depending on whether the surge impedance increases or decreases. In networks with many branches the repeated splitting of the traveling pulses quickly reduces them to harmless amplitudes.

In contrast with this, the pulses are doubled in amplitude at the end of a line. Thus end

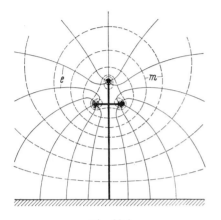

Fig. 28.4

stations, which have only one line entering, are stressed twice as much as intermediate stations, which have two lines, one entering and the other leaving. This danger to end or terminal stations is fully borne out in practice.

If a network has several local regions or circuits that have equal natural frequencies, then, even if a large distance separates them,

Fig. 28.5

it is possible that coupling between them can lead to oscillations excited by a switching process, which may work up to dangerously high voltages. Figure 28.5 shows a short length of cable Z_1 connected to a strong current source of voltage E. It is connected over a long open line Z_2 to a transformer winding Z_3, which is protected by a choke coil L and bypass condenser C. After the switch is closed the cable Z_1 oscillates at its end between voltages of $2E$ and 0 and thus a periodic pulse

train of this amplitude enters the open-line Z_2. The frequency ν_1 is determined by the length of the cable Z_1, and the pulse train decays gradually owing to the dissipation of energy in various impedances. The pulse train impinges at the transformer on the oscillatory LC circuit and if its frequency ν_1 is equal or close to the natural frequency ν_{LC} of the circuit, it will excite this circuit in resonant fashion to produce voltages that are governed by the surge impedances and the resistances of the components and lines involved. The voltage may become several times the normal operating voltage if the oscillatory-circuit impedance is much higher than the surge impedance of the feeding line. The line then can act as a fairly heavy source. Such cases have been observed in networks under short-circuit and ground faults, which give rise to pulse trains owing to spark formation.

Traveling pulses of all kinds can to a certain extent be transferred between the windings of a transformer in both directions, as calculated in Chapters 11 and 22. If, ideally, the wires of the primary and secondary windings were continuously parallel, the transfer would not distort the pulse shape. However, in power transformers the primary and secondary windings are subdivided into separate coils, so that the linkage of the two windings is far from the ideal and in fact is quite irregular. Thus in practice the pulses undergo a reduction in amplitude and a smoothing out of shape during transfer between windings. Pancake windings act more favorably in this respect than concentric cylindrical windings, since they have closer coupling. However, they have other disadvantages.

If pulses travel over long distances on a homogeneous line, the resistance of the line conductors and the ground return, as well as the glow currents in the insulation, cause losses that reduce its energy content and thus its voltage amplitude. All rapid or step pulses are confined by their high-frequency nature to the outer skin of conductors, so that they encounter much higher effective resistances than the dc resistance of the conductors. Thus fast pulse trains, and especially step pulses, are always attenuated much more heavily than the slowly alternating operating currents during propagation in the network.

As long as the front-rise length of a step pulse is shorter than a turn of the coil winding, as shown in Fig. 28.6, the actual length k does not affect the stress on the insulation between turns. Instead, the voltage between

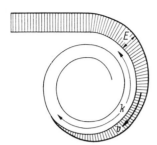

Fig. 28.6

turns becomes the full pulse voltage E after the rise. Only a front-rise length greater than the turn length begins to reduce the stress on the turn insulation. Thus a flattening of the pulse fronts due to skin or other effects will be useful only if it spreads the rise over lengths

Fig. 28.7

considerably greater than a turn. For turns 5 m long this is the case only after a passage over a 20-km two-wire line or a 1-km ground return.

The skin effect also considerably reduces the voltage amplitude of pulses. For example,

a ground distance of 20–30 km reduces it to about half its initial value. If the voltage exceeds the corona-voltage limit, a further reduction will occur that will halve the amplitude again over a distance of 15 km.

The insulators and towers of open lines and junction boxes in underground cables represent a localized discontinuity in the capacitance

Fig. 28.8

or surge impedance and thus give rise to weak transmission and reflection. In this fashion any sharp initial step rise is split into series of small steps, which reduces its abruptness considerably. Any other irregularities in the lines, which will always be present in actual construction, will produce a distortion and flattening of the pulse front that will increase with the distance traversed.

Lines supported by wooden poles are somewhat less liable to ground faults, owing to the added insulation introduced by the towers. However, strong lightning strokes can splinter an entire row of towers or poles, as shown in Fig. 28.7. This is because the high voltage, of several million volts, vaporizes the small amount of moisture in the wood and thus explodes the wood fibers.

For lines supported by steel towers, the major damage is done by the electric arc that follows an insulator flashover. For high voltages it can be kept away from the insulator surface by ring electrodes or horn gaps, as shown in Fig. 28.8. This protects the insulators from thermal damage by ground fault or short-circuit arcs. At the same time the horn or ring conductors on the line produce a more uniform voltage distribution across the insulator chains, which increases the breakdown voltage for step pulses. If the arc is not extinguished as quickly as possible by switching or inductive fault suppression, it can transfer to the actual line conductors. It may then

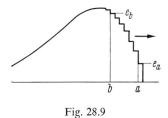

Fig. 28.9

travel along the lines away from the generator, driven by electrodynamic forces, and may do damage by melting over considerable stretches of the line.

For line insulators the spark delay is made as small as possible so that when the line is struck by lightning the flashover may occur during the initial step e_a shown in Fig. 28.9, and not some time later, when the full voltage e_b has developed. The smaller the delay the shorter are the pulse fronts that are propagated along the lines. These advantages are especially applicable to lines on steel towers connected by ground conductors because they will conduct the subsequent main portion of the pulse rapidly to ground. Ideally, it would be even more desirable to make the delay or flashover voltage so high that even the highest voltages in the tower or ground conductors due to lightning could not cause a breakdown. However, this is economically not feasible. For insulators in stations of all kinds the spark delay is made reasonably large

to obtain some resistance against even high but short lightning pulses.

(*b*) *Impulse insulation of windings.* If traveling pulses impinge on the windings of transformers, rotating machinery, or other apparatus, they can cause serious damage, partly because their voltage is usually increased in amplitude by reflection at the high-impedance entrance of the winding and partly because the space within the winding restricts the amount of insulation that can be used. Depending on their shape and voltage amplitude, the traveling pulses may cause flashover

Fig. 28.10

or breakdown between a winding and a grounded conductor or another winding on the high- or low-voltage side and also between coils, layers, or turns within a winding. Figure 28.10 shows the insulation surrounding a group of conductors in the slot of a generator that has broken down to the grounded walls of the slot in many places.

If the front-rise length of a step pulse is shorter than the length of the first turn of a winding, the insulation between the first and second turns could break down if it were merely proportioned to sustain its share of the normal operating voltage and the pulse had an amplitude roughly equal to this network operating voltage. If the pulse front is not so abrupt but is distributed over a greater length and hence rises more gradually, as shown in Fig. 28.11 where it is distributed over eight turns, the stress on the insulation between adjacent turns is much lower. The voltage drop across a turn is equal to the change in voltage along the pulse front over an equivalent length and thus the maximum value can be obtained from the pulse shape if it is known, as shown in Fig. 28.11.

In windings with short turn lengths the voltage drop in the pulse front is often small

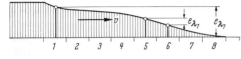

Fig. 28.11

enough not to rupture the insulation between turns. However, if the coil is wound in layers, a breakdown between layers can occur more easily. If, as shown in Fig. 28.12, each layer has four turns, the voltage difference between turns 1 and 8, which are adjacent as shown, is very much higher than that between successive turns. It can be equal to the full amplitude

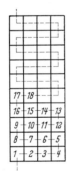

Fig. 28.12

of the impinging voltage step pulse if its front-rise length is not more than the length of seven turns, as shown in Fig. 28.11, that is, less than the length of wire between turns 1 and 8. For even longer, that is, more flattened pulse fronts, the insulation between layers may become safe but a breakdown between the first and second coils of a winding may occur if the insulation between them cannot withstand the full pulse voltage.

While step pulses most frequently cause breakdown in the entrance turns, layers, or coils of windings, periodic pulse trains can cause breakdown also well within the winding.

This is because step pulses cannot penetrate into the windings on account of flux linkage in the case of transformer, relay, and similar windings and the repeated internal reflections between the slot-embedded and end-turn portions of rotating-machinery windings. However, periodic pulse trains up to quite high frequencies can penetrate the entire winding with relatively little attenuation. If the insulation between any turns that are a half-wavelength of the periodic pulse train apart is not sufficient to withstand its voltage amplitude, a breakdown will occur between them. Such breakdowns usually occur between coils because in the main they offer the desired combination of small spacing and long wire length.

If the normal operating voltage across the two points between which a breakdown has

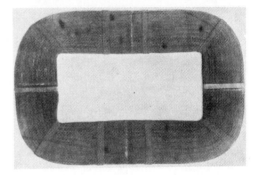

Fig. 28.13

occurred is sufficient to maintain an arc across the breakdown gap, the arc will be maintained even after the traveling pulse has passed. This produces an internal short-circuited turn which can very quickly cause the turn or coil to burn out, particularly as the short circuit occurs suddenly and thus will give rise to a large short-circuit current impulse. However, if the normal operating voltage cannot maintain an arc across the breakdown gap, the insulation is only punctured by the traveling pulse and the coil can still be useful for a long time. This is usually the case for breakdown between successive turns.

Figure 28.13 shows the entrance coil of a small transformer in which a number of breakdowns between layers can be seen. We note only puncture and charring because the regular operating voltage was low enough to cause no further damage. In contrast, Fig. 28.14 shows the entrance coil of a large

Fig. 28.14

transformer in which traveling pulses initiated an arc that was sustained by the regular operating voltage and resulted in a very substantial burn-out. Figure 28.15 shows the

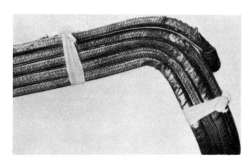

Fig. 28.15

end winding of the first coil of a medium-sized generator that has undergone marked melting on all turns. Here again an arc was not maintained. In contrast, Fig. 28.16 shows the

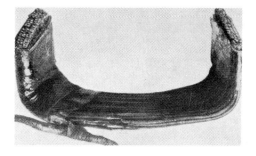

Fig. 28.16

entrance coil of an induction motor that was burned out in the first few turns by a pulse-initiated breakdown. Figure 28.17 shows a relay coil that has also had several turns burned out by a traveling pulse.

Because it is impossible in practice to avoid completely the impact of traveling pulses

Fig. 28.17

upon windings of transformers and other machinery, it is good practice to provide extra insulation on the turns comprising the first 5–10 percent of a winding. In cases where traveling pulses can arrive between the lines and ground, and thus especially in the case of transformers, it is customary to provide this strengthening of the insulation also at the star point and thus at both ends of the windings. It is usual to make the insulation between these terminal turns sufficient to withstand the full operating voltage of the network for short periods. To provide the necessary space, the end coils are made to contain fewer turns. Figure 28.18 shows the cross section of an inner coil I and an end coil E of a high-voltage transformer. In the end coil the individual conductors are surrounded by several layers of paper insulation. Figure 28.19 shows the cross section of an inner coil *I* and an end coil *E* of a high-voltage generator. In the end coil the turns are separated by thick insulating layers. All intervening spaces must be filled by an insulating compound to eliminate air, which would be decomposed in the strong electric field, leading to chemically active and hence damaging compounds. This increased insulation must of course be carried all the

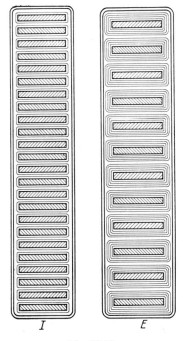

Fig. 28.18

way in the overhang portions of the coils as well as in the slot portions. Figure 28.20 shows the cross section of an inner and an end coil of a medium-sized high-voltage motor. The many turns are arranged in layers, of which there are fewer in the end coil because extra insulation is inserted between layers.

In strengthened end coils the stress decreases only inversely as the square root of the thickness of the insulation, as shown in Chapters 20 and 21. This is because the thicker insulation reduces the important internal capacitance, which increases the voltage drop across an end turn in proportion to the square root of the thickness of the insulation. Thus there are

certain limits to the extent to which it is possible to strengthen end coils.

It is possible to produce more effective results by using at the ends of windings insulating materials of higher dielectric constant between turns, layers, and coils to maintain

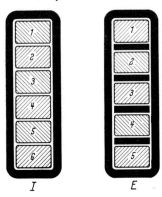

Fig. 28.19

high internal capacitance but still with a relatively low capacitance to ground by the use of materials of lower dielectric constant in these paths. This reduces the voltages developed by step pulses without other side effects at the end coils.

To obtain the smallest possible voltage difference between layers, it is useful to wind

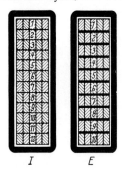

Fig. 28.20

coils of many turns not in long layers, as in Fig. 28.21, but in short layers, as in Fig. 28.12. The stress on the insulation is already reduced to a fraction by this simple measure. It is totally unsuitable to use random windings, as shown in Fig. 28.22, when traveling pulses are to be encountered. This method was formerly used for small transformers, machinery, voltage dividers, and coils of relays and other control gear. In such irregular coils two adjacent wires may span a length of several

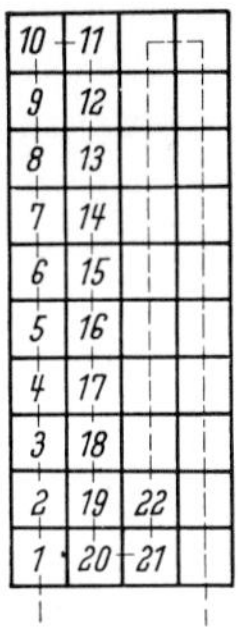

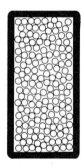

Fig. 28.21

turns and thus can easily experience a breakdown, which also may become a sustained short circuit.

Another way to protect the entrance coils of transformers against high voltages caused by incident step pulses is to use shields or rings. These distribute the incident step-pulse voltage capacitively over a large number of coils instead of causing the full impact at the end. Figure 28.23 shows several methods of

Fig. 28.22

implementing this principle as used in American transformer construction (J. M. Weed 1915; L. F. Blume and A. Boyajian 1919).

The upper plate above the first coil is connected directly to the line and balances the capacitance of the layers of this coil to ground, as described in Chapter 27. The protective shield rings around the upper coils similarly balance the ground capacitance of each of the

coils and the lower rings will have lower effective capacitance as shown to be necessary in Chapter 27. Finally, the very lowest coils are enclosed by a single conical shield, which again tends to give conditions close to the equalization discussed in Chapter 27. We see that several additional elements are involved

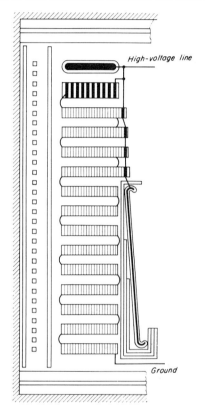

Fig. 28.23

in such a construction that require careful arrangement of the insulation as the high voltage is taken all along the outer surface.

The practical application of the findings developed in Chapter 19 led to much progress in the construction of transformers. As soon as the effect of internal capacitance was discovered (R. Rüdenberg 1914, 1923; K. W. Wagner 1915), round conductors were replaced by rectangular ones, and flat-wound coils (K. Kurda 1915) led to constructions of much greater strength against step pulses. Soon thereafter strengthened end coils were introduced for transformers and rotating machines (W. Zederbohm 1921). Favorable layer arrangements have been used for some time in shell-type transformers (J. K. Hodnette 1930; H. V. Putnam 1932) and when their advantageous capacitance relations were recognized they were also used in core-type transformers (J. Biermanns 1937; R. Elsner 1939).

Figure 28.24 shows such a layer winding that is suitable for the highest voltages. The low-voltage winding is close to the core of the transformer. It is surrounded by cylindrical

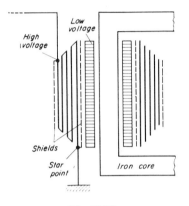

Fig. 28.24

high-voltage coils that are arranged concentrically and thus have the largest possible coil-to-coil capacitance. The ends of the coils are staggered, with the shorter coils at the outside to reduce the proximity of the higher-voltage coils to the core. Slit shields around the end layers, which are connected to the line and ground, are used to distribute incident voltage step pulses over these layers, as was shown earlier. Owing to the high capacitance between layers, the winding acts as a bypass condenser for such pulses anyway with fairly uniform voltage distribution and thus little chance for trouble between the coils or windings. Only along the cylindrical coils is it possible to encounter problems due to nonuniform voltage distribution; however, they can be handled by suitable internal capacitance distribution.

To be sure that the individual coils as well as the complete winding when assembled have sufficient strength in their insulation to withstand traveling pulses encountered in practice, it is usual to conduct a step-pulse or impulse test by means of a condenser discharge, which exposes the winding to several times the normal operating voltage (G. Stern 1915; E. Marx 1924).

This test can be performed with the sharp front of an artificially produced step pulse, by means of the arrangements described in Chapter 23, Sec. (*c*), or it is possible to use a discharge step initiated by the back portion of a test pulse, that is, an interrupted pulse whose cut-off slope now tests the windings. This method is similar to the insulator flashovers that will occur on account of incident lightning strokes in their vicinity or in a station.

(*c*) *Protective devices against traveling pulses.* Dangerous pulses are produced in high-voltage networks, sometimes by normal switching operations, more frequently by short-circuit faults, and above all by ground faults and the associated capacitive effects. Particularly dangerous, however, are the direct and indirect effects of lightning strokes which occur frequently in open lines of low, medium, and high voltage rating. For all parts of the network the voltage amplitude of the pulse constitutes a dangerous stress on the insulation. For windings of all types the rate of voltage rise and decay also determines the severity of the stress on the winding insulation. The short duration of most traveling pulses is very beneficial because the delay time for breakdown of the insulation is such that serious damage is often avoided.

Apparatus, transformers, and rotating machinery can be protected against excess voltage and traveling pulses by resistances, inductive chokes, bypass condensers, and spark gaps of various types. They can be concentrated at one location or distributed over some distance in the system. They can be made to have parameters that are constant or that vary with either current or voltage.

Current transformers, relay coils, and similar devices connected in series with power lines are hard to insulate between turns to withstand pulses of a voltage amplitude several times the system operating voltage. These will occur if these windings are in the vicinity of high capacitances such as bus bars or feedthrough bushings and form oscillatory circuits with these. As Fig. 28.25 shows, it is possible to connect a shunt resistance that bypasses the traveling pulses and even attenuates them somewhat. At the same time it has little

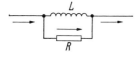

Fig. 28.25

influence on the regular operating current owing to its lower frequency. Figure 28.26 shows a current transformer so protected.

If the slight shunt operating-current losses are to be avoided completely, a sufficiently large shunt bypass condenser can be used (Fig. 28.27) instead of the resistance. This

Fig. 28.26

bypasses all high-frequency components and thus keeps them out of the winding where they might be dangerous. This arrangement is very useful in the case of series transformers to avoid the constant loss of energy. It would be still more effective to distribute this capacitance uniformly across the winding (Fig. 28.28)

to obtain a high winding capacitance, except that the many tap connections necessary prevent the use of this arrangement save in rare cases. However, in the case of step-up and regulating transformers, which have a number of taps anyway, this arrangement is the usual protective one (Fig. 28.29).

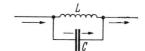

Fig. 28.27

Fig. 28.28

In end stations, transformers are often connected to a single incoming transmission line. To avoid a doubling of the voltage due to the high impedance of the transformer winding, a sufficiently large bypass condenser

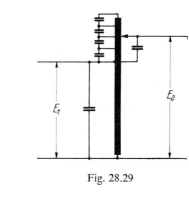

Fig. 28.29

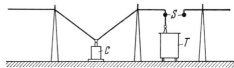

Fig. 28.30

or other protective device can be connected to the bus bar. To afford protection against very steep pulse fronts, the connection to the condenser and ground should be kept as short as possible. For open-air installations one can connect the incoming line first to the protective device and then directly to the transformer (Fig. 28.30). For installation in a building (Fig. 28.31), the protective device can be installed close to the transformer and

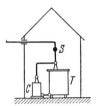

Fig. 28.31

connected by short leads to its entrance terminal and grounded tank.

Choke coils and bypass condensers have the same effect on the front of a traveling pulse but by means of quite different action. Formerly it was customary to use choke coils freely at junctions where pulses traveled into

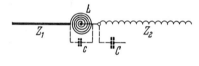

Fig. 28.32

Fig. 28.33

higher impedances, which would lead to voltage increases up to doubling. This is shown in Fig. 28.32 at the terminals of a transformer. Figure 28.33 shows such a high-voltage choke coil, which is wound in the form of a spiral of flat copper tape. This type of winding has a large internal capacitance γ, which reduces the effectiveness of the coil

below that implied by the size of its self-inductance. In addition the choke coil with its capacitance c and the terminal capacitance C of the transformer may produce oscillations owing to unfavorable parameters, which frequently produce damaging rather than protective effects.

Condensers can be wound with less distributed inductance if the conductive layers are connected in parallel. In this way dangerous oscillations can be avoided and the use of

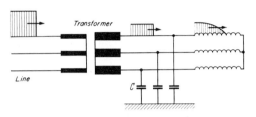

Fig. 28.34

bypass condensers at the terminals of transformer and machinery windings is safer and thus frequently encountered. Such condensers are particularly effective when connected between the secondary winding of a high-voltage transformer and a generator (Fig. 28.34). They then flatten the steep pulse fronts arriving from the line via the transformer as

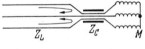

Fig. 28.35

shown and thus reduce the stress on the insulation of the generator winding.

This protection is most generally used in the case of single transformer-generator combinations because the secondary pulses are not split up among several generators. To limit also the voltage amplitude of the pulses flattened by the condenser, a shunt discharge resistance across it is frequently used.

Protective cables of at least 20-m length, as shown in Fig. 28.35, also offer good protection against traveling pulses and are frequently used for high-voltage motors because these have insulation problems owing to the large number of conductors in a single slot. Such cables should have low surge impedance, in which case they reflect step pulses, both between lines and between lines and ground, almost completely, with a strong reduction in amplitude. Thus they act essentially as

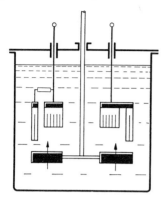

Fig. 28.36

bypass condensers but have the advantage that their performance, particularly with respect to high-frequency components, is not hampered by lengthy connections.

The generation of large traveling pulses during switching of machines, transformers,

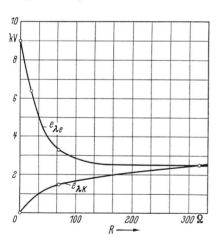

Fig. 28.37

and lines can be avoided by the use of circuit breakers, as shown in Fig. 28.36, which use a protective resistance connected to a preliminary contact that provides the initial and final switching action. When the circuit is

closed it reduces the amplitude of the step pulses, which is shown in Fig. 28.37 for an induction motor by means of the voltage across the winding $e_{\lambda e}$ during initial closing and $e_{\lambda k}$ when finally closed. When the circuit is disconnected it reduces the step pulses produced by reignition and the magnitude and thus the danger of the electric arc formed between the opening contacts. Figure 28.38

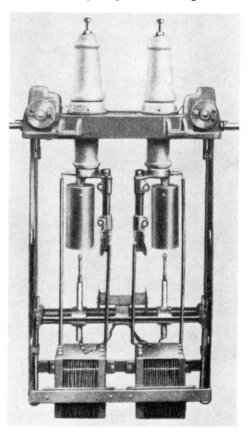

Fig. 28.38

shows the interior mechanism of a single-pole protective circuit breaker that is to operate immersed in oil.

The resistance and preliminary contact occupy considerable space, which makes the container of the circuit breaker rather large. In addition, they represent a considerable complication in construction when built to withstand the short-circuit currents that must be interrupted in large modern networks.

On account of these factors such circuit breakers were avoided for many years, but they are now coming into general use because they lead to the suppression of switching pulses at the point of origin.

On high-voltage open lines the danger from traveling pulses induced indirectly by the influence of lightning strokes frequently does not exceed that from pulses produced by switching and ground faults. In contrast, a direct lightning stroke, with its high amplitude and steep-sided impulses, leads to the highest stress on the insulation of the line. The excess voltages produced by it can be kept down during their build-up at the origin by the following two means. One is to use ground conductors so as to avoid direct strokes on the line conductors and to interconnect the towers. The other is to make the ground resistance of the tower footings as low as possible. Each by

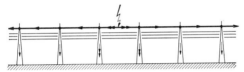

Fig. 28.39

itself already produces a significant reduction in voltages due to lightning. The use of both, together with the lowest surge impedances obtainable, can make the lines proof against even quite heavy lightning strokes.

As shown in Fig. 28.39, the ground conductors essentially produce a ground potential with a strong field concentration above the line conductors and thus considerably reduce the likelihood that a stroke will enter the line conductors directly. They further distribute the lightning current over a number of adjacent towers so that the sustained voltage at the point struck is governed by the lower impedance of all these towers connected by the ground conductors in parallel. Finally, the capacitive coupling between the ground and line conductors reduces the voltage between the towers and line conductors, which in turn reduces the stress on the line insulators. All of these effects are enhanced if several ground conductors in parallel are used.

However, it is not advisable to increase the coupling between the ground and line conductors above 50 percent, because the excess-voltage pulses on the line conductors are propagated far along the lines while those on the ground conductors are confined by the adjacent towers to the vicinity of the point struck. Thus, while a stronger coupling would give more protection in the vicinity of the lightning stroke, it would produce large stresses on the line insulators at distant towers. Because the different line conductors have different coupling to the ground conductors, lightning flashover is most frequently observed for those farthest from the ground conductors on account of their lower coupling. Usually these are the conductors lowest on the tower. However, because insulator flashover depends on the instantaneous voltage record, which is also in part due to the operating voltage, other conductors may also sometimes be affected at the instant of the stroke.

The ground resistance of the towers becomes lower the larger the dimensions of the footing. Thus high-voltage lines with large towers have lower lightning voltages generated in them than medium- or low-voltage lines. Moreover, since their insulators are built to withstand the higher operating voltages, it is clear that the susceptibility of lines to damage by lightning decreases with increasing operating voltage. Every improvement in grounding by additional grounding rods or counterpoises buried near the footing reduces the lightning voltage produced and thus the danger of breakdown. If ground conditions vary so that some single towers have high ground resistance, this can be compensated by better ground conductors to adjacent towers in such areas. If a low ground resistance cannot be achieved, as in very rocky terrain, the capacitive coupling to the underlying strata provided by a wire grid can still be useful.

In the case of especially high towers for very high-voltage lines, the time delay at the top preceding the influence of the well-grounded footing may be so large as to cause insulator flashover. A larger number of ground conductors with a low combined surge impedance can help in this case.

Ground conductors suspended above open-air switching and other installations connected to large-mesh buried grids can protect against lightning strokes. If the first few kilometers of line adjacent to such stations are protected by a few extra ground conductors above the number used on the more distant stretches of the lines in general, this will further protect the station against the strong effects of lightning strokes that occur nearby. The pulses produced by strokes farther out along the lines are reduced in strength and abruptness by the effects of corona and ground resistance before they reach the stations.

To avoid flashover in the insulation of outdoor installations, the strength of the insulators must be adequate with respect to

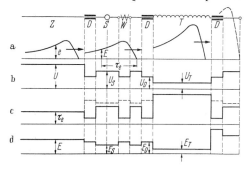

Fig. 28.40

both the flashover voltage and the time lag of discharge so as to withstand the traveling pulses encountered. Since the voltage amplitude and time duration of pulses depend on the flashover voltage and time lag on the lines, it is necessary to proportion the insulation in the stations with due regard to that of the lines in these two respects. Figure 28.40a shows a shock wave produced by a distant lightning stroke arriving along the open line Z. We shall examine the most dangerous case of termination of the line in an end station.

The capacitance of the feed-through bushings D, bus bars S, and instrument transformers W flatten the steep pulse fronts. However, the relatively smooth pulses shown are not significantly changed. Only upon entering the transformer winding T does the pulse experience a significant change, which is an increase in amplitude to almost twice its value, and at the star point it is further increased to about three or four times its incident voltage amplitude. To simplify matters,

we shall disregard the reflected pulses, which do not change matters much in principle.

The flashover voltage U of the line insulators is usually made fairly high to keep the number of breakdowns along the line low. The flashover voltage of the bushings and instrument transformers is often lower and the breakdown voltage of the transformers and windings lower still because the space available for insulation is limited by the construction features. Figure 28.40b shows the value for the different elements. In contrast, the discharge time delay τ_e in transformer oil is very large and can be made large for bus bars and other connectors with some available space and suitable insulating materials. However, for bushings and instrument transformers it is sometimes smaller than the value on line insulators, on account of space limitations in construction. Figure 28.40c shows the values for the different elements of the installation for a pulse of the same shape and amplitude.

Figure 28.40d now shows the voltage E in the pulse after the discharge time lag τ_e which applies to each of the elements. We see that in spite of the high voltage in the transformer and the limited long-term strength of the insulation, the large time lag of the oil prevents a breakdown. In contrast, the star-point connection in air is in great danger owing to both high voltage and low time lag unless it is made extremely short. Considerable danger also exists at the instrument transformers and for all feed-through bushings in transformers and circuit breakers, as well as walls and ceilings for indoor installations. Thus it is good design practice to make these units as nearly as possible equal to the open line with respect to breakdown voltage and time lag to avoid breakdowns. We also note that increasing the line insulation in these two respects will on the one hand reduce the number of line breakdowns but on the other hand considerably increase the danger of the few, now larger and longer, pulses due to line breakdown in the stations.

Conditions are particularly unfavorable with respect to indoor stations. The insulators of the open line have a much lower strength at the operating frequency when their surfaces are wetted by rain. Thus they must be designed to be safe during rain and then in dry weather they acquire an unnecessarily high flashover voltage, which may be 30–100 percent higher than that during rain. The impulse strength for short traveling pulses is then higher by another 20–30 percent. For the insulation in indoor stations the impulse strength is also usually 20–30 percent above that of the dry

Fig. 28.41

operating voltage strength at 60 cy/sec because there is no need to provide additional strength against rain. We see that it will be very useful to increase the time lag from the original value, shown in Fig. 28.40c by a solid line, to the higher value indicated by a dashed line. This is possible by the careful choice and disposition in space of the insulating materials, and by keeping the design such that the highest field concentrations occur in air away from the insulator surface discharges.

All these points are most important in the design of end stations because of the voltage doubling due to reflection. For a simple junction station they are still of considerable importance but become less so for bus bars at which three or more lines join because the transmission due to the lower parallel surge impedance produces less severe voltages in such stations from pulses arriving along a line. Such stations then experience far fewer difficulties due to insulation flashover or breakdown.

To prevent the occurrence of excessive voltages on transmission lines protective spark gaps have been in use for a long time, connected between a line and ground or between lines. The air gap is usually adjusted to break down when the voltage is 1.5 to 2 times the peak operating voltage. It should always be connected with a damping resistance to prevent almost complete reduction of the voltage, which would amount to a short circuit. Such a short circuit would itself give rise to high and dangerous currents and by its sudden formation to discharge voltage pulses that could cause dangerous conditions similar to the excess voltages against which protection was sought by the use of the spark gap. Figure 28.41 shows an early form of spark gap for a three-phase installation. The electrodes are shaped in the form of horns along which the arc formed travels upward to become longer and finally to be interrupted.

Such spark-gap arresters are effective safety valves for all longer-duration excess voltages in the network such as are produced by charging processes of any kind, resonance phenomena, or similar processes. However, they cannot suppress the basic processes that give rise to these excess voltages. Traveling pulses due to discharges and hence producing reduced voltage cannot of course activate the spark-gap arresters, and even short voltage impulses fail to do so on account of the discharge time lag. Thus spark arresters must be designed to have the shortest possible time lag, which is accomplished by the use of homogeneous electric fields in the gap and suitable materials at the electrode surfaces.

Figure 28.42 shows the voltage characteristic of a spark arrester at the junction of two lines. The transmitted voltage e_{v2} depends in addition to the incident voltage e_{v1} on the construction of the lines and the damping resistance at the junction. Prior to spark ignition it rises steeply to the ignition voltages e_{z2} and e_{z1} and thereafter less steeply. If the traveling-pulse voltage exceeds the ignition voltage e_z, current will flow in the damping resistance and reduce the voltage at the junction to e_k. By making the damping resistance

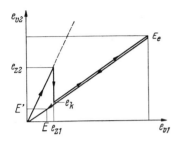

Fig. 28.42

a suitably small fraction of the surge impedance, a large reduction can be produced. However, the junction voltage will now rise further on account of linear behavior of the damping resistance if the incident pulse has a high voltage. It will rise to E_e as shown, and may exceed the ignition voltage considerably. During the decay of the pulse the voltage retraces along this same line, of lower slope, which characterizes the behavior of the arc

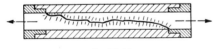

Fig. 28.43

formed. The arc must be extinguished quickly because the voltage drop caused by it reduces the voltage at the junction well below the operating voltage E'. Horn gaps, such as the one shown in Fig. 28.41, take a relatively long time for extinction and are practical only for currents up to 10 to 20 A because of the long open-air arcs formed. If the extinction is transferred to automatic switches immersed in oil, it is improved considerably, though at the expense of this more complicated construction of the arrester.

On long lines tubular arresters of the type shown in Fig. 28.43 are frequently used. Here

the arc, once ignited, vaporizes gases out of the walls of the tubular insulation, which by convection and turbulence are sufficiently cooled to prevent reignition of the arc after passage of the current through zero.

Spark arresters that are used repeatedly require good cooling of the electrodes. These electrodes are then built as a stack of disks in series and the spark and arc are drawn across the circular edges between disks. To extinguish the arc current rapidly and surely, it is possible to use a pulsed crossed magnetic field, which will also change the terminal points of the arc on the electrode surfaces and produce better cooling.

More favorable conditions for extinction are obtained by the use of nonlinear arresters. The voltage characteristic at a junction for such an arrester is shown in Fig. 28.44. Here

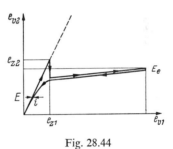

Fig. 28.44

a series spark gap is also used for ignition and the actual characteristic is shown in its rounded form as observed in practice. During the arrival of the pulse the behavior is similar to that of the spark gap with series resistance. However, owing to the nonlinearity the effective resistance can be made very small, so that, as shown, the characteristic after ignition rises only very slowly and so even large incident voltage pulses e_{v1} produce only a slight excess voltage E_e at the junction. During the decay of the pulse the knee in the nonlinear characteristic produces only a very small current in the arrester when the voltage at the junction has returned to the operating voltage E. Thus no serious voltage reduction occurs at the junction and the arc is rather easily extinguished as soon as the voltage at the junction is lower than the ignition voltage.

Actual nonlinear arresters have characteristics somewhat different from those shown in idealized form in Fig. 28.44. Figure 28.45 shows the voltage-current characteristic for an arrester made of a semiconductor material whose resistance varies as a function of voltage. Figure 28.46 shows the voltage-current characteristic for an arrester consisting of short corona-type sparks between carborundum electrodes. Because both have a small current after the voltage has returned to its normal

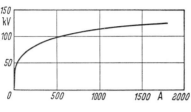

Fig. 28.45

operating value, it is necessary to connect a spark gap in series for complete extinction in spite of the nonlinear action. The total combined characteristic then also has an ignition peak, as shown in Fig. 28.44.

The ignition time delay of such nonlinear arresters must be small to protect a bus bar,

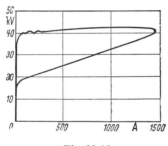

Fig. 28.46

as shown in Fig. 28.40, during the rise time of the incident pulse. An actual arrester is shown in Fig. 28.47; it consists of a spark gap f and a stack of nonlinear resistance elements p and is connected to the line at u and to the ground at e. For a voltage three to five times the normal operating voltage it has a time lag of only a fraction of a microsecond. As the rise time of pulses produced by lightning is a few microseconds after progress along a few kilometers of line, such arresters act to suppress such pulses before they have risen very much. Thus it is usually sufficient to connect an

arrester at the bus bar of a station, as in Fig. 28.40, to protect the entire station sufficiently.

Figure 28.48 shows the installation of such arresters, using short connections, to the high- and low-voltage windings of the entrance

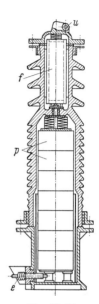

Fig. 28.47

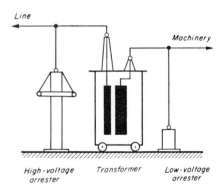

Fig. 28.48

transformer to obtain as much excess-voltage protection as possible for the transformer and the connected machinery and other equipment. Together with ground conductors above the line and station to reduce the lightning voltage and bypass condensers or lengths of cable connection on the primary and secondary sides of the transformer, such arresters afford the best protection available nowadays against flashover or breakdown at undesirable points in the network or installation.

Lightning strokes that hit a line close to a station must often produce a flashover if the voltage strength or time lag of the insulators or the tower grounding and ground-conductor protection are not sufficient. In this case, then, it is desirable to let this flashover take place in a sphere gap at a chosen location. While by this method we cannot avoid an interruption of normal operation due to the flashover, we can avoid damage to the most important portions of the installation at the station provided the relations explained in Chapter 7, Sec. (g), can be satisfied with respect to the spacing used.

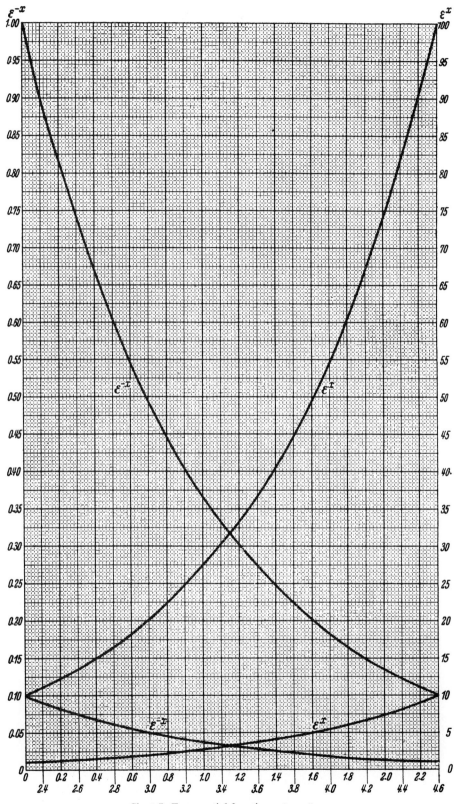

Chart I. Exponential functions ε^x, ε^{-x}

316

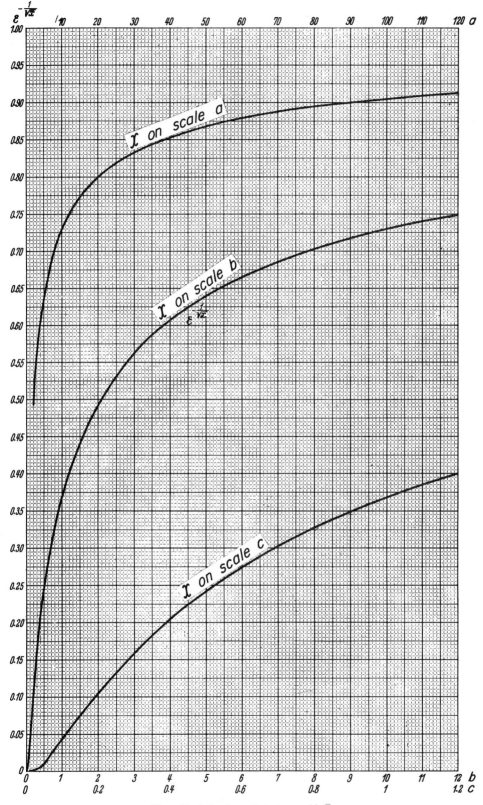

Chart II. Pulse-front function $\varepsilon^{-1/\sqrt{x}}$

317

REFERENCES

Introduction; General

Rüdenberg, R., *Transient Performance of Electric Power Systems* (New York: McGraw-Hill, 1950); cited as Ref. 1.

Thomson, J. J., *Recent Researches on Electricity and Magnetism* (Oxford: Clarendon Press, 1893).

Heaviside, O., *Electromagnetic Theory*, 3 vols. (London, 1899; 3rd ed., New York: Dover, 1950).

Abraham, M., and A. Föppl, *Einführung in die Maxwellsche Theorie der Elektrizität* (Leipzig: Teubner, 1907).

Steinmetz, C. P., *Theory and Calculation of Transient Electric Phenomena and Oscillations* (New York: McGraw-Hill, 1908; 3rd ed., 1920).

Wagner, K. W., *Elektromagnetische Ausgleichsvorgänge in Freileitungen und Kabeln* (Leipzig: Teubner, 1908).

Linke, W., "Über Schaltvorgänge bei elektrischen Maschinen und Apparaten." Dissertation, Techn. Hochschule, Hannover, 1911.

Petersen, W., *Hochspannungstechnik* (Stuttgart: F. Enke, 1911).

Peek, F. W., Jr., *Dielectric Phenomena in High-voltage Engineering* (New York: McGraw-Hill, 1915; 3rd ed., 1929).

Schumann, W. O., *Elektrische Durchbruchsfeldstärke von Gasen* (Berlin: Springer, 1923).

Roth, A., *Hochspannungstechnik* (Berlin: Springer, 1927; 4th ed., 1959).

Whitehead, J. B., *Lectures on Dielectric Theory and Insulation* (New York: McGraw-Hill, 1927).

Binder, L., *Die Wanderwellenvorgänge auf experimenteller Grundlage* (Berlin: Springer, 1928).

Bush, V., *Operational Circuit Analysis* (New York: Wiley, 1929).

Carson, J. R., *Elektrische Ausgleichsvorgänge und Operatorenrechnung*, trans. F. Ollendorff and K. Pohlhausen (Berlin: Springer, 1929).

Engel, A. v., and M. Steenbeck, *Elektrische Gasentladungen* (Berlin: Springer; vol. I, 1932; vol. II, 1934).

Küpfmüller, K., *Einführung in die theoretische Elektrotechnik* (Berlin: Springer 1932; 6th ed., 1959).

Willheim, R., *Das Erdschlussproblem in Hochspannungsnetzen* (Berlin: Springer, 1936).

Skilling, H. H., *Transient Electric Currents* (New York: McGraw-Hill, 1937).

Hak, J., *Eisenlose Drosselspulen* (Leipzig: Hohler, 1938).

British Electrical Research Assoc., *Surge Phenomena* (London, 1941).

Westinghouse Electric Corp., *Electrical Transmission and Distribution Reference Book*, 3rd ed. (East Pittsburgh, Pa., 1944).

Dwight, H. B., *Electrical Coils and Conductors* (New York: McGraw-Hill, 1945).

Carter, G. W., *The Simple Calculation of Electrical Transients* (Cambridge, England: University Press, 1945).

Biermanns, J., *Hochspannung und Hochleistung* (Munich: Hanser, 1949).

Clarke, E., *Circuit Analysis of a.c. Power Systems* (New York: Wiley, vol. 1, 1943; vol. 2, 1950).

Wagner, K. W., *Operatorenrechnung nebst Anwendungen in Physik und Technik* (Leipzig: J. A. Barth, 1940; 2nd ed., 1950).

Lewis, W. W., *The Protection of Transmission Systems Against Lightning* (New York: Wiley, 1950).

Bewley, L. V., *Traveling Waves on Transmission Systems* (New York: Wiley, 1933; 2nd ed., 1951).

Byers, H. R., *Thunderstorm Electricity* (Chicago: University of Chicago Press, 1953).

Rüdenberg, R., *Elektrische Schaltvorgänge* (Berlin: Springer, 1923; 4th ed., 1953).

Strigel, R., *Elektrische Stossfestigkeit* (Berlin: Springer, 1938; 2nd ed., 1955).

Hayashi, S., *Surges on Transmission Systems* (Kyoto: Denki-Shoin, 1955).

Burstyn, W., *Elektrische Kontakte und Schaltvorgänge* (Berlin: Springer, 4th ed., 1956).

Küchler, R., *Transformatoren* (Berlin, Göttingen, Heidelberg: Springer, 1956).

Baatz, H., *Überspannungen in Energieversorgungsnetzen* (Berlin, Göttingen, Heidelberg: Springer, 1956).

Willheim, R., and M. Waters, *Neutral Grounding in High-Voltage Transmission* (New York, London: Elsevier, 1956).

1. Laws of Propagation of Traveling Waves

Sommerfeld, A., "Über die Fortpflanzung elektrodynamischer Wellen längs eines Drahtes," *Ann. Physik*, 67 (1899), 233.

Mie, G., "Elektrische Wellen an zwei parallelen Drähten," *Ann. Physik*, 2 (1900), 201.

Poincaré, H., "Étude de la propagation du courant en période variable," *Eclair. électr.*, 40 (1904), 121.

Steinmetz, C. P., "The general equations of the electric circuit," *Proc. Amer. Inst. Elec. Engr.* (1908), 1121, and (1919), 249.

Wagner, K. W., "Der Verlauf telegraphischer Zeichen in langen Kabeln," *Physik. Z.* (1909), 865.

Nicholson, L. C., "A practical method of protecting insulators from lightning and power arc effects," *Proc. Amer. Inst. Elec. Engr.* (1910), 241.

Wagner, K. W., "Die Fortpflanzung von Strömen

in Kabeln mit unvollkommenem Dielektrikum," *Göttinger Nachr., Math.-Phys. Klasse* (1910), 425.

Malcolm, H. W., "The theory of the submarine telegraph cable," *Electrician*, **68** (1912), 876.

Wagner, K. W., "Elektromagnetische Wellen in elementarer Behandlungsweise," *Elecktrotech. Z.* (1913), 1053.

Hiecke, R., "Der Einfluss des Ohmschen Widerstandes auf den Verlauf der Wanderwellen," *Elektrotechn. and Maschinenb.* (1919), 125.

Carson, J. R., "Radiation from transmission lines," *J. Amer. Inst. Elec. Engr.* (1921), 789.

Karapetoff, V., "A graphical theory of traveling electric waves between parallel conductors," *Trans. Amer. Inst. Elec. Engr.* (1929), 508; *J.* (1929), 113.

Blondel, A., "L'Évolution des méthodes de calcul des phénomènes transitoires," *Rev. Gén. Élec.*, **41** (1937), 227, 259, 298, 327, 579, 650 (contains lists of references).

Gates, B. G., "Theory of surge propagation," in *Surge Phenomena*. London: British Electrical Research Assoc., 1941.

Roadhouse, C. S., C. A. Streifus, and R. B. Gow, "Measured electrical constants of 270-mile 154-kV transmission line," *Trans. AIEE*, **63** (1944), 538.

Pélissier, R., "La propagation des ondes transitoires et périodiques le long des lignes électriques," *Rev. Gén. Élec.*, **59** (1950), 379, 437.

2. The Generation of Simple Traveling Pulses

Hertz, H., "Über sehr schnelle elektrische Schwingungen," *Gesammelte Werke*, vol. 2 (Leipzig: J. A. Barth, 1894), p. 32.

Thomas P. H., "Static strains in high tension circuits and the protection of apparatus," *Trans. Amer. Inst. Elec. Engr.* (1902), 189.

Wagner, K. W., "Freie Schwingungen in langen Leitungen," *Elektrotech. Z.* (1908), 707.

Thomas, P. H., "Static strains in high-tension circuits," *Elec. J.* (1910), 228.

Pfiffner, E., "Die Eigenschwingungen elektrischer Stromkreise," *Elektrotech. u. Maschinenb.* (1916), 209.

3. The Generation and Shape of Pulses

Petersen, W., "Überspannungen und Überspannungsschutz," *Elektrotech. Z.* (1913), 167.

Küpfmüller, K., "Über Einschwingvorgänge in Wellenfiltern," *Elek. Nachr.-Tech.* (1924), 141.

Küpfmüller, K., and H. F. Mayer, "Über Einschwingvorgänge in Pupinleitungen und ihre Verminderung," *Wiss. Veröff. Siemens-Konz.*, **5** (1926), no. 1, p. 51.

Ledoux, C., "Ondes mobiles: propagation, formation et protection," *Rev. Gén. Électr.*, **22** (1927), 815.

Krutzsch, J., "Berechenbare Wanderwellenformen unter Zugrundelegung des Toeplerschen Funkengesetzes," *Elektrotech. Z.* (1928), 607.

Fortescue, C. "La foudre et ses effets sur les lignes aériennes," *Internat. Electrical Congress, Paris*, **6** (1932), 399; *Trans. Amer. Inst. Electr. Engr.* **49** (1930), 1503.

Peek, F. W., Jr., "La foudre," *Internat. Electrical Congress, Paris*, **6** (1932), 465; *Elec. World*, **100** (1932), 247.

Kopeliowitch, J., and P. Fourmarier, "Études à l'oscillographe cathodique relatives à la propagation des ondes dans les réseaux et à leur pénétration dans les enroulements des machines électriques," *Rev. Gén. Elec.*, **40** (1936), 163, 197.

Bewley, L. V., "Traveling waves initiated by switching," *Trans. AIEE*, **58** (1939), 18.

McCann, G. D., and C. F. Wagner, "Induced voltages on transmission lines," *Trans. AIEE*, **61** (1942), 916.

Stevens, R. F., and T. W. Springfield, "A transmission line fault locator using fault generated surges," *Trans. AIEE*, **67** (1948), 1168.

Froidevaux, J., and C. Rossier, "Impulse testing of large transformers: The maintenance of the wave shape," *CIGRE* (1956), No. 118.

Langlois-Berthelot, R., M. Monnet, J. Derippe, and R. Favié, "Chopped wave tests of transformers with a wave of reduced steepness," *CIGRE* (1956), No. 138.

4. The Influence of Line Terminations

Hiecke, R., "Über Schwingungen mit hoher Spannung und Frequenz in Gleichstromnetzen," *Elektrotech. Z.* (1907), 334.

Wagner, K. W., "Elektromagnetische Ausgleichsvorgänge in Freileitungen und Kabeln," *Elektrotech. Z.* (1911), 899.

Cunningham, J. H., and C. M. Davis, "Propagation of impulses over a transmission line," *Proc. Amer. Inst. Elec. Engr.* (1912), 649.

Siegbahn, M., "Über die Ausbreitung der Spannung und des Stromes beim Einschalten eines Kabels an eine Wechselstromquelle," *Arch. Elektrotech.*, **2** (1913), 155.

Franklin, W. S., "Some simple examples of transmission line surges," *Proc. Amer. Inst. Elec. Engr.* (1914), 547.

Weed, J. M., "Theory of electric waves in transmission lines," *General Electric Rev.* (1915), 1148; (1916), 141.

Brylinski, E., "Propagation sur une ligne a circuit ouvert," *Bull. Soc. Intern. Élec.* (1917), 217.

Bush, V., "Transmission line transients," *J. Amer. Inst. Elec. Engr.* (1923), 1155.

Rogowski, W., and E. Flegler, "Die Wanderwelle nach Aufnahmen mit dem Kathodenoszillographen, *Arch. Elektrotech.*, **14** (1925), 529.

Clem, J. E., K. W. Miller, and H. Halperin, "Transient voltages on bonded cable sheaths," *Trans. AIEE*, **54** (1935), 73.

Baatz, H., "Vorgänge beim Abschalten leerlaufender Hochspannungsleitungen," *VDE-Fachber.*, **7** (1935), 35.

Halperin, H., and G. B. Shanklin, "Impulse strength of insulated power-cable circuits," *Trans. AIEE*, **63** (1944), 1190.

Dorsch, H., "Ausgleichsspannungen auf einer 700 km langen 380-kV Drehstromübertragung bei Lastabwurf," *ETZ*, **71** (1950), 685.

Fischer, U., "Analyse und Synthese der Vorgänge beim Abschalten leerlaufender Hochspannungsleitungen," *VDE-Fachber.*, **15** (1951), 36.

5. Distortion of Pulse Shape

Pleijel, H., "Vandringsvågor och deras Formförändringar under Fortplantningen utefter Ledningar," *Tekn. T.* (1918), no. 11, p. 129.

Ryan, H. J., and H. H. Henline, "The hysteresis character of corona formation," *Trans. Amer. Inst. Elec. Engr.* (1924), 1118; *J.* (1924), 825.

Whitehead, J. B., "The corona as lightning arrester," *Trans. Amer. Inst. Elec. Engr.* (1924), 1172; *J.* (1924), 914.

Carson, J. R., "Die Behandlung der Telegraphengleichung unter Berücksichtigung der Stromverdrängung nach der Operatorenmethode," *Elek. Nachr.-Tech.* (1925), 359.

Holm, R., "Die Theorie der Korona an Hochspannungsleitungen," *Wiss. Veröff. Siemens-Konz.*, **4** (1925), no. 1, p. 14.

Moeller, F., "Die Abflachung steiler Wellenstirnen unter Berücksichtigung der Stromverdrängung im Leiter," *Arch. Elektrotech.* **15** (1926), 547.

Moeller, F., "Über den Einfluß der Wanderwellenlänge auf die Abflachung steiler Stirnen," *Arch. Elektrotech.*, **18** (1927), 399.

Lewis, W. W., "Surge voltage investigation on transmission lines," *Trans. Amer. Inst. Elec. Engr.* (1928), 1111; *J.* (1928), 795.

Frühauf, G., "Dämpfung von Wanderwellenschwingungen auf Freileitungen nach Aufnahmen mit dem Kathodenoszillographen," *Elektrotech. Z.* (1929), 892.

Jacottet, P., "Einfluß der Stromverdrängung auf die Stirnform von Sprungwellen," *Wiss. Veröff. Siemens-Konz.*, **8** (1929), no. 3, p. 54.

McEachron, K. B., J. G. Hemstreet, and W. J. Rudge, "Traveling waves on transmission lines with artificial lightning surges," *Trans. Amer. Inst. Elec. Engr.* (1930), 885; *J.* (1930), 377.

Brune, O., and J. R. Eaton, "Experimental studies in the propagation of lightning surges on transmission lines," *Trans. Amer. Inst. Elec. Engr.* (1931), 1132.

Jacottet, P., "Dämpfung und Verzerrung kurzer Sprungwellen durch Stromverdrängung im Erdreich," *Wiss. Veröff. Siemens-Konz.*, **10** (1931), no. 1, p. 42.

Skilling, H. H., "Corona and line surges," *Electr. Engng.* (1931), 798.

Flegler, E., and J. Röhrig, "Die Dämpfung von Wanderwellen auf Hochspannungsleitungen," *Arch. Elektrotech.*, **27** (1933), 637.

Voerste, F., "Die Verformung und Dämpfung von Wanderwellen durch Coronaverluste nach Aufnahmen mit dem Kathodenstrahl-Oszillographen," *Elektrotech. Z.* (1933), 452.

Carroll, J. S., B. Cozzens, and T. M. Blakeslee, "Corona losses from conductors of 1.4 inch diameter," *Elec. Eng.*, **52** (1933), 854.

Skilling, H. H., and P. de K. Dykes, "Distortion of traveling waves by corona," *Trans. AIEE*, **56** (1937), 850.

Böckmann, M., N. Hyltén-Cavallius, and S. Rusck, "Propagation of surge generator waves up to 850 kV on a 132 kV line," *CIGRE* (1950), No. 314.

Moritz, K., "Rechnerische Ermittlung der Dämpfung und Verzerrung von Wanderwellen," *Arch. Elektrotech.*, **41** (1953), 160.

Chura, V., "Der Anteil des Erdwiderstandes an der Dämpfung von Wanderwellen" *Elektrotech. u. Maschinenbau*, **71** (1954), 467 (abstract).

Wagner, C. F., and B. L. Lloyd, "Effects of corona on traveling waves," *Trans. AIEE*, **74**, part III (1955), 858.

6. Transmission and Reflection

Wagner, K. W., "Eine neue künstliche Leitung zur Untersuchung von Telegraphierströmen und Schaltvorgängen," *Elektrotech. Z.* (1912), 1289.

Biermanns, J., "Beiträge zur Frage des Überspannungsschutzes," *Arch. Elektrotech.*, **2** (1914), 217.

Gewecke, H., "Überspannungsschutz bei Stromwandlern," *Elektrotech. Z.* (1914), 386.

Müller-Hillebrand, D., "Der Kathodenfallableiter als Gewitterschutz," *VDE-Fachber.* (1929), 51.

Röhrig, J., "Wanderwellenaufnahmen an zusammengesetzten Betriebsleitungen," *Arch. Elektrotech.*, **25** (1931), 411.

Buller, F. H., W. J. Rudge, and H. G. Brinton, "Traveling wave voltages in cables," *Trans. AIEE*, **52** (1933), 121.

Teszner, S., "La propagation des ondes électromagnétiques dans les circuits heterogènes," *Rev. Gén. Élec.*, **38** (1935), 695, 723.

7. Spark Discharges on Lines During Switching

Finckh, F., Über einen bemerkenswerten Fall einer schädlichen Spannungserhöhung bei einem Drehstromgenerator," *Elektrotech. Z.* (1903), 198.

Cunningham, J. H., "Design, construction and test of an artificial transmission line," *Proc. Amer. Inst. Elec. Engr.* (1911), 87.

Berger, K., "Die Abschaltung von Kurzschlüssen am Ende unverzweigter Leitungen und die sich dabei ergebenden Überspannungen, nach Versuchen mit dem Kathodenstrahl-Oszillographen," *Bull. Schweiz Elektrotech. Verein* (1929), 681.

Berger, K., "Untersuchungen mittels Kathodenstrahl-Oszillograph der durch Erdschluss hervorgerufenen Überspannungen in einem 8-kV Verteilnetz," *Bull. Schweiz. Elektrotech. Verein* (1930), 756.

Hawley, W. G., and H. M. Lacey, "Essais de choc sur une ligne à 33 kV," *CIGRE* (1937), No. 309, 21; discussion, D–71.

Witzke, R. L., and T. J. Bliss, "Surge protection of cable-connected equipment," *Trans. AIEE,* **69** (1950), 527.

Hagenguth, J. H., A. F. Rohlfs, and W. J. Degnan, "Sixty-cycle and impulse sparkover of large gap spacings," *Trans. AIEE,* **71**, part III (1952), 455 (contains lists of references).

Witt, H., and L. R. Bergström, "Restrikes and microsecond-transients in capacitance switching," *CIGRE* (1960), No. 101.

8. Natural Frequencies of Three-phase Circuits

Boehne, E. W., "Voltage oscillations in armature windings," *Trans. AIEE,* **49** (1930), 1587.

Hameister, G., "Untersuchungen über die Frequenz der wiederkehrenden Spannung als Unterlage für die Prüfbestimmungen," *VDE-Fachber.,* **7** (1935), 42.

Wanger, W., and K. Brown, "Berechnung der Schwingungen der wiederkehrenden Spannung nach Kurzschlüssen," *B.B.C.-Mitteil.,* **24** (1937), 283.

Rüdenberg, R., "Natural frequencies of three-phase windings," *J. Franklin Inst.,* **231** (1941), 157, 269.

ter Horst, T. J., "Les fréquences naturelles dans le réseau de transmission à 150 kV des Pays-Bas," *CIGRE* (1948), No. 123.

Kurth, F., "Fréquences de l'oscillation de la tension de rétablissement dans les réseaux à courant alternatif," *CIGRE* (1948), No. 139.

Belot, R., "Les fréquences propres dans les réseaux de transport d'énergie des Unions des Centrales Électriques du Hainaut," *CIGRE* (1950), No. 317.

Someda, G., "Détermination expérimentale des fréquences propres dans un réseau italien," *CIGRE* (1950), No. 329.

Gosland, L., and J. S. Vosper, "Network analyzer study of inherent restriking voltage transients on the British 123-kV Grid," *CIGRE* (1952), No. 120.

Böcker, H., "Einschwingfrequenzen in Netzen," *ETZ-A,* **76** (1955), 792 (contains lists of references).

Kern, B., "Die Reflexionen von Stosswellen in Maschinenwicklungen," *Elektrotech. u. Maschinenb.,* **76** (1959), 415.

9. Protective Resistances for Line Switching

Rüdenberg, R., "Der Einschaltvorgang bei elektrischen Leitungen," *Elektrotech. u. Maschinenb.* (1912), 157.

Kesselring, F., "Betrachtungen über den Hörnerableiter," *Elektrotech. Z.* (1924), 819.

Slepian, J., "Theory of the autovalve arrester,"

Trans. Amer. Inst. Elec. Engr. (1926), 167; *J.* (1930), 3.

Westermann, O., "Beurteilung des Schutzwertes von Funkenableitern," *Elektrotech. Z.* (1926), 217.

McEachron, K. B., and H. G. Brinton, "Performance of thyrite arresters for any assumed form of traveling wave and circuit arrangement," *General Electric Rev.* (1930), 350.

Slepian, J., R. Tanberg, and C. E. Krause, "Theory of a new valve type lightning arrester," *Trans. Amer. Inst. Elec. Engr.* (1930), 257; *J.* (1930), 34.

Meyer, H., "Der Ausgleichsvorgang beim Ansprechen von Überspannungsableitern und seine Berechnung unter Berücksichtigung der Spannungsabhängigkeit der Widerstände," *Bull. Schweiz. Elektrotech. Verein,* **33** (1942), 94.

Breuer, G. D., R. H. Hopkinson, I. B. Johnson, and A. J. Schultz, "Arrester protection of high-voltage stations against lightning," *Trans. AIEE,* **79**, part III (1960), 414.

10. Distributed Impedance Changes

Rüdenberg, R., "Der Verlauf elektrischer Wellen auf Leitungen mit räumlich veränderlicher Charakteristik," *Elektrotech. u. Maschinenb.,* **31** (1913), 421.

Wallot, J., "Bemerkungen zu vorstehender Arbeit," *Elektrotech. u. Maschinenb.,* **32** (1914), 607.

Wallot, J., "Der senkrechte Durchgang elektromagnetischer Wellen durch eine Schicht räumlich veränderlicher Dielektrizitätskonstante," *Ann. Physik,* **60** (1919), 734.

Barthold, L. O., "An approximate transient solution of the tapered transmission line," *Trans. AIEE,* **76,** part III (1957), 1556.

Baur, E., "Beitrag zur Transformation mit inhomogenen Leitungen," *Arch. elek. Übertrag.,* **13** (1959), 114 (contains lists of references).

11. Interaction Between Adjacent Lines

Wagner, K. W., "Induktionswirkungen von Wanderwellen in Nachbarleitungen," *Elektrotech. Z.* (1914), 639.

Finckh, F., Spannungsstösse auf Freileitungen bei Blitzentladungen, ihre Hochtransformierung an den Transformatoren der Überlandnetze und ihre Bekämpfung," *Elektr-Wirtsch.* (1926), 314.

Piloty, H., "Wanderwellenreflexion und Schutzwert von Überspannungsableitern bei einphasigem Ansprechen, *Elektrotech. Z.* (1927), 1755.

Satoh, Y., "Electric oscillations in the double-circuit three-phase transmission line," *J. Amer. Inst. Elec. Engr.* (1927), 868.

Bewley, L. V., "Resolution of surges into multi-velocity components," *Elec. Eng.,* **54** (1935), 1199.

MacLane, G. L., C. F. Wagner, and G. D.

McCann, "Shielding of transmission lines," *Trans. AIEE*, **60** (1941), 313.

McCann, G. D., "The effect of corona on coupling factors between ground wires and phase conductors," *Trans. AIEE*, **62** (1943), 818.

Cotte, M., "Théorie de la propagation d'ondes de choc sur deux lignes parallèles," *Rev. Gén. Élec.*, **56** (1947), 343.

Lewis, L. L., "Traveling wave relations applicable to power-system fault locators," *Trans. AIEE*, **70** (1951), 1671 (contains lists of references).

Hayashi, S., and Y. Hattori, "Analytical investigation of surge impedances and admittances of multiconductor transmission lines," *Tech. Repts. Eng. Res. Inst., Kyoto University*, **5** (1955), No. 19, p. 1.

Schultz, A. J., F. C. Van Wormer, and A. R. Lee, "Surge performance of aerial cable," *Trans. AIEE*, **76,** part III (1957), 923, 930.

Delson, J. K., "Electromagnetic field phenomena in shielded aerial cables under surge conditions," *Trans. AIEE*, **77,** part III (1958), 247.

Schneider, W., "Über den Einfluss von Schirmleitern auf die Wellengeschwindigkeit in gestreckten Leitern bei Starkstromfrequenzen," *Elektrotech. u. Maschinenb.*, **75** (1958), 613.

Adams, G. E., "Wave propagation along unbalanced high-voltage transmission lines," *Trans. AIEE*, **78,** part III (1959), 467.

Mitropoulos, T. N., R. J. Fogel, and C. J. Tang, "Characteristics of power cable shielding," *Trans. AIEE*, **79,** part III (1960), 192.

12. Pulse Shaping

Léauté, A., "Surintensités dues à la fermeture des interrupteurs de tableau," *Lum. élec.*, **12** (1910), 227.

Pfiffner, E., "Theorie und Praxis des Überspannungsschutzes," *Elektrotech. u. Maschinenb.* (1912), 953; (1913), 45.

Wagner, K. W., "Über Reflexion und Brechung von Wanderwellen mit steiler Front an Schaltungen mit Kondensatoren und Drosselspulen," *Arch. Elektrotech.*, **2** (1914), 299.

Faccioli, G., "Rectangular waves," *General Electric Rev.* (1914), 742.

Schumann, W. O., "Beiträge zur Frage der Wellenformen und Deformationen bei Ausgleichsvorgängen längs gestreckter Leiter," *Elektrotech. u. Maschinenb.* (1914), 345.

Traverse, P., and G. Silon, "Note sur la protection des réseaux contre les ondes rectangulaires," *Rev. Gén. Élec.*, **18** (1925), 9.

Schilling, W., and J. Lenz, "Über die Stirnform und die Absenkung der Stirnsteilheit durch Kondensatoren bei durch Funken in Luft ausgelösten Wanderwellen," *Elektrotech. Z.* (1930), 1138.

—— "Die Umbildung der Wellenform durch Kapazitäten und Induktivitäten bei durch Funken ausgelösten Wanderwellen," *Arch. Elektrotech.*, **25** (1931), 97.

Berger, K., "Vom Blitzschlag bedingter Spannungsverlauf in einer am Ende einer Freileitung angeschlossenen Kapazität," *Bull. Schweiz. Elektrotech. Verein*, **35** (1944), 14.

13. Effects of Change of Shape

Wagner, K. W., "Die Oberschwingungen elektrischer Schwingungskreise," *Arch. Elektrotech.*, **1** (1912), 47.

Riepl, W., "Messungen über die Verschleifung von Wanderwellen an Freileitungen," *Arch. Elektrotech.*, **18** (1927), 416.

Schwenkhagen H., "Die Einwirkung der Mastkapazitäten auf die Ausbreitung von Wanderwellen auf Leitungsbündeln," *Arch. Elektrotech.*, **31** (1937), 73.

Gates, B. G., "Effect of line insulators on wave front," in *Surge Phenomena* (British Electrical Research Assoc., London, 1941), p. 113.

14. Protective Value of Coils and Condensers

Jackson, R. P., "Recent investigation of lightning protective apparatus," *Proc. Amer. Inst. Elec. Engr.* (1906), 843.

Döry, J., "Freie Schwingungen in langen Leitungen," *Elektrotech. u. Maschinenb.* (1909), 105.

Capart, G., "Die atmosphärischen Erscheinungen und die Störungen, welche durch dieselben in den elektrischen Verteilungsnetzen hervorgerufen werden," *Elektrotech. u. Maschinenb.* (1913), 782.

Böhm, O., "Beiträge zur Frage der Schutzwirkung von Drosselspulen," *Elektrotech. u. Maschinenb.* (1918), 377.

Schilling, W., "Einschaltvorgang der kapazitiv belasteten endlichen Leitung bei endlicher Stirnsteilheit der Schaltwelle nach der Operatorenrechnung," *Arch. Elektrotech.*, **25** (1931), 241.

Boll, G., "Schutz gegen Überspannungen durch Kondensatoren und Kabel," *B.B.C.-Nachricht.*, **18** (1931), 47.

Meyer, H., Überspannungsschutz mit Kapazitäten," *Bull. Schweiz. Elektrotech. Verein*, **31** (1940), 597.

Szpcr, S., "Inductance coils for the protection of substations against the effects of lightning," *CIGRE* (1958), No. 316.

15. Surge Impedance of Lumped Circuit Elements

Petersen, W., "Wanderwellen als Überspannungserreger," *Arch. Elektrotech.*, **1** (1912), 233.

Rogowski, W., "Eine Erweiterung des Reflexionsgesetzes für Wanderwellen," *Arch. Elektrotech.*, **4** (1916), 204.

Biermanns, J., "Über Wanderwellen-Schutzeinrichtungen," *Arch. Elektrotech.*, **5** (1917), 215.

Warren, A. G., "The transmission of electric waves along wires," *J. Inst. Elec. Engr.*, **59** (1921), 330.

McEachron, K. B., J. G. Hemstreet, and H. P. Seelye, "Study of the effect of short lengths of cable on traveling waves," *Trans. Amer. Inst. Elec. Engr.* (1930), 1432; *J.* (1930), 760.

Fucks, W., "Dämpfung einer Stosswelle auf einem Kabel," *Arch. Elektrotech.*, **26** (1932), 118.

Brinton, H. G., F. H. Buller, and W. J. Rudge, "Traveling-wave voltages in cables," *Trans. Amer. Inst. Elec. Engr.* (1933), 121; *Elec. Eng.* (1932), 660.

Dejuhasz, K. J., "Graphical analysis of transient phenomena in electric circuits," *J. Franklin Inst.*, **228** (1939), 339.

Satche, P., and V. Grosse, "The calculation of recovery voltages and internal voltage surges by means of Bergeron's method," *CIGRE* (1950), No. 128.

Bulla, W., "Das Bergeron-Diagramm für Wanderwellen," *Elektrotech. u. Maschinenb.*, **71** (1954), 37.

Boehne, E. W., "Traveling wave protection problems," *Trans. AIEE*, **73**, part III (1954), 920; **74** part III (1955), 880 (contains lists of references).

Waste, W., "Surge stresses in sub-stations connected to overhead lines through cables," *CIGRE* (1960), No. 314, p. 36 (contains lists of references).

16. Joint Action of Coils and Condensers

Linke, W., "Überspannungserscheinungen bei Schaltvorgängen," *Arch. Elektrotech.*, **1** (1912), 163.

Rüdenberg, R., "Eine neue Schutzanordnung für elektrische Stromkreise gegen Überspannungen und ähnliche Störungen," *Elektrotech. Z.* (1914), 610.

Faccioli, G., and H. G. Brinton, "High frequency absorbers," *General Electric Rev.* (1921), 444.

Gabor, D., "Einige Untersuchungen mit dem Kathodenstrahloszillographen zur Aufklärung von Überspannungserscheinungen," *Elek.-Wirtsch.* (1926), 307.

Peek, F. W., "Lightning," *General Electric Rev.* (1929), 602.

Katzschner, M., "Über die Gefahr von Wanderwellen-Resonanzschwingungen," *Arch. Elektrotech.*, **27** (1933), 57.

Woodruff, L. F., "Transmission line transients in motion pictures," *Trans. AIEE*, **57** (1938), 391.

Wedmore, E. B., *et al.*, "Report on the E.R.A.—Surge Filter," in *Surge Phenomena* (London, 1941), p. 401.

Kaneff, S., "Some aspects of lightning surges on power systems and their suppression," *Trans. AIEE*, **69** (1950), 1544.

17. Propagation Velocity and Surge Impedance in Simple Windings

Abetti, P. A., "Survey and classification and bibliography of published data on the surge performance of transformers and rotating machines," *Trans. AIEE*, **77,** part III (1958), 1150, 1403.

Seibt, G., "Vorführung von Experimenten über schnelle elektrische Schwingungen," *Elektrotech. Z.*, **24** (1903), 105.

Biermanns, J., "Elektrische Schwingungen in Maschinenwicklungen," *Arch. Elektrotech.*, **4** (1916), 211.

Boehne, E. W., "Voltage oscillations in armature windings under lightning impulses," *Trans. AIEE*, **49** (1930), 1587.

Willheim, R., "Die Gewitterfestigkeit des Drehstromtransformators," *Elektrotech. u. Maschinenb.*, **50** (1932), 16.

Hunter, E. M., "Transient voltages in rotating machines," *Elec. Eng.*, **54** (1935), 599.

Neuhaus, H., and K. Strigel, "Der Verlauf von Wanderwellen in elektrischen Maschinen und deren Schutz beim Anschluss an Freileitungen," *Arch. Elektrotech.*, **29** (1935), 702.

Norinder, H., "Versuche mit Wanderwellen," in I. L. la Cour and K. Faye-Hansen, *Die Transformatoren* (Berlin: Springer, 1936), p. 195.

Kroemer, H., and A. Wallraff, "Einschaltschwingungen auf leitungsähnlichen Spulen," *Arch. Elektrotech.*, **30** (1936), 780.

Fielder, F. D., and E. Beck, "Effects of lightning voltages on rotating machines and methods of protecting against them," *Trans. AIEE*, **49** (1936), 1577 (contains lists of references).

Calvert, J. F., and F. D. Fielder, "Switching surges in rotating machines," *Elec. Eng.*, **55** (1936), 376.

Pirenne, J., "Théorie générale des phénomènes oscillatoires dans les enroulements des transformateurs," *Rev. Gén. Élec.*, **47** (1940), 19.

Rüdenberg, R., "Surge characteristics of two-winding transformers," *Trans. AIEE*, **60** (1941), 1136.

Lerstrup, K., "Single-layer coils under electrical impacts: an experimental investigation with traveling waves," thesis, Harvard University, Cambridge, Mass., 1942.

Friedländer, E., "Traveling waves in high-voltage alternator windings," *J. Inst. Elec. Engr.*, **89**, part II (1942), 492.

Wellauer, M., "Surge voltage stresses in the windings of machines and insulation coordination of rotating machinery," *CIGRE* (1946), No. 117.

Abetti, P. A., I. B. Johnson, and A. J. Schultz, "Surge phenomena in large unit-connected steam turbine generators," *Trans. AIEE*, **71**, part III (1952), 1035.

von Hippel, A. R., *Dielectric Materials and Applications* (New York: Wiley, 1954).

Abetti, P. A., and F. J. Maginniss, "Fundamental oscillations of coils and windings," *Trans. AIEE*, **73,** part III (1954), 1.

Heller, B., and A. Veverka, "Die Modelltheorie der Stosserscheinungen in Transformatoren," *Elektrotech. u. Maschinenb.*, **74** (1957), 248.

Abetti, P. A., "Pseudo-final voltage distribution in impulsed coils and windings," *Trans. AIEE*, **79** (1960), 87.

18. Analysis of Pulse Propagation in Transformer Windings by Maxwell's Equations

Lenz, W., "Berechnung der Eigenschwingungen einlagiger Spulen," *Ann. Phys.*, **43** (1914), 749.

Steidinger, W., "Das elektromagnetische Verhalten der einlagigen Zylinderspule," *Arch. Elektrotech.*, **13** (1924), 237.

Zuhrt, H., "Eine quasistationäre Berechnung der Eigenwellen einlagiger Flach- und Zylinderspulen," *VDE-Fachber.* (1931), 36; *Arch. Elektrotech.*, **27** (1933), 613.

Rüdenberg, R., "Electromagnetic waves in transformer coils treated by Maxwell's equations," *J. Appl. Phys.*, **12** (1941), 219.

Brillouin, L., *Spiraled coils as waveguides* (Office of Naval Research, Contract N5-ori-76, Task Order No. 1, 1947).

Kaden, H., "Allgemeine Theorie des Wendelleiters," *Arch. elek. Übertrag.*, **5** (1951), 534.

Poritsky, H., P. A. Abetti, and R. P. Jerrard, "Field theory of wave propagation along coils," *Trans. AIEE*, **72**, part III (1953), 930.

Sensiper, S., "Electromagnetic wave propagation on helical structures," *Proc. Inst. Rad. Engr.*, **48** (1955), 149 (contains lists of references).

Piefke, G., "Wendelleitung als Verzögerungsleitung," *Arch. elek. Übertrag.*, **14** (1960), 15.

19. Linkages in Coil Windings

Steinmetz, C. P., "Underground transmission and distribution of electrical energy," *Proc. Amer. Inst. Elec. Engr.* (1907), 196.

Rüdenberg, R., "Entstehung und Verlauf elektrischer Sprungwellen," *Elektrotech. u. Maschinenb.*, **32** (1914), 729.

Wagner, K. W., "Das Eindringen einer elektromagnetischen Welle in einer Spule mit Windungskapazität," *Elektrotech. u. Maschinenb.*, **33** (1915), 89.

Weed, J. M., "Abnormal voltages in transformers," *Proc. Amer. Inst. Elec. Engr.* (1915), 1621.

Wagner, K. W., "Beanspruchung und Schutzwirkung von Spulen bei schnellen Ausgleichsvorgängen," *Elektrotech. Z.* (1916), 425.

Böhm, O., "Rechnerische und experimentelle Untersuchung der Einwirkung von Wanderwellen-Schwingungen auf Transformatorenwicklungen," *Arch. Elektrotech.*, **5** (1917), 383; Dissertation, Technische Hochschule, Darmstadt (1916).

Rogowski, W., "Spulen und Wanderwellen," *Arch. Elektrotech.*, **6** (1918), 265, 377, **7** (1918), 33, 161; **7** (1919), 320.

Dreyfus, L., "Einschaltspannungen der Spule aus zwei Windungen," *Arch. Elektrotech.*, **7** (1918), 175.

Blume, L. F., and A. Boyajian, "Abnormal voltages within transformers," *Trans. Amer. Inst. Elec. Engr.*, **38** (1919), 577.

Rogowski, W., "Überspannungen und Eigenfrequenzen einer Spule," *Arch. Elektrotech.*, **7** (1919), 240.

Gothe, A., "Kritische Frequenz und Eigenfrequenzen einlagiger Spulen," *Arch. Elektrotech.*, **9** (1920), 1.

Schröder, W., "Berechnung der Eigenschwingungen der doppellagigen langen Spule," *Arch. Elektrotech.*, **11** (1922), 203.

Weed, J. M., "Prevention of transient voltage in windings," *J. Amer. Inst. Elec. Engr.*, **41** (1922), 14.

Rüdenberg, R., "Windungsverkettung in Spulen," in *Elektrische Schaltvorgänge* (Berlin: Springer, 1st ed., 1923), pp. 446–463.

Trage, H., "Messungen über den Durchgang von Wanderwellen durch Schutzdrosselspulen," *Arch. Elektrotech.*, **15** (1925), 345.

Fallou, J., "Surtensions dans les transformateurs, essais contre les ondes à front raid," *Rev. Gén. Élec.*, **20** (1926), 772.

—— "Contribution expérimentale à l'étude des surtensions dans les transformateurs," *Bull. Soc. Franç. Élec.* (1926), 237.

Ledoux, C. "Propagation des ondes dans les enroulements," *Bull. Soc. Franç. Élec.* (1928), 1026.

Rücklin, R., "Ein experimenteller Beitrag zum Spulenproblem," *Arch. Elektrotech.*, **20** (1928), 507.

Palueff, K. K., "Effect of transient voltages on power transformer design," *Trans. Amer. Inst. Elec. Engr.*, **48** (1929), 681.

Reimann, E., "Sprungwellenversuche an Strom- und Spannungswandlern," *Wiss Veröff. Siemens-Konz.*, **7** (1929), no. 2, 31; **8** (1930), no. 3, 1.

Klein, R., "Die Grösse der Gesamt-Windungskapazität von Schutzdrosselspulen," *Elektrotech. u. Maschinenb.* (1930), 337.

Bewley, L. V., "Transient oscillations in distributed circuits," *Trans. AIEE*, **50** (1931), 1215.

Bewley, L. V., J. H. Hagenguth, and F. R. Jackson, "Methods of determining natural frequencies in coils and windings," *Trans. AIEE*, **60** (1941), 1145.

Meador, J. R., "Power transformers for high voltage transmission," *CIGRE* (1950), No. 134.

Hochrainer, A., "Die Spannungsverteilung in einer Transformatorwicklung bei beliebiger Form der Stossspannung," *ETZ-A*, **74** (1953), 153.

Vitins, I., "Die Schwingungsgleichungen eines idealisierten Hochspannungstransformators," *Arch. Elektrotech.*, **41** (1954), 196, 301.

Abetti, P. A., "Méthodes pour l'étude des tensions anormales dans les transformateurs," *Bull. Scientifique de l'Institut Électrotechnique Montefiore, Liège* (1957), No. 12 (contains lists of references).

Rabins, L., "A new approach to the analysis of impulse voltages and gradients in transformer windings," *Trans. AIEE*, **78,** part III (1959), 1784.

Jayaram, B. N., "Bestimmung der Stossspannungsverteilung in Transformatoren mit Digitalrechner," *ETZ-A*, **82** (1961), 1.

20. Entry of Step Pulses into Windings

Frühauf, G., "Phénomènes transitoires dans les enroulements de transformateurs abordés par les ondes mobiles," *Congr. Internat. d'Électricité, Paris* (1932), **5**, 939, *CIGRE* (1932), No. 28.

Calvert, J. F., "Protecting machines from line surges," *Trans. AIEE*, **53** (1934), 139.

Allibone, T. E., D. B. McKenzie, and F. R. Perry, "The effects of impulse voltages on transformer windings," *J. Inst. Elec. Engr.*, **18** (1937), 128.

Hallén, E., "Über das Eindringen von Wanderwellen in Wicklungen," *Arch. Elektrotech.*, **32** (1938), 515.

Rüdenberg, R., "Performance of traveling waves in coils and windings," *Trans. AIEE*, **59** (1940), 1031.

—— "Standing or traveling waves," discussion, *Trans. AIEE*, **59** (1940), 1260.

Elsner, R., "Die Gewittersicherheit moderner Hochspannungstransformatoren," *Elektrotech. u. Maschinenb.*, **61** (1943), 493.

Wellauer, M., "Die Spannungsbeanspruchung der Eingangsspulen von Wicklungen beim Auftreffen von Stossspannungen verschiedener Steilheit," *Bull. Assoc. Suisse Elec.*, **38** (1947), 655.

Jahnke, E., and F. Emde, *Tafeln Höherer Funktionen* (4th ed., Leipzig: Teubner, 1948), p. 1.

Heller, B., J. Hlavka, and A. Veverka, "Die Eigenfrequenz der einlagigen Zylinderspule bei Spannungsstössen," *Bull. Assoc. Suisse Elec.*, **40** (1949), 951.

Degoumois, C., and W. Zoller, "The stresses due to surges in modern transformers," *CIGRE* (1950), No. 124.

Heller, B., and A. Veverka, "Electrical stresses in a winding produced by a unit voltage impulse," *CIGRE* (1952), No. 140.

Gänger, B., "Surge phenomena arising from impulse tests on transformers," *CIGRE* (1954), No. 125.

Biorci, G., "Sulla propagazione delle onde a fronte ripido negli avvolgimenti dei trasformatori," *Energia Elettrica.*, **32** (1955), 899.

Kern, B., "Über das Eindringen von Stossspannungen in Wicklungen elektrischer Maschinen und die Beanspruchungen der Isolierung," *ETZ-A*, **78** (1957), 849.

Heller, B., "Surge phenomena in high voltage equipment with voltage regulation by means of series transformers," *CIGRE* (1958), No. 136.

Lee, T. H., and A. Greenwood, "The effect of current chopping in circuit breakers on networks and transformers," *Trans. AIEE*, **79** part III (1960), 535, 545.

21. Oscillations and Pulses in Subdivided Windings

Wagner, K. W., "Wanderwellen-Schwingungen in Transformatorwicklungen," *Arch. Elektrotech.*, **6** (1918), 301.

Reiche, W., "Messungen über die Spannungsverteilung auf Transformatorwicklungen unter dem Einfluss von Sprungwellen," *Arch. Elektrotech.*, **15** (1925), 216.

Ohkohchi, J., "Untersuchung des Eindringens von Wanderwellen in eine Transformatorwicklung," Dissertation, Technische Hochschule Dresden (1931).

Bewley, L. V., "Transient oscillations in distributed circuits," *Trans. AIEE*, **50** (1931), 1215.

Einhorn, H., "Modellversuche zur Ermittlung der Sprungwellenbeanspruchung von Transformatorwicklungen," *Elektrotech. u. Maschinenb.*, **52** (1934), 309.

Rüdenberg, R., "Electric oscillations and surges in subdivided windings," *J. Appl. Phys.*, **11** (1940), 665.

Bellaschi, P. L., and A. J. Palermo, "Analysis of transient voltages in networks," *Trans. AIEE*, **59** (1940), 973.

Norris, E. T., "The lightning strength of power transformers," *J. Inst. Elec. Engr.*, **95** (1948), 389.

Newton, J. P., "The natural frequency spectra of coils and windings," thesis, Harvard University, Cambridge, Mass. (1952), pp. 152–184.

Abetti, P. A., "Transformer models for the determination of transient voltages," *Trans. AIEE*, **72**, part III (1953), 468.

—— and F. J. Maginniss, "Natural frequencies of coils and windings determined by equivalent circuit," *Trans. AIEE*, **72**, part III (1953), 495.

Waldvogel, P., and Rouxel, "Predetermination by calculation of the electric stresses in a winding subjected to a surge voltage," *CIGRE* (1956), No. 125.

22. Pulse Transfer Between Windings

Fallou, J., "Surtensions dans les transformateurs; propagation des ondes de l'enroulement primaire vers l'enroulement secondaire," *Rev. Gén. Élec.*, **27** (1930), 5.

Krug, W., "Über die Umbildung einer Wanderwelle beim Auflaufen auf eine Transformatorwicklung," *Bull. Schweiz. Elektrotech. Verein*, **22** (1931), 277.

Röhrig, J., "Untersuchungen an Betriebstransformatoren mit dem Kathodenoszillographen," *Arch. Elektrotech.*, **25** (1931), 420.

Bewley, L. V., "Transient oscillations of mutually coupled windings," *Trans. AIEE*, **51** (1932), 299; *Elec. Eng.* (1932), 388.

Rorden, H. L., "Transient oscillations of mutually coupled windings," discussion, *Trans. AIEE*, **51** (1932), 324.

Palueff, K. K., and J. H. Hagenguth, "Transition of lightning waves from one circuit to another through transformers," *Trans. AIEE*, **51** (1932), 601.

Neuhaus, H., and R. Strigel, "Modellversuche zur Wanderwellenübertragung auf die Unterspannungswicklung von Transformatoren,"

Wiss. Veröff. Siemens-Konz., **15** (1936), no. 1, p. 51 (contains lists of references).

Rebhan, I., in the preceding reference, p. 52.

Elsner, R., "Neuere Untersuchungen zur Frage der Stossbeanspruchung von Transformatoren," *Arch. Elektrotech.*, **30** (1936), 368.

—— "Zur Frage der Übertragung von Stossspannungen auf die Unterspannungsseite von Drehstromtransformatoren," *Wiss. Veröff. Siemens-Konz.*, **16** (1937), no. 1, p.1.

Wellauer, M., "Die Übertragung von Überspannungen von der Oberspannungs- auf die Unterspannungswicklung von Transformatoren," *Bull. Schweiz. Elektrotech. Verein*, **30** (1939), 124.

Rüdenberg, R., "Surge characteristics of two-winding transformers," *Trans. AIEE*, **60** (1941), 1136.

Bellaschi, P. L., "Lightning surges transferred from one circuit to another through transformers," *Trans. AIEE*, **62** (1943), 731.

Meyer, H., "Untersuchungen über die Übertragung von Stossvorgängen in Transformatoren," *Bull. Schweiz. Elektrotech. Verein*, **36** (1945), 416.

Boyajian, A., "Oscillations of a high-voltage secondary winding," *Trans. AIEE*, **65** (1946), 1010.

Stenkvist, E., "Surge voltage transmission through transformer windings," *CIGRE* (1946), No. 130.

Beardsley, K. D., W. A. McMorris, and H. C. Stewart, "Voltage stresses in distribution transformers due to lightning currents in low-voltage circuits," *Trans. AIEE*, **67** (1948), 1632.

Abetti, P. A., and H. F. Davis, "Surge transfer in 3-winding transformers and electrostatic voltage distribution," *Trans. AIEE*, **73**, part III (1954), 1395, 1407.

Vitins, J., "Der Schwingungsvorgang in unbelasteter Hochspannungswicklung von Transformatoren bei plötzlicher niederspannungsseitiger Einschaltung," *Arch. Elektrotech.*, **41** (1954), 196, 301.

Abetti, P. A., G. E. Adams, and F. J. Maginniss, "Oscillations of coupled windings," *Trans. AIEE*, **74**, part III (1955), 12.

Hayward, A. P., J. K. Dillard, and A. R. Hileman, "Lightning protection of unit-connected turbine generators," *Trans. AIEE*, **75**, part III (1956), 1370 (contains lists of references).

Hileman, A. R., "Surge transfer through 3-phase transformers," *Trans. AIEE*, **77**, part III (1958), 1543.

Chang, K. H., and T. B. Thompson, "Surge protection of unit-connected generators," *Trans. AIEE*, **78**, part III (1959), 1580 (contains lists of references).

23. Spark Discharges

Algermissen, J., "Verhältnis von Schlagweite und Spannung bei schnellen Schwingungen," *Ann. Physik*, **19** (1906), 1016.

Hayden, J. L. R., and C. P. Steinmetz, "Disruptive strength with transient voltages," *Proc. Amer. Inst. Elec. Engr.* (1910), 747.

Binder, L., "Messungen über die Form der Stirn von Wanderwellen," *Elektrotech. Z.* (1915), 241.

Peek, F. W., "The effect of transient voltages on dielectrics," *Proc. Amer. Inst. Elec. Engr.* (1915), 1695; (1919), 717; (1923), 623.

Peters, J. F., "The klydonograph," *Elec. Wld.*, **83** (1924), 769.

Cox, J. H., and J. W. Legg, "The klydonograph and its application to surge investigation," *Trans. Amer. Inst. Elec. Engr.* (1925), 857; *J.* (1925), 1094.

Laue, M. v., "Bemerkung zu K. Zubers Messung der Verzögerungszeiten bei der Funkenentladung," *Ann. Physik*, **76** (1925), 261.

McEachron, K. B., and C. J. Wade, "Study of time lag of the needle gap," *J. Amer. Inst. Elec. Engr.* (1925), 622.

Toepler, M., "Funkenkonstante, Zündfunken und Wanderwelle," *Arch. Elektrotech.*, **14** (1925), 305.

Burawoy, O., "Die Funkenverzögerung bei Spannungsstössen von sehr kurzer Dauer," *Arch. Elektrotech.*, **16** (1926), 186.

Rogowski, W., "Townsends Theorie und der Durchschlag der Luft bei Stossspannungen," *Arch. Elektrotech.*, **16** (1926), 496.

Gabor, D., "Oszillographieren von Wanderwellen," *Arch. Elektrotech.*, **16** (1926), 296; **18** (1927), 48.

Lee, E. S., and C. M. Foust, "The measurement of surge voltages," *J. Amer. Inst. Elec. Engr.* (1927), 149.

Matthias, A., and D. Gabor, *Kathodenoszillograph* (Forschungshefte der Studiengesellschaft für Höchstspannungsanlagen, no. 1; Berlin, 1927).

Mayr, O., "Eine neue Schaltung zur Messung der Durchschlagsverzögerung elektrischer Isolatoren," *Arch. Elektrotech.*, **19** (1927–28), 108.

Rogowski, W., E. Flegler, and R. Tamm, "Über Wanderwelle und Durchschlag. Neue Aufnahmen mit dem Kathodenoszillographen," *Arch. Elektrotech.*, **18** (1927), 479.

—— and R. Tamm, "Stossspannungen und Funkenbilder," *Arch. Elektrotech.*, **20** (1928), 625.

Slepian, J., "Breakdown of spark gaps," *Elec. Wld.*, **91** (1928), 761.

Torok, J. J., "Surge impulse breakdown of air," *Trans. Amer. Inst. Elec. Engr.* (1928), 349; *J.* (1928), 177.

Fielder, F. D., and P. H. McAuley, "Time lag of breakdown," *Elec. Wld.*, **94** (1929), 1019.

Lissmann, M. A., "High-voltage phenomena in thunderstorms," *Trans. Amer. Inst. Elec. Engr.* (1929), 146; *J.* (1929), 45.

Slepian, J., and J. J. Torok, "Streamer currents in high voltage sparkover," *Elec. J.* (1929), 108.

Berger, K., "Überspannungen in elektrischen Anlagen erläutert an Hand von Untersuchungen mit dem Kathodenoszillographen," *Bull. Schweiz. Elektrotech. Verein* (1930), 77.

Krug, W., "Neuere Aufnahmen von Funken-durchbrüchen mit dem Kathodenstrahl-Oszillographen," *Z. tech. Physik* (1930), 153.

—— "Eine Sprungschaltung für Sperr- und Zeitkreise für Kathodenstrahl-Oszillographen," *Elektrotech. Z.* (1930), 605.

Peek, F. W., "Law of impulse sparkover and time lag," *Trans. Amer. Inst. Elec. Engr.* (1930), 1456; *J.* (1930), 868.

Rüdenberg, R., "Die Kopfgeschwindigkeit elektrischer Funken und Blitze," *Wiss. Veröff. Siemens-Konz.*, 9 (1930), no. 1, p. 1.

Schilling, W., and J. Lenz, "Der Spannungsverlauf bei der Stossprüfung nach Aufnahmen mit dem Kathodenstrahloszillographen," *Elektrotech. Z.* (1931), 107.

Bellaschi, P. L., "Characteristics of surge generators for transformer testing," *Trans. Amer. Inst. Elec. Engr.* (1932), 936; *Elec. Eng.* (1932), 407.

Holzer, W., "Optische Untersuchung der Funken-zündung in Luft von Atmosphärendruck mittels des unterdrückten Durchbruchs," *Z. Physik.*, 77 (1932), 676.

Rogowski, W., "Die Zündung einer Gasentladung," *Physik. Z.* (1932), 797.

Strigel, R., "Über den Entladeverzug in homogenen elektrischen Feldern und in Luft von Atmosphärendruck," *Wiss. Veröff. Siemens-Konz.*, 11 (1932), no. 2, p. 52.

Hippel, A. v., "Die Entwicklungsgeschichte des elektrischen Funkens und seiner Vorentladungen," *Z. Physik* (1933), 19.

Strigel, R., "Zur Frage der Materialabhängigkeit des Entladeverzuges im homogenen elektrischen Felde in Luft von Atmosphärendruck," *Arch. Elektrotech.*, 27 (1933), 137.

—— "Über den Entladeverzug im inhomogenen elektrischen Feld bei kleinen Schlagweiten," *Arch. Elektrotech.*, 27 (1933), 377.

Foust, C. M., H. P. Kuehni, and N. Rohats, "Impulse testing technique," *General Electric Rev.*, 35 (1932), 358.

Strigel, R., "Über den Entladeverzug im gleich-förmigen Feld bei grösseren Schlagweiten," *Wiss. Veröff. Siemens-Werke*, 15 (1936), no. 3. p. 1.

—— "Über die Aufbauzeit des Entladever-zuges im Spitzenfelde," *Wiss. Veröff. Siemens-Werke*, 15 (1936), no. 3, 13.

Hagenguth, J. H., "Short-time spark-over of gaps," *Trans. AIEE*, 56 (1937), 67.

Raske, W., "Messteiler für hohe Spannungen; Der Widerstandsteiler," *Arch. Elektrotech.*, 31 (1937), 653.

Hardy, D. R., and J. D. Craggs, "The irradiation of spark gaps for voltage measurement," *Trans. AIEE*, 69 (1950), 584 (contains lists of references).

Dawes, C. L., C. H. Thomas, and A. B. Drought, "Impulse measurements by repeated-structure networks," *Trans. AIEE*, 69 (1950), 571.

Beldi, F., "Impulse testing of transformers,

measuring procedures and test circuits," *CIGRE* (1952), No. 112.

Hagenguth, J. H., A. F. Rohlfs, and W. J. Degnan, "Sixty-cycle and impulse sparkover of large gap spacings," *Trans. AIEE*, 71, part III (1952), 455.

Hagenguth, J. H., and J. R. Meador, "Impulse testing of power transformers," *Trans. AIEE*, 71, part III (1952), 697.

Provoost, P. G., "Impulse testing of transformers," *CIGRE* (1952), No. 123; (1954), No. 115.

Liao, T. W., and J. G. Anderson, "Propagation mechanism of impulse corona and breakdown in oil," *Trans. AIEE*, 72, part I (1953), 641 (contains lists of references).

Elsner, R., "Detection of insulation failures during impulse testing of transformers," *CIGRE* (1954), No. 101.

Gänger, B., "Auslegung von Stossgeneratoren zum Prüfen von Transformatoren," *ETZ-A*, 75 (1954), 17.

Tetzner, V., "Der Stossdurchschlag der Anordnung Spitze—Platte unter Öl," *Arch. Elektrotech.*, 44 (1958), 56.

Kind, D., "Die Aufbaufläche bei Stossbeanspruchungsspannung technischer Elektrodenanordnungen in Luft," *ETZ-A*, 79 (1958), 65.

Baatz, H., H. Böcker, and M. Oezkaya, "Die Eichung von Stossspannungsmesskreisen mittels Rechteck-Entladestoss," *ETZ-A*, 79 (1958), 553.

Kern, B., "Fehlererkennung bei der Stossspan-nungsprüfung elektrischer Maschinen," *ETZ-A*, 79 (1958), 957.

Vondenbusch, A., "Beitrag zur Berechnung von Stossschaltungen mit zwei Energiespeichern," *ETZ-A*, 80 (1959), 617.

Früngel, F., "Stosstransformator zur Erzeugung von Hochspannungsimpulsen," *ETZ-A*, 81 (1960), 355.

Hortopan, G., "A novel expression concerning transformer impulse tests," *CIGRE* (1960), No. 125 (contains lists of references).

Baatz, H., "Überspannungsregistrierung in Netzen," special number of *ETZ-A*, 81 (1960), 581.

24. Lightning Discharges

Emde, F., "Die Schwingungszahl des Blitzes," *Elektrotech. Z.* (1910), 675.

Peek, F. W., "Lightning," *General Electric Rev.* (1916), 586; (1919), 900.

Toepler, M., "Gewitter und Blitze," *Verb.-Mitt. Dresden. Bez-V. d. I.* (1917), 43.

Walter, B., "Über die Ermittlung der zeitlichen Aufeinanderfolge zusammengehöriger Blitze sowie über ein bemerkenswertes Beispiel dieser Art von Entladungen," *Physik. Z.* (1918), 273.

Norinder, H., "Untersuchungen über das luftelektrische Feld bei Gewittern," *Elektrotech. Z.* (1921), 764 (abstract).

Wilson, C. T. R., "Investigations on lightning discharges and on the electric field of thunderstorms," *Phil. Trans. Roy. Soc. (London)*, 221A (1921), 73.

Norinder, H., "Gewitterforschung und Überspannungsschutz," *Elektrotech. Z.* (1924), 935 (abstract).

Peek, F. W., "Lightning and other transients on transmission lines," *J. Amer. Inst. Elec. Engr.* (1924), 696.

Appleton, E. V., R. A. Watson Watt, and J. F. Herd, "On the nature of atmospherics," *Proc. Roy. Soc.* (*London*), **111** (1926), 615.

Toepler, M., "Gewitter, Blitze und Wanderwellen auf Leitungsnetzen," *Hescho-Mitteil.* (1926), 743.

Fallou, J., "La forme probable des surtensions induites par les décharges orageuses sur les lignes aériennes," *Rev. Gén. Électr.*, **23** (1928), 957.

Simpson, G. C., "Die Theorie der Gewitter," *Meteorol. Z.* (1928), 321.

Peek, F. W., "Progress in lightning. Research in the field and in the laboratory," *Trans. Amer. Inst. Elec. Engr.* (1929), 436; *J.* (1929), 303.

Pupp, W., "Über Funkenerregung kurzer elektrischer Wellen unter 1 m Wellenlänge und einen neuartigen Stossfunkensender," *Ann. Physik*, **2** (1929), 865.

Simpson, G. C., "Lightning," *J. Inst. Elec. Engr.*, **67** (1929), 1269; *Nature* (*London*), **124** (1929), 801.

Wilson, C. T. R., "Some thunderstorm problems," *J. Franklin Inst.*, **208** (1929), 1.

Fortescue, C. L., "Transmission tower design for maximum lightning protection," *Elec. J.* (1930), 640.

Matthias, A., *Gewitterforschungen und Blitzschutz* (Gesamtbericht der zweiten Weltkraftkonferenz, Berlin, 1930), vol. 14, p. 518.

Norinder, H., "Surges and over-voltage phenomena on transmission lines, due to lightning," *J. Inst. Elec. Engr.*, **68** (1930), 525; *Bull. Soc. Franç. Élec.* (1930), 594.

—— "Recherches oscillographiques sur le mécanisme de décharge des éclairs," *Internat. Elektrizitäts-Kongress, Paris*, **12** (1932), 217.

Foust, C. M., and H. P. Kuehni, "The surge-crest ammeter," *General Electric Rev.*, **35** (1932), 644.

Schonland, B. F. J., "The development of the lightning discharge," *Trans. South-African Inst. Elec. Engr.*, **24**, part VI (1933), 145 (contains lists of references).

Rüdenberg, R., "Die Influenzwirkung von Blitzschlägen auf benachbarte Freileitungen," *Wiss. Veröff. Siemens-Konz.*, **13** (1934), no. 2, p. 1.

Aigner, V., "Induzierte Blitzüberspannungen und ihre Beziehung zum rückwärtigen Überschlag," *ETZ*, **56** (1935), 497.

Berger, K., "Resultate der Gewittermessungen in den Jahren 1932–33 und 1934–35," *Bull. Schweiz. Elektrotech. Verein*, **25** (1934), 213; **27** (1936), 145.

Simpson, G. C., and F. J. Scrase, "The distribution of electricity in thunderclouds," *Proc. Roy. Soc.* (London), **161A** (1937), 309.

Watson Watt, R. A., J. F. Herd, and F. E. Lutkin, "On the nature of atmospherics," *Proc. Royal. Soc.* (London), **162A** (1937), 267.

Schonland, B. F. J., "The diameter of the lightning channel," *Phil. Mag.* [7], **23** (1937), 503.

Norinder, H., "Indirekte Blitzüberspannungen in Kraftleitungen," *ETZ.*, **59** (1938), 105.

McEachron, K. B., "Lightning to the Empire State Building," *J. Franklin Inst.*, **227** (1939), 149 (contains lists of references).

Lewis, W. W., and C. M. Foust, "Lightning investigation on transmission lines, part VII," *Trans. AIEE*, **59** (1940), 227.

Wagner, C. F., G. D. McCann, and E. Beck, "Field investigations on lightning," *Trans. AIEE*, **60** (1941), 1222.

McEachron, K. B., "Lightning to the Empire State Building," *Trans. AIEE*, **60** (1941), 885.

Bruce, C. E. R., and R. H. Golde, "The lightning discharge," *J. Inst. Elec. Eng.*, **88** (1941), 242 (contains lists of references).

Workman, E. J., R. E. Holzer, and G. T. Pelsor, *The Electrical Structure of Thunderstorms* (National Advisory Committee for Aeronautics, Tech. Note 864; Washington, D.C., 1942).

Berger, K., "Neuere Resultate der Blitzforschung in der Schweiz," *Bull. Schweiz. Elektrotech. Verein*, **38** (1947), 813.

Gunn, R., "Electrical field intensity inside of natural clouds," *J. Appl. Phys.*, **19** (1948), 481.

Schonland, B. F. J., *et al.*, "Progressive lightning," *Proc. Royal Soc.* (*London*), **143A** (1934), 654 to **206A** (1950), 145 (contains lists of references).

McEachron, K. B., "Lightning, a hazard to electric systems," *Trans. AIEE*, **71**, part III (1952), 977 (contains lists of references).

Hagenguth, J. H., and J. G. Anderson, "Lightning to the Empire State Building, part III," *Trans. AIEE*, **71** part III (1952), 641.

Schlomann, R. H., W. S. Price, I. B. Johnson, and J. G. Anderson, "Lightning field investigations of the OVEC. 345-kV system," *Trans. AIEE*, **76**, part III (1957), 1447 (contains lists of references).

Müller-Hillebrand, D., "Rückwärtiger Blitzeinschlag in Häuser und Energieumsatz im eingeengten Blitzkanal," *ETZ-A*, **78** (1957), 548.

Hösl, A., "Vorgänge beim Blitzeinschlag in Gebäude und einzelnstehende Bäume," *ETZ-A*, **79** (1958), 291.

Griscom, S. B., "The prestrike theory and other effects in the lightning stroke," *Trans. AIEE*, **77**, part III (1958), 919 (contains lists of references).

Lampe, W., "Der Blitzschutz an Hochhäusern," *ETZ-A*, **80** (1959), 201.

Golde, R. H., "Measurement of lightning current amplitudes (by magnetic links)," *CIGRE* (1960), No. 314, p. 3 (contains lists of references).

Wagner, C. F., "Determination of the wave front of lightning stroke currents from field measurements," *Trans. AIEE*, **79**, part III (1960), 581.

Müller-Hillebrand, D., "Zur Physik der Blitzentladung," *ETZ-A*, **82** (1961), 232 (contains lists of references).

Berger, K., "Gewitterforschung auf dem Monte San Salvatore," *ETZ-A*, **82** (1961), 249.

Frühauf, G., "Modellmessungen mit Blitzströmen," *ETZ-A*, **82** (1961), 265.

25. Lines Struck by Lightning

Berger, K., "Die ersten Beobachtungen des Verlaufes von durch Gewitter verursachten Spannungen in Mittelspannungsnetzen mittels des Kathodenstrahl-Oszillographen des S. E. V.," *Bull. Schweiz. Elektrotech. Verein* (1929), 321.

Fortescue, C. L., A. L. Atherton, and J. H. Cox, "Theoretical and field investigations of lightning," *Trans. Amer. Inst. Elec. Engr.* (1929), 449; *J.* (1929), 277.

Matthias, A., "Der gegenwärtige Stand der Blitzschutzfrage," *Elektrotech. Z.* (1929), 1469.

Biermanns, J., "Blitzschutz von Freileitungen," *Forsch. u. Techn.* (Berlin, 1930), p. 234.

Cox, J. H., and E. Beck, "Lightning on transmission lines, cathode ray oscillograph studies," *Trans. Amer. Inst. Elec. Engr.* (1930), 857.

Lewis, W. W., and C. M. Foust, "Lightning investigations on transmission lines," *General Electric Rev.* (1930), 185.

Bewley, L. V., "Critique of ground wire theory," *Trans. Amer. Inst. Elec. Engr.* (1931), 1; *J.* (1930), 780.

Müller-Hillebrand, D., "Die Einwirkung unmittelbarer Blitzentladungen auf Hochspannungsnetze und ihre Bekämpfung," *Elektrotechn. Z.* (1931), 722.

Pittman, R. R., and J. J. Torok, "Lightning investigation on a wood pole transmission line," *Trans. Amer. Inst. Elec. Engr.* (1931), 568.

Sporn, P., "1929 lightning experience on the 132 kV transmission lines of the American Gas and Electric Company," *Trans. Amer. Inst. Elec. Engr.* (1931), 574.

—— and W. L. Lloyd, Jr., "1930 lightning investigations on the transmission system of the American Gas and Electric Company," *Trans. Amer. Inst. Elec. Engr.* (1931), 1111.

Matthias, A., "Modellversuche über Blitzeinschläge," *ETZ*, **58** (1937), 881, 928, 973.

AIEE Committee Report, "Lightning performance of 220-kV transmission lines," *Trans. AIEE*, **65** (1946), 70.

Harder, E. L., and J. M. Clayton," Transmission line design and performance based on direct lightning strokes," *Trans. AIEE*, **68** (1949), 439.

Baatz, H., "Blitzeinschlagsspannungen in Freileitungen," *ETZ*, **72** (1951), 191.

Langrehr, H., "Der Schutzraum der Erdseile," *AEG-Mitteil.*, **41** (1951), 295.

Golde, R. H., "Lightning surges on overhead distribution lines caused by indirect and direct lightning strokes," *Trans. AIEE*, **73,** part III (1954), 437.

Johnson, I. B., and A. J. Schultz, "A hypothesis concerning lightning phenomena and transmission line flashover," *Trans. AIEE*, **76,** part III (1957), 1470.

Stolte, E., "Spannungsgrenzen für die Verwendung von Erdseilen bei Freileitungen," *ETZ-A*, **79** (1958), 797.

26. Impulse Characteristics of Towers and Ground Rods

Fortescue, C. L., "La foudre et ses effects sur les lignes aériennes," *Congrès International d'Électricité, Paris* (1932), vol. 6, p. 399.

Aigner, V., "Das Verhalten gestreckter Erder bei Stossbeanspruchung," *ETZ*, **54** (1933), 1233.

Fortescue, C. L., "Counterpoises for transmission lines," *Trans. AIEE*, **53** (1934), 1781.

Bewley, L. V., "The counterpoise," *General Electric Rev.*, **37** (1934), 73.

Jordan, C. A., "Lightning computations for transmission lines with overhead ground wire," *General Electric Rev.*, **37** (1934), 130.

Dwight, H. B., "Calculation of resistances to ground," *Trans. AIEE*, **55** (1936), 1319.

Bewley, L. V., "Flashovers on transmission lines," *Trans. AIEE*, **55** (1936), 342.

Ryle, P. J., "Direct lightning strokes to towers and effect of wave fronts, footing resistance and tower height," *CIGRE* (1937), No. 307.

Baatz, H., "Über den wirksamen Widerstand von Erdern bei Stossbeanspruchung," *ETZ*, **59** (1938), 1263.

Sunde, E. D., "Surge characteristics of a buried bare wire," *Trans. AIEE*, **59** (1940), 987.

Bellaschi, P. L., "Impulse and 60-cycle characteristics of driven grounds," *Trans. AIEE*, **60** (1941), 123; **61** (1942), 349.

Rüdenberg, R., "Fundamental considerations on ground currents," *Elec. Eng.*, **64** (1945), 1.

Berger, K., "Das Verhalten von Erdungen unter hohen Stossströmen," *Bull. Schweiz. Elektrotech. Verein*, **37** (1946), 197.

Rüdenberg, R., "Comparative properties of grounding electrodes," *Elec. World*, **129** (1948), 72.

Norinder, H., and G. Petropoulos, "Characteristics of pointed electrodes and direct strokes in surge currents to ground," *CIGRE* (1948), No. 310.

Grünewald, H., "Über den Stossausbreitungswiderstand von Erdern," *ETZ.*, **70** (1949), 505.

Norinder, H., and O. Salka, "Stosswiderstände der verschiedenen Erdelektroden und Einbettungsmaterialien," *Bull. Schweiz. Elektrotech. Verein*, **42** (1951), 321.

McCrocklin, A. J., and C. W. Wendlandt, "Determination of resistance to ground of grounding grids," *Trans. AIEE*, **71,** part III (1952), 1062.

Gross, E. T. B., B. V. Chitnis, and L. J. Stratton, "Grounding grids for high-voltage stations," *Trans. AIEE*, **72,** part III (1953), 799.

Koch, W., *Erdungen in Wechselstromanlagen über 1 kV* (Berlin: Springer, 1948; 2nd ed., 1955).

Miller, C. J., "Anomalous flashovers on transmission lines," *Trans. AIEE*, **75,** part III (1956), 897 and discussion.

Johnson, I. B., and A. J. Schultz, "Analytical studies on lightning phenomena involving towers, insulator strings, and transmission lines," *Trans. AIEE*, **76**, part III (1957), 1310.

Breuer, G. D., R. J. Schultz, R. H. Schlomann, and W. S. Price, "Field studies of the surge response of a 345 kV transmission tower and ground wire," *Trans. AIEE*, **76**, part III (1957), 1392.

AIEE Committee Report, "Voltage gradients through the ground under fault conditions," *Trans. AIEE*, **77**, part III (1958), 669 (contains lists of references).

Rüdenberg, R., "Discussion on Ground resistance and voltage gradients," *Trans. AIEE*, **77**, part III (1958), 685.

Lundholm, R., R. B. Finn, and W. S. Price, "Calculation of transmission line lightning voltages by field concepts," *Trans. AIEE*, **77**, part III (1958), 1271, and C. J. Miller, discussion, 1281.

Hagenguth, J. H., and J. G. Anderson, "Factors affecting the lightning performance of transmission lines," *Trans. AIEE*, **77**, part III (1958), 1379, (contains lists of references).

Johnson, I. B., W. S. Price, and A. J. Schultz, "Lightning current distribution in towers and ground wires," *Trans. AIEE*, **77**, part III (1958), 1414.

Fisher, F. A., J. G. Anderson, and J. H. Hagenguth, "Determination of lightning response of transmission lines by means of geometrical models," *Trans. AIEE*, **78**, part III (1959), 1724.

Wagner, C. F., and A. R. Hileman, "A new approach to the calculation of the lightning performance of transmission lines," *Trans. AIEE*, **75**, part III (1956), 1233; **78**, part III (1959), 996; **79**, part III (1960), 589.

Armstrong, H. R., and L. J. Simpkin, "Grounding electrode potential gradients from model tests," *Trans. AIEE*, **79**, part III (1960), 618.

Wilpernig, H., "Wanderwellen auf dem Mast bei graduellem Spannungsanstieg beim Einschlag in die Spitze," *ETZ-A*, **81** (1960), 14.

Baatz, H., "Gewitterschutz und Wirksamkeit direkter Schutzmassnahmen in Netzen," *ETZ-A*, **82** (1961), 278.

27. Impulse Strength of Insulation

Orlich, E., "Über einen Spannungsteiler für Hochspannungsmessungen," *Arch. Elektrotech.* **1** (1912), 1.

Rüdenberg, R., "Die Spannungsverteilung an Kettenisolatoren," *Elektrotech. Z.*, **35** (1914), 412.

Ollendorff, F., "Modelltheorie der Hängeisolatoren," in *Potentialfelder der Elektrotechnik* (Berlin: Springer, 1932), p. 33.

Schönenberg, R., "Der Einfluss von Isolatorenform und Schutzarmaturen auf die Stosskennlinien von Freileitungsisolatoren," *ETZ-A*, **74** (1953), 485.

Maass, H. F., "Steuerung des elektrischen Randfeldes von Hochspannungsdurchführungen und Kondensatoren durch Widerstandsschichten," *ETZ-A*, **79** (1958), 128.

Jacottet, P., "Durchschlagsverhalten von Anordnungen in Luft und unter Öl bei Schaltüberspannungen," *ETZ-A*, **79** (1958), 337.

Widmann, W., "Einfluss der Erdelektroden auf die Stehstossspannung von Pegelfunkenstrekken," *ETZ-A*, **79** (1958), 965.

Hermstein, W., "Einfluss von Vorentladungen auf das Überschlagsverhalten grundsätzlicher Stützeranordnungen in Luft," *ETZ-A*, **81** (1960), 413.

Berger, K., and A. Asner, "Neue Erkenntnisse über das Verhalten und die Prüfung von Spannungsteilern zur Messung sehr hoher, rasch veränderlicher Stossspannungen," *Bull. Schweiz. Elektrotech. Verein*, **51** (1960), 769.

Denholm, A. S., "The electric stress grading of insulator strings," *Trans. AIEE*, **79**, part III (1960), 167.

Lantz, A. D., "Measuring the lightning strength of high-voltage insulators," *Trans. AIEE*, **79**, part III (1960), 298.

28. Performance of Networks and Windings

Steinmetz, C. P., "Abnormal strains in transformers," *General Electric Rev.* (1912), 737.

Rüdenberg, R., "Entstehung und Verlauf elektrischer Sprungwellen," *Elektrotech. u. Maschinenb.*, **32** (1914), 729.

Wirz, E., "Überspannungserscheinungen bei Stromwandlern," *Bull. Schweiz. Elektrotech. Verein* (1915), 121.

Wagner, K. W., "Das Eindringen einer elektromagnetischen Welle in eine Spule mit Windungskapazität," *Elektrotech. u. Maschinenb.*, **33** (1915), 89.

Weed, J. M., "Abnormal voltages in transformers," *Proc. Amer. Inst. Elec. Engr.* (1915), 1621.

Blume, L. F., and A. Boyajian, "Abnormal voltages within transformers," *Trans. Amer. Inst. Elec. Engr.*, **38** (1919), 577.

Sarolea, J., "Les accidents de surtension," *Rev. Gén. Élec.*, **1** (1917), 215.

Grünewald, F., "Das Verhalten der Freileitungs-Isolatoren unter der Einwirkung hochfrequenter Spannungen," *Elektrotech. Z.* (1921), 1377.

Zederbohm, W., "Fortschritte in der Isolierung von Wechselstrom-Hochspannungswicklungen," *Siemens-Z.*, **1** (1921), 15.

Courvoisier, G., "Über Sprungwellenbeanspruchung von Transformatoren," *Bull. Schweiz., Elektrotech. Verein* (1922), 437.

Rump, S., "Statistische Untersuchungen über Störungen in elektrischen Anlagen durch Blitzschläge," *BBC-Mitt. Baden* (1922), 234.

Bucksath, W., "Elektrische Stossprüfung von Porzellan-Isolatoren," *Elektrotech. Z.* (1923), 943.

Rüdenberg, R., "Windungsverkettung in Spulen," in *Elektrische Schaltvorgänge* (Berlin: Springer, 1st ed., 1923), pp. 446–463.

Meyer, G. J., "Wanderwellenschäden," *Mitt. Ver. Elek.-Werk.* (1923), 407.

Marx, E., "Versuche über die Prüfung von Isolatoren mit Spannungsstössen," *Elektrotech. Z.*, **45** (1924), 652.

Slepian, J., "Theory of the autovalve arrester," *J. Amer. Inst. Elec. Engr.*, **45** (1926), 3.

Flegler, E., "Die Wirkungsweise von Überspannungsschutzvorrichtungen nach Untersuchungen mit dem Kathodenoszillographen," *Arch. Elektrotech.*, **19** (1928), 527.

Wade, E. J., and G. S. Smith, "Time lag of insulators," *Elec. Wld.*, **92** (1928), 309.

Torok, J. J., and W. Ramberg, "Impulse flashover of insulators," *Trans. Amer. Inst. Elec. Engr.* (1929), 239; *J.* (1928), 864.

Fortescue, C. L., "Rationalization of station insulating structures with respect to insulation of the transmission line," *Trans. Amer. Inst. Elec. Engr.* (1930), 1450; *J* (1930), 674.

McMorris, W. A., and J. H. Hagenguth, "The non-resonating transformer," *General Electric Rev.*, **33** (1930), 558.

Hodnette, J. K., "Effect of surges on transformer windings," *Trans. AIEE*, **49** (1930), 68.

Melsom, S. W., A. N. Arman, and W. Bibby, "Surge investigations on overhead lines and cable systems," *J. Inst. Elec. Engr.*, **68** (1930), 1476.

Sporn, P., "Rationalization of transmission insulation strength," *Trans. Amer. Inst. Elec. Engr.* (1930), 1470; *J.* (1930), 662.

Torok, J. J., "Surge characteristics of insulators and gaps," *Trans. Amer. Inst. Electr. Engr.* (1930), 866; *J.* (1930), 276.

Stern, G., *Hochspannungsforschung und Hochspannungspraxis* (Berlin: Springer, 1931), p. 175.

Fielder, F. D., "Surge testing of suspension insulators," *Elec. J.* (1931), 436.

Kopeliovitch, J., "À propos de l'essai de choc des isolateurs," *Bull. Schweiz. Elektrotech. Verein* (1931), 461.

Müller, H., and W. Weicker, "Zur Frage der Stossüberschlagsspannungen verchiedener Isolatorformen in Abhängigkeit von der Stossspannung," *VDE-Fachber.* (1931), 98.

Neuhaus, H., "Überspannungsmessungen mit dem Klydonographen in deutschen Hochspannungsnetzen," *Arch. Elektrotech.*, **25** (1931), 333.

Biermanns, J., "Sprungwellen- und Stossprüfung von Transformatoren," in *Hochspannungsforschung und Hochspannungspraxis* (Berlin: Springer, 1931), 173.

Torok, J. J., "Protection against lightning surges," *Elec. Eng.*, **50** (1931), 498.

—— and C. G. Archibald, "Suspension insulator assemblies, their design and economic selection," *Trans. Amer. Inst. Elec. Engr.*, **51** (1932), 682; *Elec. Eng.* (1932), 513.

Putman, H. V., "Surge proof transformers," *Trans. AIEE*, **51** (1932), 579.

Melvin, H. L., "Lightning experience on wood pole lines," *Elec. Eng.*, **52** (1933), 36.

McEachron, K. B., I. W. Gross, and H. L. Melvin, "The expulsion protective gap," *Trans. AIEE*, **52** (1933), 884.

Müller-Hillebrand, D., "Die neuzeitliche Entwicklung von Überspannungs-Schutzgeräten in Hochspannungs-Anlagen," *Elektrotech. Z.*, **55** (1934), 733, 765, 782.

Hunter, E. M., "Tests on lightning protection for a–c rotating machines," *Elec. Eng.*, **55** (1936), 137.

Bewley, L. V., "Flashover on transmission lines," *Elec. Eng.*, **55** (1936), 342.

Biermanns, J., "Fortschritte im Transformatorenbau," *Elektrotech. Z.*, **58** (1937), 659.

Elsner, R., "Die Gewitterfestigkeit moderner Reguliertransformatoren," *CIGRE* (1937), No. 115.

—— "Zur Theorie des schwingungsfreien Drehstromtransformators," *Wiss. Veröff. Siemens. Konz.*, **18** (1939), no. 1, p. 1.

Miller, J. L., and J. M. Thomson, "The surge protection of power transformers," *J. Inst. Elec. Engr.*, **84** (1939), 187 (contains lists of references).

McCann, G. D., E. Beck, and L. A. Finzi, "Lightning protection for rotating machines," *Trans. AIEE*, **63** (1944), 319.

Meyerhans, A., "Neue Bauweisen bei Transformatoren und Drosselspulen," *Bull. Schweiz. Elektrotech. Verein*, **35** (1944), 632.

Vogel, F. J., "A study of the relative severity of steep front waves and chopped waves on transformers," *Trans. AIEE*, **66** (1947), 64.

Meador, J. R., "360,000-Volt power transformer," *General Electric Rev.*, **51** (December 1948), 19.

Carpenter, T. J., I. B. Johnson, and L. E. Saline, "Evaluation of lightning-arrester lead length and separation in coordinated protection of apparatus against lightning," *Trans. AIEE*, **69** (1950), 933.

Heller, B., J. Hlavka, and A. Veverka, "Surge phenomena in transformers (non-oscillating windings)," *CIGRE* (1950), No. 143.

Blume, L. F., and A. Boyajian, "Transient voltage characteristic of transformers," in *Transformer Engineering* (New York: Wiley, 2nd ed., 1951), p. 475.

White E. L., "An experimental investigation of surge voltage distribution in open-circuited sections of a transformer winding," *CIGRE* (1952), No. 111.

Aeschlimann, H., and J. Amsler, "Schutz von Transformatoren gegen Überspannungen durch Ableiter oder Stabfunkenstrecken," *Bull. Schweiz. Elektrotech. Verein*, **44** (1953), 88.

Itschner, M., "Transformatoren für 380 kV Übertragungsleitungen," *Bull. Schweiz. Elektrotech. Verein*, **44** (1953), 145.

Aeschlimann, H., "Insulation stresses on transformer winding coils due to chopped waves," *CIGRE* (1954), No. 126.

Rorden, H. L., and R. S. Gens, "Protective practices as a criterion for high-voltage transmission design," *Trans. AIEE*, **73**, part III (1954), 465.

Griscom, S. B., J. K. Dillard, and A. R. Hileman, "Direct-stroke protection of high-voltage switching stations and transformers," *Trans. AIEE*, **74**, part III (1955), 354.

Gänger, B., "Stossprüfung von Transformatoren," *ETZ-A*, **76** (1955), 177.

Helmchen, G., "Fortschritte auf dem Gebiet der Stossprüfung von Transformatoren," *ETZ-A*, **77** (1956), 193.

Liebscher, F., and H. Meyer, "Die Stossspannungsprüfung der Isolation elektrischer Maschinen," *ETZ-A*, **78** (1957), 481.

Short, C. M., T. J. Bliss, and M. K. Enns, "Surge protection of cable-connected power trans-formers at Scattergood steam plant," *Trans. AIEE*, **76**, part III (1957), 1464.

Rabus, W., "Isolationsprobleme bei Transformatoren höchster Spannung (380 bis 500 kV)," *ETZ-A*, **81** (1960), 35 (contains lists of references).

Bader, J., and H. Czagacbanian, "Schutz von Höchstspannungsanlagen gegen äussere und innere Überspannungen," *ETZ-A*, **81** (1960), 549.

Greve, A. W., "Aktuelle Probleme im Zusammenhang mit dem Überspannungsschutz bei hohen Betriebsspannungen," *ETZ-A*, **82** (1961), 103.

Auth, W., "Stossspannungsprüfung der mit einem Parallelableiter geschützten Reihenwicklung von Spartransformatoren," *ETZ-A*, **82** (1961), 641 (contains lists of references).

INDEX

335